Externalizing Migration Management

The extension of border controls beyond a country's territory to regulate the flows of migrants before they arrive has become a popular and highly controversial policy practice. Today, remote control policies are more visible, complex and widespread than ever before, raising various ethical, political and legal issues for the governments promoting them.

The book examines the externalization of migration control from an interdisciplinary and comparative perspective, focusing on 'remote control' initiatives in Europe and North America, with contributions from the fields of politics, sociology, law, geography, anthropology and history. This book uses empirically rich analyses and compelling theoretical insights to trace the evolution of 'remote control' initiatives and assesses their impact and policy implications. It also explores competing theoretical models that might explain their emergence and diffusion. Individual chapters tackle some of the most puzzling questions underlying remote control policies, such as the reasons why governments adopt these policies and what might be their impact on migrants and other actors involved.

Ruben Zaiotti is Director at European Union Centre of Excellence and Associate Professor in the Department of Political Science at Dalhousie University, Canada.

Routledge Research in Place, Space and Politics Series

Series Edited by Professor Clive Barnett
Professor of Geography and Social Theory, University of Exeter, UK

This series offers a forum for original and innovative research that explores the changing geographies of political life. The series engages with a series of key debates about innovative political forms and addresses key concepts of political analysis such as scale, territory and public space. It brings into focus emerging interdisciplinary conversations about the spaces through which power is exercised, legitimized and contested. Titles within the series range from empirical investigations to theoretical engagements and authors comprise of scholars working in overlapping fields including political geography, political theory, development studies, political sociology, international relations and urban politics.

Published

Urban Refugees
Challenges in Protection, Service and Policy
Edited by Koichi Koizumi and Gerhard Hoffstaedter

Space, Power and the Commons
The Struggle for Alternative Futures
Edited by Samuel Kirwan, Leila Dawney and Julian Brigstocke

Externalizing Migration Management
Europe, North America and the spread of 'remote control' practices
Edited by Ruben Zaiotti

Forthcoming

Nation Branding and Popular Geopolitics in the Post-Soviet Realm
By Robert A. Saunders

Political Street Art
Communication, Culture and Resistance in Latin America
By Holly Ryan

Geographies of Worth: Rethinking Spaces of Critical Theory
By Clive Barnett

Externalizing Migration Management

Europe, North America and the spread of 'remote control' practices

Edited by Ruben Zaiotti

 Routledge
Taylor & Francis Group

LONDON AND NEW YORK

First published 2016
by Routledge
2 Park Square, Milton Park, Abingdon, Oxon OX14 4RN

and by Routledge
711 Third Avenue, New York, NY 10017

First issued in paperback 2018

Routledge is an imprint of the Taylor & Francis Group, an informa business

British Library Cataloguing in Publication Data
A catalogue record for this book is available from the British Library

Library of Congress Cataloging in Publication Data
Names: Zaiotti, Ruben, editor.
Title: Externalizing migration management : Europe, North America and the spread of 'remote control' practices / edited by Ruben Zaiotti.
Description: New York, NY : Routledge, 2016. | Series: Routledge research in place, space and politics series | Includes bibliographical references.
Identifiers: LCCN 2015032561| ISBN 9781138121591 (hardback) | ISBN 9781315650852 (e-book)
Subjects: LCSH: United States--Emigration and immigration--Government policy.
| Europe--Emigration and immigration--Government policy. |
Canada--Emigration and immigration--Government policy. |
Mexico--Emigration and immigration--Government policy. | Border security--United States. | Border security--Europe. | Border security--Canada. | Border security--Mexico.
Classification: LCC JV6483 .E87 2016 | DDC 325.4--dc23
LC record available at http://lccn.loc.gov/2015032561

ISBN 13: 978-1-138-54649-3 (pbk)
ISBN 13: 978-1-138-12159-1 (hbk)

Typeset in Times New Roman
by GreenGate Publishing Services, Tonbridge, Kent

Contents

Figures and tables

Figures

Tables

Contributors

Roberto Dominguez is an Associate Professor of Government and International Relations at Suffolk University, Boston (US). His research interests include the European Union, International Security, Regional Governance, Trans-Atlantic Relations and Latin American Politics and Economy. He is the author of *The OSCE: Soft Security for a Hard World* (Peter Lang 2013), *Security Governance of Regional Organizations* (Routledge 2011; with Emil Kirchner) and articles on regionalism in Europe and North America.

Stephan Dünnwald works at the Bavarian Refugee Council, Munich. Until 2014, he was researcher at the Centro de Estudos Africanos at ISCTE, Lisbon University Institute, and research fellow at the Lab for Critical Migration and Borderregime Studies, Institute for Cultural Anthropology and European Ethnography, University of Göttingen. He studied anthropology, sociology and geography at the Ludwig Maximilians University at Munich, and holds a PhD in sociology from the University of Augsburg. He conducted research on refugees and migrants in Central and Southeastern Europe (Kosovo) as well as in West Africa (Mali, Mauritania, Cape Verde), focusing on social impacts of migration and (forced) return to countries of origin, the externalization of European migration management, and bordering processes. Dünnwald is member of the editorial boards of *Hinterland-Magazin* and *Movements – Journal of Critical Migration and Border Regime Studies*.

Bruno Dupeyron is Associate Professor and Graduate Chair at the Johnson-Shoyama Graduate School of Public Policy, University of Regina (Canada). His major research interests are border and immigration issues, using political sociology and comparative perspectives. Bruno's research focuses on two streams: (1) the transformations of cross-border governance in North America and Europe, and (2) the regulation of mobility and security in North America. He is the author of *L'Europe au défi de ses frontières: expériences rhénane et pyrénéenne* (Peter Lang 2008) and articles on borders and migration in peer reviewed journals.

Karolina Follis teaches in the Department of Politics, Philosophy and Religion at Lancaster University (UK). Dr Follis holds a PhD in Anthropology from the New School for Social Research and an LLM in International Human Rights Law from Lancaster University. She is a political anthropologist working on borders, citizenship and non-citizenship, human rights and new security technologies in the European Union. Her current research encompasses projects which concern maritime migration across the Mediterranean, trafficking in human beings and surveillance technologies in border management. She is the author of *Building Fortress Europe: The Polish–Ukrainian Frontier* (University of Pennsylvania Press 2012).

Martin Geiger is Assistant Professor of Politics of Human Migration and Mobility at Carleton University. His research, teaching and writing focuses on migration and mobility from an interdisciplinary perspective. Martin Geiger is the founding editor of the 'Mobility & Politics' series with Palgrave Macmillan. He previously worked as a migration researcher for the European Migration Centre in Berlin. He has (co-)authored and co-edited several publications including *The Politics of International Migration Management* (Palgrave 2010) and *Disciplining the Transnational Mobility of People* (Palgrave 2013).

Liette Gilbert is a Professor in the Faculty of Environmental Studies at York University (Canada). Liette's research interests are articulated around two poles: *Immigration, Multiculturalism and Citizenship* (multicultural cities and identities; politics of difference in the city; neoliberalization of immigration policy; social justice, media representations of immigration and multiculturalism, and North American border politics) and *Urban and Environmental Politics* (planning, design and urbanism; exurban growth and environmental conservation; political ecology of landscapes; and environmental justice).

Maarten den Heijer is Assistant Professor of international law at the Amsterdam Center for International Law at the University of Amsterdam (The Netherlands). He is vice-chairman of the Standing Committee of Experts on International Immigration, Refugee and Criminal Law and member of the editorial boards of the caselaw journal *European Human Rights Cases* (EHRC) and the *Netherlands Yearbook of International Law*. Maarten den Heijer previously worked as policy officer for the Dutch Refugee Council and as legislative lawyer for the Dutch Ministry of the Interior and Kingdom Relations. He holds degrees in Law and Political Science from the University of Leiden, where he also received his PhD (*cum laude*).

Martín Iñiguez Ramos is a PhD candidate in the department of History at the Iberoamericana University (Mexico City, Mexico). Previously he was deputy director at Center for Migration Studies of the National Institute of Migration (2005) and consultant for the International Organization for Migration (IOM) (2007–2010). His areas of expertise include transnational gangs, unaccompanied minor migrants, the Sanctuary Movement in the USA and international migration.

Oleg Korneev is a Marie Curie Fellow at the Department of Politics, University of Sheffield (UK), and a Visiting Associate Professor at the Department of World Politics, Tomsk State University (Russia). Prior to that, he was a Jean Monnet Fellow at the Robert Schuman Centre for Advanced Studies, European University Institute. He has received his MA in Sociology and Social Anthropology from Central European University, his BA in International Relations and his PhD in History from Tomsk State University (Russia). His main research interests include global migration governance, international organizations, production and transfer of expert knowledge, EU–Russia cooperation on migration issues, migration policies in Eastern Europe and Central Asia, EU and Russian migration governance strategies in Central Asia.

Rey Koslowski is Associate Professor of Political Science and Public Policy, Rockefeller College of Public Affairs and Policy, University at Albany (SUNY), and Director of the Center for Policy Research Program on Border Control and Homeland Security. His primary teaching and research interests are in the field of international relations dealing with international organization, European integration, international migration, information technology and homeland security. Koslowski is the author of *Migrants and Citizens: Demographic Change in the European States System* (Cornell University Press 2000) and *Real Challenges for Virtual Borders: The Implementation of US-VISIT* (Washington: Migration Policy Institute 2005). His articles have appeared in *International Organization, International Studies Quarterly, The Journal of European Public Policy, Journal of Common Market Studies, The Journal of Ethnic and Migration Studies, International Migration, The Cambridge Journal of International Studies* and *The Brown Journal of World Affairs*.

Joshua Labove is a PhD candidate in geography at Simon Fraser University (Canada) and holds a BA (Hons) from the University of Chicago and an MA from Dartmouth College. His recent work focuses on the way legal traditions are performed and (re)constituted along the Canada–US border. Engaging with the border as both a geopolitical production and a means of investigating methods of inclusion and exclusion, he continues to do research about transnational mobility, particularly in the post-9/11 North American context.

Andrey Leonov is Assistant Professor at the Department of European and International Law, Nizhny Novgorod State University (Russia). In 2003–2006, as a recipient of an EGIDE/Eiffel Research Fellowship (French Ministry of Foreign Affairs), he was an invited Research Fellow at the Centre of International Security and European Cooperation, Pierre Mendes France University of Grenoble (France). He has received his Specialist

Diploma in Law and completed his postgraduate studies in International and European Law at Nizhny Novgorod State University. His research interests focus mainly on European Human Rights Law, ECHR Case-law in the Russian Legal Order, international criminal law and procedure, EU Human Rights legal framework and policies, EU action against human trafficking, interaction between the EU and the Council of Europe legal frameworks, Responsibility to Protect.

Alison Mountz is Professor and Canada Research Chair in Global Migration in Geography and the Balsillie School of International Affairs at Wilfrid Laurier University. She researches human migration, with a particular interest in boat travel, interception, detention, asylum and borders. Her recent work explores the role of islands in migration management. Mountz teaches and advises students in migration studies, political, urban and feminist geography, and is conducting collaborative research with Keegan Williams on the relationship between losses and enforcement at sea. She is the author of *Seeking Asylum: Human Smuggling and Bureaucracy at the Border* (University of Minnesota Press 2010). *Seeking Asylum* was awarded the 2011 Meridian Book Prize from the Association of American Geographers.

Keegan Williams is a doctoral candidate in the Waterloo-Laurier Graduate Program in Geography at Wilfrid Laurier University. His research focuses on international migration, political geography and statistical methodology. Keegan's thesis explores how states are reconfiguring spaces at sea to manage the movement of people by boat and shift borders in the Central Mediterranean Sea under the supervision of Dr. Alison Mountz. He has worked as a research assistant with the International Migration Research Centre in Waterloo, Ontario, since 2012.

Sarah Wolff is Lecturer at the School of Politics and International Relations at Queen Mary University of London (UK). Dr Wolff has extensive research, teaching and consultancy experience in EU public policies, Justice and Home Affairs (JHA), migration and border management policies, as well as EU–Arab Mediterranean relations. Her monograph *The Mediterranean Dimension of the European Union's Internal Security* (Palgrave 2012) builds upon fieldwork in Europe, Morocco, Egypt and Jordan. She received the LISBOAN Research Award 2012 for her book *Freedom, Security and Justice after Lisbon and Stockholm* (Asser 2012; co-editors F. Goudappel and J. de Zwaan). In 2014/2015 she was a Fulbright-Schuman fellow at the Transatlantic Academy (Washington DC) to work on the role of religion in Transatlantic Foreign policy. She was also awarded a 2014/2015 Leverhulme Research grant for a research on EU Engagement with Islamist political parties in Morocco and Tunisia. Dr Wolff holds a PhD in International Relations and an MSc in European Politics and Governance from the London School of Economics and a BA in Public Administration from Science Po Grenoble.

Ruben Zaiotti is Director of the European Union Centre of Excellence and Associate Professor in the Department of Political Science at Dalhousie University (Canada). His main areas of interest are international security, border control and European Union politics. He is the author of *Cultures of Border Control: Schengen and the Evolution of European Frontiers* (University of Chicago Press) and articles for *Review of International Studies, European Security, Journal of European Integration, Journal of Borderland Studies* and *International Journal of Refugee Law.* Ruben is currently working on two research projects. The first looks at the transatlantic partnership over issues of homeland security. The second examines the European Union's troubled quest to define a stable and coherent identity on the international stage.

Lyubov Zhyznomirska is an Assistant Professor in the Department of Political Science at Saint Mary's University (Canada). She works at the intersection of such fields as migration studies, the politics of post-Soviet countries and European Union politics. She is interested in the politics of international migration governance, issues of citizenship and belonging, and in transformations in governance of migration in Europe. Dr Zhyznomirska is also interested in the European Union's foreign relations with its eastern neighbours – specifically, with Russia and Ukraine, and the EU's impact on these countries. The EU's external migration relations, the European Neighbourhood Policy and the EU's relations with Russia are among her current research interests.

Preface

The extension of border controls beyond a country's territory has become a popular – and highly controversial – policy practice that governments around the world, and especially in the west, have adopted to manage the political, social and economic challenges posed by contemporary international migration. This book examines the array of 'remote control' initiatives (i.e. visas, overseas immigration officers, border pre-clearance, externalized asylum processing, offshore interdiction and detention) elaborated in Europe and North America since the turn of the millennium. Relying on an original interdisciplinary and comparative approach, the book seeks to map these initiatives, trace their recent evolution and assess their impact and policy implications. It also explores competing arguments that might explain their emergence and diffusion.

The volume examines the externalization of migration control from an interdisciplinary and comparative perspective, and it is based on a combination of empirically rich analyses and compelling theoretical insights. It includes contributions from various scholarly fields in the social sciences (politics, sociology, law, geography, anthropology and history). These different disciplinary perspectives offer a highly nuanced and textured understanding of the multifaceted, complex and ever-evolving nature of the phenomenon under investigation and highlights its wide-ranging political, social, legal and economic implications. To date this multifocal approach has not been used to study remote control policies.

The book adopts a comparative approach to examine the externalization of migration management. It focuses on two regions, Europe and North America, that represent two of the most popular destinations of global migratory flows and that are at the vanguard in the development of remote control policies. The book foregrounds the similarities and the differences between the two regions in terms of migratory dynamics and the policy responses deployed to manage them. The book also addresses the transatlantic dimensions of these phenomena, taking into consideration issues such as inter-regional mobility, the ongoing policy cooperation between European and North American governments, and the policy diffusion taking place between the two regions.

The volume blends new empirical material and theoretically driven arguments in an engaging and accessible manner. The contributions are based on fieldwork experiences and on other primary and secondary data about border policies collected and analyzed by the authors. Relying on theoretical approaches and hypotheses drawn from the social sciences, the individual chapters tackle some of the most puzzling questions underlying remote control policies, such as the reasons why governments adopt these policies and what might be their impact on migrants and other actors involved. This combination of empirical and theoretical insights provides a solid analytical foundation for the policy-relevant prescriptions drawn from the European and North American experiences that are developed in the volume.

The chapters included in this volume were originally presented at a symposium organized by Dalhousie University's European Union Centre of Excellence (EUCE) in Halifax in September 2014. This event was generously supported by the European Union and the EU–Canada Transatlantic Dialogue. Special thanks go to Andrea D'Silva, EUCE Coordinator, for her work in organizing the symposium, Carolyn Ferguson for the editing of the text, and all the students at Dalhousie University who helped make this event a success.

Part I

Introduction

1 Mapping remote control

The externalization of migration management in the 21st century

Ruben Zaiotti

Introduction

Since the end of the cold war, countries of immigration have introduced a raft of increasingly restrictive measures to discourage the inflow and permanence of foreign citizens on their territory (Gibney 2005). This is especially the case in the western world, still the most sought after destination for individuals who live in turbulent parts of the globe and are seeking refuge or a better life. With hopes for a peaceful and stable new world order quickly fading after the fall of the Iron Curtain, panicked governments rushed to fend off political pressure coming from domestic constituencies less and less willing to open up their doors to newcomers and to address what was presented as an impending 'global migration crisis' (Weiner 1995; Hollifield 1994; Loescher 1993).

The 'crisis' narrative provided the backdrop for the tough policy responses deployed in this period (Huysmans 2006). Governments could in fact justify their actions on the grounds that times of turmoil call for exceptional measures. This narrative is premised on the assumption that the state of emergency under which draconian policy measures are taken should be temporary, and lifted once the problem is under control and more long-term solutions are introduced. Well into the new millennium, however, these 'exceptional' conditions are still prevalent in most countries of immigration (Solimano 2010). Political and economic instability in western countries' neighborhoods, especially in North Africa and Central America, has meant that migratory pressures have not abated. The policy response has not changed either. Migration is still defined in terms of emergency, and that the dominant approach remains that of restriction, with only minimal – and, when provided, only temporary – respite is offered to those in desperate need (e.g. refugees). What was supposed to be an exceptional phenomenon has turned into a *permanent* state of crisis.

This situation poses a serious challenge to policy-makers. Managing migration is a complicated endeavor under normal circumstances (Castles 2004), but even more so when policy responses are couched in terms of emergency. Short-term results cannot be guaranteed, and exceptional measures cannot be justified ad infinitum. Short of enacting long-term solutions, the issue thus becomes that of managing domestic audiences' expectations and finding more politically acceptable solutions. Countries of immigration have introduced various initiatives to limit regular and irregular migratory movements at and within their borders (e.g. restrictions on family reunifications and access to welfare benefits, clamp downs on workers without work permits, strengthening of border checks). Managing migratory movements is, however, difficult. Various political, legal, budgetary and ethical constraints limit the ability of policy-makers to effectively regulate mobility within and across borders, especially in liberal democracies.

It is in this context that we can understand the reasons why a growing number of governments have tried to stop incoming migrants *before* they reach their final destination. Externalizing border management is not a new phenomenon. Indeed some of what Zolberg (1999) calls *remote control* practices (e.g. the imposition of visas) date back to the origins of immigration policy at the turn of the 20th century (Torpey 2000). Other initiatives (e.g. the 'offshoring' of border checks, asylum processing, interdictions and migrant detention) were introduced in the post cold war era. Since the millennium, however, these practices, and the context in which they unfold, have experienced significant transformations. These changes have affected the constellation of policy actors involved in the management of migration and the nature of their relations, the technological tools they employ, and the magnitude of the legal and political challenges they confront. As a result, remote control policies have become more complex, widespread and prominent in migration strategies around the world than ever before, and, due to their popularity, the object of greater scrutiny and criticism.

It is for these reasons that revisiting remote control practices today, highlighting their transformations and challenges, is a timely and worthy endeavor. This is especially the case for policies developed in and around Europe and North America, for the two regions are still the primary targets of global migration flows and governments across the Atlantic are leading policy innovators in this domain. By exploring theoretical approaches that can make sense of these practices, this book also provides an important contribution to the existing literature on migration policy, border control and international security.

This introductory chapter is organized as follows. The first section presents current trends defining the externalization of migration management phenomenon. The second section examines the concept of 'remote control' and some of its theoretical underpinnings, and then outlines a typology of remote control instruments that will guide the contributions to this volume. The third section maps recent developments characterizing individual remote control policies, focusing on events unfolding in Europe and North America. The concluding section previews the content of the chapters included in the present volume.

The externalization of migration management: emerging trends

The most notable development in the externalization of migration policy in the 21st century has occurred on the institutional front, with a reconfiguration of the constellation of actors responsible for this policy area. The major restructuring of the governance of migration took place in North America and Europe. In the United States, the Immigration Act of 2001 created the Department of Homeland Security (DHS), which has combined immigration functions with anti-terrorism coordination, and new methods of identification and surveillance. These changes are reflected in the organization and mandate of units within this department which have responsibility for external controls, namely the Immigration and Customs Enforcement Agency (ICE) and the US Coast Guard. ICE, for instance, now has responsibility not only for preventing irregular migration but also for preventing terrorism, while the Coast Guard (a unit of the military) has increased its role in the migration field and expanded the geographical scope of its activities beyond US territorial waters. Canada has followed a similar path with the establishment of a dedicated ministry of interior, Public Safety Canada, whose remit includes border control matters.

Across the Atlantic, Europeans have been active as well in reorganizing the border control field. In 2004 the European Union (EU) created a single border-monitoring agency, Frontex. The Warsaw-based agency is a unique hybrid structure with both supra-national and intergovernmental features whose mandate is to coordinate cooperation among EU member states, and to work alongside national immigration agencies, navies and border guards. Frontex has also the ability to engage with border authorities of third states, and to coordinate EU member states' international engagements. The restructuring of Europe's border control governance has also led to the proliferation of operational agencies which support the implementation of remote control policies in the region.[1]

Besides national governmental agents, international actors have over time acquired a more prominent role in the externalization of migration management (Geiger and Pécoud 2014). This has been the case for agencies linked to the United Nations (UN) such as the United Nations High Commissioner for Refugees (UNHCR) and the International Organization for Migration (IOM), or for hybrid entities such as the above mentioned Frontex. Their actions have typically been complementary to those of national governments, but in some circumstances these organizations have taken a more proactive profile. The presence and scope of the involvement of non-state actors have also expanded. Private companies have increasingly taken on the role of service providers on behalf of governments, such as in the case of airlines for document controls at airports or the defense industry for equipment (Bloom 2015).

The transformations affecting the governance of migration in North America and Europe and the greater geographical scope of their activities are reflected in new dynamics characterizing their international engagements.

The push to externalize border controls have led European and North American governments to establish a thick network of cooperative arrangements between them and with sending and transit countries of migration around the world. Since the 2001 terrorist attacks on American soil, transatlantic cooperation on migration matters has intensified and become more institutionalized, expanding to areas such as intelligence and data gathering. Migration issues, for instance, have frequently appeared on the agenda of various security-related transatlantic bilateral forums that were set up in this period (e.g. EU–US Justice and Home Affairs Ministerial Meetings; High Level Political Dialogue on Border and Transportation Security; Buonanno et al. 2015, 99–102). This enhanced cooperation has contributed to greater policy diffusion and, to a certain extent, policy convergence between North America and Europe, setting the foundations for the emergence of a 'transatlantic homeland security field' (Zaiotti 2014).

European and North American governments have also made greater efforts to engage sending and transit countries. These moves are a reflection of the contribution that these countries offer in managing international migration. Examples of these engagements include the *EU*–Horn of Africa Migration Route Initiative (Khartoum Process) and the US-backed Mexico–Guatemala–Belize Border Region Program, launched in 2014 and 2011, respectively. Through a combination of threats and incentives, European and North American governments have tried to convince both sending and transit countries to change domestic legislation to crack down on irregular migration on their own (Kimball 2007). These concerted actions have had limited success in stemming migration flows, although they did have an impact on the routes migrants have chosen for their journeys. The more barriers governments have introduced, the more creative – and dangerous – the attempts to get around them, as evidenced by the popularity of migration sea routes across the Mediterranean and the Caribbean. There are also variations in these flows' trajectories in response to policy changes in receiving and transit countries, such as in the case of the switch from north-western African routes directed to Spain's Canary Islands to further south; or Libya becoming a launching pad for Europe-bound human smuggling after the fall of the Qaddafy's regime. Similar arguments can be made about Mexico's southern border, which has become a new hotspot of activity due to the pressure of US officials on their Mexican counterparts to stem the flow of north-bound migrants.

The externalization of border controls has not only expanded geographically, but also become more sophisticated. Their formulation and implementation is more professional, with greater focus on management techniques borrowed from the security realm such as risk analysis (Geiger, this volume). This development is reflected in the array of 'smart' border technologies introduced or upgraded in this period to increase the reliability and efficiency of control measures beyond national borders (Kenk et al. 2013; Broeders 2007; Ajana 2013; Salter 2004). These devices are deployed for the collection and sharing of travelers' information (e.g. large databases), verification of identity (e.g. biometric scanners) and detection of suspicious movements (e.g. drones, sensors and satellite tracking systems; Kenk et al. 2013). The European Border Surveillance System (EUROSUR), an EU-led program that since 2013 supports EU member states' authorities in carrying out border surveillance, is the most ambitious example of 'smart' remote control initiatives.

The growing reliance on technologies borrowed from the intelligence and military realms is consistent with the 'security turn' in migration policy that took place in the wake of the series of high profile terrorist attacks that shook North America and Europe in the early 2000s (Guild and Bigo 2008; Guild 2009; Khursheed and Lazaridis 2015 Andreas and Biersteker 2003). Early discussions about the security dimension of migration tended to emphasize the domestic impact of large numbers of foreign nationals on a country's social and economic stability (Weiner 1993). Since the 2000s, the link between migration and security outside national borders and, more specifically, on the connection between terrorism and population movements, have become commonplace in official statements and policy plans elaborated by countries of immigration (Lahav 2010; Maguire 2015). The result has been the blurring of distinctions between national and international security and the creeping militarization of migration policy.

The expansion and greater complexity of externalization policies have raised their public profile, and with it, the level of scrutiny they face. Critics have highlighted various ethical, political and legal issues that governments promoting these policies have to grapple with. The main criticism levelled against the externalization of migration control is that it has a negative impact on migrants' lives and their rights (see Mountz and Williams, this volume). These mechanisms do not address the root causes of migration, leaving migrants in a state of limbo and exposing them to abuses (Taylor 2011, 2). The use of new technologies in remote control also raises thorny legal questions (Donohue 2013). The most serious problem has to do with data privacy. Rules about the storing and sharing of information about migrants are often opaque, inconsistent and weakly enforced, particularly when exchanges between foreign governments are involved. Remote control policies have also been criticized because they are used to circumvent domestic legal obligations in liberal democracies (Gibney 2005). Domestic and international courts have delved into these practices and emphasized their problematic nature. This is especially the case in Europe. Through its jurisprudence, the European Court of Human Rights has evaluated whether acts by the European Union and its member states are consistent with the Charter of Fundamental Rights, and, in the process, it has led some degree of policy standardization on matters of extraterritorial jurisdiction (McNamara 2013, 322).[2] This period has also witnessed the growing presence and activism of transnational non-state actors focusing on the plight of migrants. Through direct action, support of migrants and monitoring of government activities, a number of vocal non-government organizations (NGOs) have been putting pressure on governmental officials to limit the use of remote control policies. Yet, despite the mounting critiques and number of challenges governments are facing, these policies remain a popular approach to manage international migration in Europe and North America and, following these regions' examples, in other countries of immigration around the world as well.

Conceptualizing remote control: a multidimensional approach

Before exploring further how the trends described in the previous section are reflected in actual practice, some conceptual clarification is in order. The term 'remote control' was first introduced by Aristide Zolberg (2003) in an analysis of the United States' early attempts to manage migration at its source. As noted, since the 1990s the relevance and visibility of the externalization of migration policy has increased considerably. These developments are reflected in the growth of academic interest in this phenomenon (McNamara 2013; Ryan and Valsamis Mitsilegas 2010; Gammeltoft-Hansen 2011; Taylor 2011; Geiger and Pécoud 2014; Gibney 2005; Lavenex 2006; Collinson 1996; Guiraudon 2003, Bosswell 2003). Albeit using different language (i.e. 'offshoring,' 'extraterritoriality,' 'outsourcing,' 'buffering'), migration scholars have paid attention to the externalization of migration policy and explored its dynamics in different geographical contexts. Besides references to the fact that policies occur at a distance, however, most of these works lack explicit theoretically informed elaborations of the phenomenon they are examining. As a result, existing accounts of remote control policies tend to be either so broad as to include all types of foreign policies aimed at managing international migration (see, for instance Boswell 2003), or so narrow as to focus only in specific policy areas (e.g. 'offshoring' for interdiction on the high seas). These contributions also tend to shun explicit examinations of the power dynamics defining the relations among the policy actors involved in the externalization of migration management. In order to provide a more compelling understanding of this phenomenon, the next paragraphs elaborate an analytical model of remote control as applied to migration policy. This model is built on four interrelated dimensions: 1) spatial; 2) relational; 3) functional; and 4) operational. Taken together, these dimensions provide the analytical foundations for the construction of a typology of remote control policies that will guide the contributions to this volume.

1) Spatial dimension

The first and most distinguishable feature that characterizes remote control is its spatial dimension. The term 'remote' hints at the existence of some geographical distance between the location of the object that needs to be protected (typically, the territory of a state receiving large numbers of migrants) and the location where a specific migration policy is implemented. Distance in remote control policies can vary. Some occur just outside the borders of the externalizing state; others take place further away from that territory. Externalization thus consists of 'moving borders out' of one's territory and their redeployment elsewhere.[3] Yet externalization does not need to entail an actual expansion of territory. As Lord Bingham of Cornhill put it in a recent legal opinion on the legality of extraterritorial border checks conducted by the British government, the newly created frontier is not 'real' (in a legal sense) but 'metaphorical.'[4] It is also not necessarily material, as in the case of a fence at a land border. It can be a bureaucratic 'paper wall,' such as the one built on visas, or the 'virtual wall' created by data information systems tracking incoming travelers. Although counterintuitive, the externalization of migration management does not require the expansion of borders either. Parts of a state territory can in fact be 'excised,' thus creating zones in which officials are not obliged to provide foreign individuals with the same protections available to those officially on state territory (Coutin 2010). The most notorious example of territorial excision is represented by the Guantanamo Bay base, a US-held territory in Cuba, which the US government has used for the processing of Haitian and Cuban asylum claims since the 1980s (see more on this point below). Guantanamo Bay provided the inspiration for what is arguably the most radical experiment in territorial excision, namely the Australian government's redefinition of the status of some of its island territories for immigration purposes in 2001 (Gibney 2005, 8). In the European context, similar arrangements can be found in the Spanish enclaves of Ceuta and Melilla in North Africa, which, if entered, do not give the right to travel to the Spanish mainland. Switzerland, France, Germany and Spain have also declared parts of their airports 'international zones,' thus creating a new category, that of the 'offshore excised person' (Adey 2004).

2) Relational dimension

The externalization of migration management involves a complex set of relationships among various policy actors. As noted above, the types of policy actors in this field can be both national and international, official and private. The most influential players in this field are national governments, acting through relevant ministries such as those of defense, foreign affairs and the interior. Transnational cooperation among these entities is common. Engagement with governments in sending and transit countries and other actors such as international organizations and the private sector have become the norm as well. The nature of these interactions can be unilateral (e.g. offshore interdiction on the high seas, deportation), bilateral between sending and receiving or transit countries, but also between receiving countries (e.g. border preclearance arrangements), multilateral (e.g. discussions involving

United Nations agencies). These modes of engagement are often combined, such as the bi-multilateral approach that the EU adopts in its relations with third countries. The purpose of these interactions is the management of human flows. The type of targeted flows can be either 'regular' (economic migrants, tourists, businesspeople), or 'irregular' (undocumented migrants, trafficked individuals, asylum seekers). Each flow involves different types of policy response, although in their practices policy-makers and border practitioners often fail to take into consideration the peculiarities that characterize each migration movement.

The ongoing interactions between sending, transit and receiving countries of immigration have led to the spread and eventual institutionalization of western-inspired border control policies in peripheral countries (Lavenex and Uçarer 2004). Policy diffusion takes place along a continuum that runs from fully voluntary adaptation to direct imposition. This process can occur through emulation or mimicking of externalized strategies (e.g. the diffusion of offshore detention policies from the United States to Australia and then Europe; Flynn 2014, 24). This type of diffusion entails a mix of voluntary and involuntary adaption in response to policies adopted by another country. In most instances, however, especially those involving, on one hand, receiving countries, and, on the other, sending and transit countries, the relationship is defined by a certain degree of imposition or even coercion. This is the case of policy transfer through conditionality as applied by the European Union with its eastern and southern neighbors (Lavenex and Uçarer 2002). These coercive dynamics can be diffused and indirect. Externalizing agents often pressure third countries' governments, supra-national and non-governmental actors to become proxies to carry out their agendas. The role of sending and transit countries' governments thus becomes ambiguous, since they are both the target of the externalizing agents' actions but also active externalizing agents themselves. Therefore, even in what seem the most asymmetrical of relationship, power can be re-distributed among policy actors in subtle and unexpected ways.

3) Functional dimension

A popular explanation for why governments opt for the externalization of border controls is based on the calculation that policy-makers make when trying to address the issue of unregulated and uncontrolled migration directed at their shores. In this 'rational' perspective, externalizing migration control is an efficient and cost effective policy tool to manage this phenomenon. This interpretation of 'externalization' echoes the use of this term in the economics literature. From this perspective, externalization describes a process through which an individual or business attempts to maximize his or her profits by off-loading indirect costs and forcing negative effects to a third party (Laffont 1991). Applied to the case under investigation, the benefits of remote control policies lie in the fact that they allow governments to offload part of the burden of implementing border control measures to sending or transit countries of migration (Lavenex and Wichmann 2009). By raising the difficulty of accessing one's territory, the costs for potential migrants increases as well, and the deterrence effect it creates helps reduce future expenses incurred by governments.

These calculations are not limited to the pecuniary aspect. Costs can also be estimated in political terms. As noted earlier, offshoring immigration controls allows liberal governments to circumvent domestic legal and political constraints. 'Ethical' calculations can also come into play. Externalizing migration can be justified because it prevents abuse (as in the case of combating smugglers), or it can relieve the burden on countries disproportionally affected by migration (Gibney 2005, 15).

Rationalist interpretations are not the only ones available. One of the most appealing alternatives is represented by accounts that take inspiration from (Freudian) psychology. From this perspective, 'externalization' is an unconscious defense mechanism to deal with stress or anxiety, where an individual attributes, or 'projects,' his or her own internal characteristics that cannot be accepted as one's own onto the outside world, particularly onto other people. To make sense of current dynamics characterizing the management of migration we should therefore consider that times of rapid economic and cultural change have contributed to the spreading of fears about one's identities in western societies, fears that have in turn been projected onto a threatening 'other,' especially if this other comes from far away and it is perceived to be culturally 'different.' In policy terms, this projection has taken the form of the externalization of policies aimed at the migrant 'other.'

As with other defense mechanisms, externalization is a protection against anxiety and is, therefore, part of the everyday life of 'normal' individuals. However, if taken to excess, it can lead to the development of 'neurosis,' a functional disorder characterized by excessive and irrational anxiety, and frequent compulsive acts. The popularity and exponential growth of externalization of migration, and especially its most extreme forms, can be understood from this angle.

Although in tension with each another, economic and psychological interpretations of externalization of migration are not mutually exclusive. Anxiety, and thus the justification for exceptional measures, can be rationally manipulated by political elites for the purpose of distracting the population from their limited ability to control the economic forces of globalization (Turton 2002; Bauman 1998). In turn, neurotic behavior can be rationalized as an acceptable form of policy output. Conceptualized in this fashion, this perspective mixing rationalist and psychological features echoes the sociological approach of securitization as it has been applied to the issue of migration (Huysmans 1995).

4) Instrumental dimension

The policy 'toolbox' that governments in Europe, North America and elsewhere have developed over the years to manage incoming migration flows contains a wide range of externalizing policy instruments. These instruments differ in terms of format, purpose, policy actors responsible for their implementation and their relations. Legal-administrative and law enforcement measures are the most common in this policy domain. Examples of the former include visas and readmission agreements, while the latter encompass extraterritorial detention, overseas immigration officers and border preclearance. While until recently deployed only in exceptional circumstances, initiatives with a military or economic focus have become more prominent modalities of remote control. Examples of such practices include offshore interdiction and the inclusion of migration clauses in international trade agreements. Each of these policy instruments is deployed for a particular purpose in mind, whether it is to facilitate, prevent, contain or stop migratory flows, although in practice these distinctions are often blurred. Visas, for instance, can be deployed both to facilitate and restrict access to a territory. Their degree of formalization also varies, ranging from instruments enshrined in domestic and international law to practices informally or discretely carried out by officials or private agents abroad (e.g. training exercises, offshore detention, anti-smuggling operations, travel document checks). These instruments' format influences the profile of the actors responsible for their implementation. These actors can be official representatives of a government (e.g. ministry of foreign affairs for visas) or international organization (e.g. International Organization for Migration's desk officers), or private actors (air carriers checking incoming passengers' documents). Relations among these actors can also take different forms. As noted earlier, these relations range from the unilateral (interception on the high seas) to the bilateral (cooperation between officials of receiving countries or between them and sending/transit countries) and the multilateral (policies agreed upon within the European Union), or a combination of these approaches. Table 1.1 maps the most important externalizing policy instruments currently in use, highlighting their format, purpose and the policy actors responsible for their implementation and their relations. This snapshot of existing externalizing policies demonstrates the complexity and wide scope of the policy toolbox available to governments to manage migration. The next section examines more in detail each of these instruments and how they have evolved in recent times.

Table 1.1 Remote control policies and their features

FEATURES	POLICIES						
	Visa	Immigration officers	Pre-clearance	Asylum processing	Re admission	Interdiction	Detention
Spatial scope	Near/Far	Near/Far	Near/Far	Near	Near/Far	Near	Near/Far
Purpose	Facilitation/prevention	Prevention	Facilitation	Facilitation/prevention	Prevention	Prevention	Prevention
Policy format	Admin	Policing	Policing	Legal/Admin	Admin/economic	Military	Policing
Policy actors	Foreign affairs/private	Interior/private	Interior	Interior/Justice	Interior/Justice/Foreign affairs	Defense	Interior
Nature of interaction	Uni-lateral	Bilateral	Bilateral	Bi/multilateral	Bilateral	Bi/multilateral	Bilateral
Migration type	Regular	Regular/Irregular	Irregular/regular	Regular/Irregular	Irregular	Irregular	Regular/irregular

Remote control policies in the 21st century: North America, Europe and beyond

While in recent times the externalization of migration policy has undergone important transformations, and the scope, pace and impact of these changes have varied depending on the particular policy areas and context in which they have unfolded. The following pages trace the recent evolution of remote control, paying particular attention to developments affecting visa policy, overseas immigration officers, border preclearance, offshore interdiction, asylum processing and detention in North America and Europe.

Visas and the regulation of mobility

Visas are documents that provide permission to enter a foreign country. The first and most notable development characterizing the recent evolution of visa policy is the shift to advanced controls (Yale-Loer et al. 2005). Historically, in western countries entry requirements have been mostly processed upon arrival. Nowadays requiring visas prior to arrival has become the norm. Even when no visa is formally needed, some upstream controls are routinely performed. Since 2003, for instance, the US Visa Waiver Program – the program that allows citizens of selected (mostly western) countries to travel to the United States without a visa – has added as requirement a pre-trip entry permission (the Electronic System for Travel Authorization – ESTA). The visa application process itself has become more burdensome. Notable changes include expanded interview requirements at consulates, additional security checks on visa applicants, new special registration programs and biometric identifiers (Taylor 2011, 5).

These developments are part of a larger trend, namely the politicization of visa policy. Although the original purpose of visas is to manage mobility in general, countries of immigration have progressively relied on this tool as a means to prevent, or at least limit, unwanted migration. Evidence of this trend is represented by the fact that the number of countries to which western states apply visa restrictions since the 1990s closely reflects rising concerns over asylum (Gibney 2005). Visas are also more frequently used as foreign policy tools. European countries, for instance, have used the promise of lifting entry requirements into the European Union as bargaining chips with their neighbors, and especially those who wish to join the EU. Visas have also been employed as 'weapons' alongside more traditional tools of coercive diplomacy such as sanctions. Exemplary in this regard is EU and US imposition of visa bans on Russian officials believed to be involved in the conflict in Eastern Ukraine, which has been brewing since 2014. Not surprisingly, the politicization of visas has been a source of tension. Officials of countries whose citizens still require visas to access Europe and North America have complained about the creation of 'paper walls,' and, more generally, the bullying attitude of western governments on the issue of cross-border mobility.

Bilateral relations have been put under strain as a result. This is not just the case of relations between sending and receiving countries, but also between sending and transit countries (e.g. Canada vs. Mexico, see Gilbert, this volume), and even between allies (e.g. EU vs. US; see Koslowski, this volume).

Another important development has been the increasing use of technology in visa processing. North American and European countries have spearheaded the establishment of electronic visas, whereby the travel document is stored in a computer and is electronically tied to the passport number. The processing of electronic visas has been reinforced with the requirement to include biometric technology. Indeed, in the post 9/11 world biometric-enabled passports have become an 'imperative' (Gold 2013). Biometrics are the core of the 'US-VISIT,' a program developed by the US government that supplies technology for collecting, storing and analysing biometric data on incoming travelers and that it has been used to verify identities of visa applicants at consulates and embassies abroad. Biometrics are also the cornerstone of the 'Visa Information System' (VIS), a common system of communication which facilitates the exchange of information about visas among EU member states and their consular services around the world.

The growing role of data information gathering to manage international travel is also behind the development of 'Passenger Name Record' (PNR) and 'Advance Passenger Information' (API) programs. These programs collect personal information about travelers when making a reservation. Although not formally visas, these systems serve a similar purpose of pre-screening travelers before they leave their destination. The existence of programs such as PNR also highlights another important aspect of visa processing, namely its outsourcing to private companies.[5] If not granted on arrival, visas are typically issued though a prior application at a country's embassy or consulate. More and more, however, visas are processed by private visa services that are granted the authority by foreign governments to issue international travel documents. In the European case, some of these applications are pooled among countries.

The trend toward the privatization of document checks builds upon the practice established from the 1990s to inflict fines or other penalties against transport companies (airlines, train and shipping companies) that fail to properly screen foreign travelers' documentation (Gilboy 1997). Once again, European and North American countries have been at the forefront of this policy development. The United States introduced carrier sanctions in the Immigration Act of 1990. This Act mentioned not only fines for incoming irregular passengers, but also the responsibility for holding passengers, paying their detention costs and for their deportation. Canada and most European countries followed suit. A directive harmonizing carrier sanctions across EU states, whose purpose was "to combat illegal immigration," was formally adopted in June 2001.[6] By imposing sanctions of non-compliant carriers, governments have therefore de facto privatized migration management, for carriers have to take on the responsibility to make decisions on the possession and authenticity of relevant documents on their behalf. One of the most problematic consequences of this development is that the use of administrative discretion has increased, while public accountability and possibility of legal recourse have been reduced (Taylor 2011, 5).

Overseas immigration officers and border preclearance

Although some aspects of border control such as travel document checks have been outsourced to private entities, national governments have not completely surrendered their sovereign power in this policy area. Indeed one of the most striking developments in the post 9/11 environment has been the expansion of the 'extraterritorial' presence of government agencies for the purpose of combating illegal immigration. By the millennium, the UK, Canada, the US, Sweden, France, the Netherlands and Norway had posted civil servants abroad (mostly in airports) to gather information, support and train airline companies, and co-operate with local authorities. In 2001, the EU established a network of Immigration Liaison Officers (ILOs) to coordinate its immigration control activities. The US followed suit in 2003 with the overseas deployment of the first Customs and Border Protection attachés. Border control officials posted abroad are a common feature of the contemporary migration management policy field. Their role and power, however, remains ambiguous. Although in theory these officers should provide only advice and support to local partners, in practice their influence extends much further than their formal mandate indicates.[7]

Besides sending officials abroad, North American and European governments have offered authorities in sending and transit countries financial incentives to strengthen their capacity to deal with migration flows directed toward their territories. These actions are controversial, especially when targeted at local enforcement practices. The Mexican government, for instance, is particularly sensitive about perceptions of direct US interference (see Dominguez and Iniguez, this volume). The United States have also established joint training centres in Latin American countries to strengthen local capacity in the fight against human smuggling (Flynt 2014: 15). Border control-related US actions abroad therefore provide support for the claim that today "the American homeland is the planet" (Miller 2015).[8] In the European context, the European Union has created a Border Assistance Mission to Moldova and Ukraine (EUBAM). In this project, launched in 2005, European border control officials, with the support of the International Organization for Migration, are posted in the two Eastern European countries with the purposes of training local border officials.[9] Whether these training activities achieve their goals is difficult to determine. There is evidence, however, that transit countries have been persuaded to change domestic legislation to crack down on irregular migration on their own (Kimball 2007).

The ambiguity that characterizes overseas immigration officers is less prominent in another extraterritorial model of border control, namely 'border preclearance.' This arrangement enables destination countries of immigration to post immigration officers at ports of entries (airports, train stations, ports) in third countries in order to screen incoming travelers before departure. Border preclearance is not a new phenomenon, but it has expanded substantially since 9/11.[10] The US government has nowadays pre-embarkation immigration checks at airports in Canada, the Caribbean, Ireland and the United Arab Emirates. In Europe, Belgium, France and the United Kingdom have introduced so-called 'juxtaposed controls' in each other's jurisdiction allowing immigration pre-checks on selected routes across the English Channel (Clayton 2010). Aside from regular pre-control activities, governments have also carried out targeted extraterritorial anti-smuggling operations (e.g. the US-led 'Operation Firm Grip' in 2001). Although formally host governments provide consent for border preclearance activities, this form of remote control has been at times contentious.[11] It is not surprising, therefore, that most preclearance agreements are signed between 'friendly' countries whose main objective is to encourage business or tourism.

Offshore interdiction

One of the side effects in the surge in the phenomenon of 'boat people' risking their lives to reach rich countries' shores has been the parallel expansion of interdiction practices on the high seas. The term 'interdiction' is taken from military jargon and refers to "the act of destroying, damaging, or cutting off … to stop or hamper an enemy" (Merriam-Webster Dictionary). Interdiction is often used in conjunction with 'interception,' namely the "the act of stopping someone or something that is going from one place to another place before that person or thing gets there" (Merriam-Webster Dictionary). Patrolling coastlines with naval or coastguard ships is a state's core security function. Although in some circumstances (e.g. during war times or search and rescue operations) states do conduct patrolling exercises beyond their sovereign boundaries, the presence of vessels on the high seas with the intended purpose of interdicting migrant boats is a relatively new phenomenon (Taylor 2011, 10; Tondini 2012, 59–60). The origins of modern interdiction policy can be traced back to the early 1980s, when the United States Coast Guard was granted authority to stop Haitian vessels on the high seas, and to return individuals who were not deemed to be refugees (Flynn 2014, 7). This policy was redeployed and further expanded during the George H.W. Bush and Clinton presidencies. The 1990s is the period when the US Coast Guard's 'extraterritorial' powers and the security component of its mandate were substantially expanded (ibid., 13).

These trends have continued into the new millennium. Following his father's footsteps, George W. Bush declared that the US "will turn back any [Haitian] refugee that attempts to reach our shore" (Cited in Marquis 2004).[12] The number of reported interdictions remained high throughout the decade, reaching a peak of 10,000 in 2005 (BCP 2014; Flynn 2014, 15). To deal with security threats at sea, the US has also established joint initiatives with third countries. This is the case, for instance of *Shiprider,* a program introduced in 2005 involving the sharing of resources and personnel with the Canadian Coast Guard.

US-led interdiction policies have influenced similar practices outside North America. The US 'imprint,' for instance, is apparent in Australia's Pacific Solution (Magner 2004.). Among European states, Spain and Italy have been among the most active in maritime interdiction. Spain's *Sistema Integrado de Vigilancia Exterior* (SIVE), one of the first of this kind in Europe (it became operational in 2002), employs radar and surveillance cameras to detect incoming vessels and intercept them if suspected to be carrying irregular migrants (Lutterbeck 2006, 2). This project was the template for other maritime surveillance and interdiction initiatives such as *Seahorse Atlantic* (Casas-Cortes et al. 2014).[13] Since its creation, Frontex has taken an active role in offshore operations led by EU member states (HRW 2010). The EU agency has coordinated joint patrols of European maritime forces along the Mediterranean and West African coasts and, in some cases, it has taken a more operational role (e.g. Joint Operation Triton).[14]

Offshore interdiction has raised questions of legal responsibility for the states conducting these types of operations (Blay et al. 2007). The issue of extraterritorial control has been examined in high profile cases before national and international courts. In *Sale vs. Haitian Centers Council* (1993), the US Supreme Court accepted the government's argument that the duty not to return refugees applied only to refugees within US territorial waters (Gibney 2005, 10). In *Hirsi Jamaa and Others v. Italy* (2012), the European Court of Human Rights also weighed in on this issue, although coming to a different conclusion than its US counterpart. In its ruling the court argued that if a person is under the 'effective' control of a state (whether on its territory or not) then that state is obliged to protect his or her fundamental rights.[15] This legal threshold is, however, quite high and difficult to apply, and for some commentators this means that for the foreseeable future governments will still be able to exert extraterritorial control while avoiding legal responsibilities (McNamara 2013, 323; see also den Heijer, this volume).

Externalized asylum processing and readmission agreements

Asylum seekers are one of the most vulnerable categories of migrants, and for this reason they have special protection under international law. Yet, governments around the world have progressively restricted access to asylum because of the belief that a number of individuals are not genuine refugees. As for migration more generally, governments of countries receiving high numbers of asylum claims have therefore tried to externalize their asylum policy to circumvent their legal obligations. One of the centerpieces of this approach is the so-called 'safe country of asylum' policy (Kneebone,

2008). The concept of 'safe country of asylum' is the key component of the Dublin Convention, which European countries signed in 1990 and became effective in 1997. According to this agreement, if an alien applies for asylum after traveling through a country that is party to the UN Convention relating to the status of refugees and thus considered to be 'safe,' he or she will be sent back to that country to file his/her application. The Dublin Convention in turn inspired a similar arrangement in North America between the US and Canadian governments (see Labove, this volume; Arbel 2013). The safe country of asylum arrangement relies on extensive use of data. In Europe, an asylum fingerprint database (EURODAC) was established for this purpose in 2003. As evidence of the ongoing securitization of migration policy, the mandate of EURODAC has recently been expanded to allow national police forces and Europol to compare fingerprints linked to criminal investigations.

In parallel to, and as a corollary measure to the safe country of asylum arrangements, European and North American governments have pushed for the establishment of external asylum processing facilities outside their borders. The purpose is to compel migrants to submit asylum claims before they reach their final destination. European officials first began suggesting the offshoring of asylum procedures in the 1980s (Flynn 2014, 23). In light of, and inspired by, Australia's 'Pacific Solution,' the proposal resurfaced at the turn of the millennium. In March 2003, Prime Minister Tony Blair proposed the creation of 'external asylum processing centres' outside Europe to its EU partners. The proposal was then shelved because it was considered to be too controversial. The interest in this project has never completely waned, however. Indeed the idea of external processing centers resurfaced again in 2015 as the number of boat people crossing the Mediterranean reached new record levels.

Since the direct management of asylum claims abroad has yet to materialize, governments in receiving countries have adopted an alternative strategy, namely strengthening the capacity of their counterparts in refugee-producing or transit regions through a mix of financial incentives and operational support. The European Union has, for instance, introduced *Regional Protection Programs*, which are elaborated in collaboration with EU member states, UNHCR and the targeted countries. The first programs of this type were implemented in mid 2000s in Eastern Europe and the African Great Lakes Region, and then extended to the Horn of Africa and North Africa.

In order to deal with failed asylum seekers and, more generally, undocumented migrants already present on their territory, and to dissuade potential new claimants, receiving countries have negotiated readmission agreements with sending and receiving countries. These agreements oblige countries to accept not only their own nationals but also third country nationals in exchange for 'compensatory measures' like visa facilitation programs or financial support. The US and Canada do not have formal agreements of this kind, and therefore undocumented migrants usually are deported to the country of origin without the latter's consent or involvement (Ellermann 2008; Kanstroom 2012). In Europe, the number of bilateral readmission agreements have instead bourgeoned since the millennium (Billet 2010; Wolff, this volume). Spain and Morocco signed such an agreement in 2003; after years of negotiations in 2008 Libya and Italy signed a 'Friendship Pact' whereby the North African country agreed to increase cooperation in tackling illegal migration across the Mediterranean and to allow the Italian navy to patrol Libyan waters in exchange for long-term direct financial support. In December 2013 the EU finalized a similar arrangement with Turkey.

If not explicitly mentioned in dedicated arrangements, a 'readmission clause' has become commonplace in economic agreements between migrants' sending and receiving countries. Again, Europe is this regard has been at the vanguard. The 2000 Cotonou Agreement, which represents the current overarching framework for economic and development cooperation between the European Union and Asian, Caribbean and Pacific (ACP) countries, included for the first time rules on readmission and the fight against irregular migration. ACP countries vocally contested these provisions and, although they were not able to eliminate them from the agreement's final text, they successfully lobbied against the obligation to readmit third country nationals, as their European counterparts had demanded (Lavenex 2006). This linkage to politics involving migration issues still colors the relationship between the EU and its southern partners. In one of the official declarations issued at the 4th EU–Africa summit that was held in Brussels in 2014, dedicated to 'migration and mobility,' readmissions are mentioned side by side with measures to promote the employment of young people as the centerpiece of Euro-African cooperation in the management of migration.

Offshore detention

The external processing of asylum is strictly linked to another type of externalizing instrument, namely offshore detention. Since asylum seekers are either forced to enter the proposed processing centers or prevented from leaving them before their claims are processed, these facilities are de facto prisons (Noll 2003). Detention centers for migrants are typically located in transit countries, although other arrangements have been devised.[16] The modern roots of this phenomenon can be traced to the early 1990s when the United States begun to extend its enforcement measures against illegal migrants beyond its borders (Welch 2002, 107; Flynn 2014, 5; Frenzen 2010, 377). As noted earlier, the most notorious offshore detention facility was the one set up in 1991 at the US naval base in Guantanamo Bay. There are speculations that in the same period the US government funded other detention centers in Guatemala for migrants from Central America and other regions (Taylor 2011, 9).

Influenced in part by the US example, European governments have considered the offshoring detention option. In light of the controversies surrounding the UK-backed external processing proposal (see discussion in previous section), the idea of directly managing processing facilities was set aside. Instead the EU and its member states have started to delegate to transit country governments the responsibility of building and managing centers for irregular migrants. Some of these efforts date back to the late 1990s. For instance, in 1998, as part of an agreement between Italy and Tunisia on the readmission of irregular migrants, Italy provided financial support for the creation of migrant detention facilities ('centri di permanenza') in the North African country. Since the millennium, however, this practice has become more widespread. In 2002, for instance, the Mauritanian government, with the financial support of the Spanish Agency for International Development Cooperation, created a detention center for irregular migrants (known as 'El Guantanamito') in the city of Nouadhibou (see Dünnwald, this volume). Similar facilities have sprung up in other locations (e.g. Ukraine, Belarus).

Another significant development has been the increasing cooperation between national governments and international organizations in the offshore detention business. The European Union, for instance, has provided funds to the International Organization for Migration to support detention efforts in Ukraine. The EU has also offered financial assistance for the creation and management of detention facilities within the framework of the 'Twinning' program, an initiative in which EU states partner with governments in Europe's neighborhood to strengthen local administrative and bureaucratic capacity.[17]

In most cases, the practice of confining migrants offshore has been shrouded in secrecy. Not all details of the arrangements regulating detention centers in third countries, and in some cases even their very existence, have been made public (Taylor 2011, 9). This dearth of information allows externalizing countries to avoid taking responsibility for the ways migrants are (mis)treated in countries with poor human rights records. More problematically, by limiting the scrutiny they are exposed to, this secretive approach shields them from the type of public outcry these facilities would receive if they were located 'on shore.'

Chapters preview

The chapters included in the present volume elaborate on the themes outlined in the previous sections, taking the reader though a tour of the main features and debates characterizing remote control policies and exploring their recent applications in and around Europe and North America.

Chapter 2 offers an empirically-based assessment of the impact of externalization policies on migrants and their migratory trajectories. In their contribution, geographers Mountz and Williams test two hypotheses that reflect claims advanced in the migration studies literature, namely that increased maritime enforcement leads to more migrant losses and that increased enforcement does not deter migrants but may lead them to take new routes. For this purpose the authors have compiled data sets on global migrant boat losses and border-related operations at sea since 1990. The data show that most reported boat incidents are concentrated around the European Union, Australia and the United States, and that there has been a notable increase in incidents since millennium. Correspondingly, reported maritime interdiction operations predominantly occurred around these places and times. Using test case data from the Europe Union between 2006 and 2013, the authors find a strong correlation between the intensity of operations and most measures of migrant and boat losses. They also find some evidence that variations in enforcement in specific areas lead to changes in the number of incidents in other areas, although these results do not lead to conclusive answers regarding this issue.

The volume then turns to the analysis of externalization of migration management around the European continent. Chapters 2 and 3 examine remote control policies in the Mediterranean Sea. In Chapter 3 den Heijer provides a legal analysis of recent developments in the externalization of maritime border controls in Europe, with a focus on the activities coordinated by the EU border agency Frontex. This contribution's starting point is the recognition of the main trend characterizing the patrolling of European coastlines, namely the geographical relocation of controls away from coastal border crossing points to the various maritime zones and the involvement of non-EU countries in these activities. The author points to the legal implications of this trend and examines the attempts by the European Union to clarify the legal regime applicable to such controls in the sphere of fundamental rights, powers of interdiction and search and rescue obligations. The contribution argues that these attempts are fundamentally flawed as they fail to harmonize Member State practice and erode the already limited human rights protections granted to migrants.

In Chapter 4 Follis relies on an ethnography approach to unearth the inherent contradictions that characterize the 'humane' rhetoric used by governments to support remote control policies. For this purpose the author scrutinizes the central nodes of EU and national politics (parliamentary chambers, governmental offices) where the expert knowledge that underpins the dynamics of the EU border regime travels and considers them as ethnographic field sites for the study of the contemporary European border regime. Using as term of reference the British government's so-called 'Let them drown policy,' Follis foregrounds the paradox of border control practitioners claiming to uphold human rights standards while engaging in violent practices against migrants. The author then shows that the British government's call to halt rescuing operations in the Mediterranean because they might encourage more migrants to take to the sea represents the moment that the "humanitarian mask came off." Follis concludes by arguing that understanding how remote control policies are formulated is a necessary condition to challenge the unequal power dynamics constituting Europe's border regime.

The next two chapters look at European remote control efforts toward its southern neighbors. In Chapter 5 Wolff examines the political dynamics shaping the negotiations between the European Union and Morocco and Turkey over the signing of bilateral readmission agreements. Focusing on the role of EU incentives and third countries' preferences, this chapter reveals that beyond the function of this instrument to co-opt third countries in the EU's fight against irregular migration, a series of obstacles forced the EU to revise the design of these agreements and to take into account domestic and regional factors. In her analysis, Wolff critically engages with the meanings and representations carried by readmission agreements in third countries and outlines their implications for the European Union external migration policy.

Attempts by European countries to externalize migration management is also the theme of Chapter 6. Here Dünnwald argues that by creating new spaces of border surveillance and disciplining of migrants, remote control practices have remodeled the relations between Europe and the African continent, and left traces in the social, political and economic life of the targeted countries, as well as the ways mobility is conceived and organized. Although the European Union and its member states have formulated a comprehensive and 'global' approach to deal with migration issues, the practical outcomes in terms of inter-state cooperation and consequences on the societal level show remarkable differences. Taking Mauritania and Mali as examples, the chapter examines these differences by first drawing on recent developments in externalization of European migration policy, then sketching out the process of how this externalization evolved in the two Western African countries and contributes to the transformation of bordering processes.

The next two chapters move the geographical focus to Eastern Europe. In Chapter 7 Zhyznomirska examines the effects of the EU's migration diplomacy on the relations with its eastern neighbors and on the international governance of migration. The author also compares the conditions under which countries co-operate with the EU on international migration management. As the analysis of the actions of two Eastern European states (Russia and Ukraine) suggests, although the EU acts as the facilitator and promoter of norms in the area of migration, ultimately, the decision to pursue – or not – certain policy directions is made and justified by the government of a given country, either in accordance with its domestic interests and/or under external pressure. The chapter therefore problematizes the role of non-EU countries in establishing and maintaining the EU's regime of migration controls.

In Chapter 8 Korneev and Leonov examine the ongoing EU–Russia cooperation on migration matters and then extend the analysis to include countries of origin and transit for migrants coming to Russia. The chapter argues that the EU–Russia cooperation on migration control has triggered significant spillover dynamics in the wider Eurasian region. It shows how Russia has played a crucial role in the diffusion of readmission mechanisms to its partners and major migrants' origin/transit countries in Central Asia and the Middle East, as well as in South and South East Asia. The authors argue that attention to what happens beyond the framework of the EU–Russia readmission agreement can therefore help to get better insights about the degree of EU success in the externalization of its migration control approaches. The chapter concludes with several normative issues and contradictions stemming from the externalization of migration management in a region with low human rights standards.

The second part of the volume addresses the evolution of remote control policies in North America. In Chapter 9 Koslowski examines the controversies that have surrounded the US Visa Waiver Program (VWP). The chapter traces the intense diplomatic exchanges between the Department of Homeland Security and its European counterparts, which led to a revision of the EU's common visa policy and reform of the US Visa Waiver Program after 9/11. Koslowski argues that the political compromise that was struck on this issue is not only tenuous (the EU is unlikely to back down on its position that all EU member states should be included in the arrangement), but also that it may have unforeseen consequences for international travel, which, in turn, will raise new thorny issues for policymaking and diplomacy. Koslowski also outlines some policy options that policy-makers could consider in order to overcome the challenges that the VWP is currently facing.

The next two contributions look at the US northern neighbor, i.e. Canada. In Chapter 10 Labove examines the role of Canadian courts in the production and expansion of borders and bordering processes in North America. Drawing upon research on performativity, and using the US–Canada Safe Third country agreement as a term of reference, Labove analyses how borders are worked in to and out of the Canadian legal system, and the tools, notably the *Canadian Charter of Rights and Freedoms*, that make such legal framings possible. Labove argues that foregrounding the discretionary power of judges to "stretch, pull, twist, and move" legal boundaries enables us to see how borders do not simply expand, but are constantly open to reinterpretation. A focus on legal decisions at the border therefore produces a more complex geography of bordering. The author suggests that this line of inquiry should be the basis for a new research agenda on law at the border, which brings socio-legal studies in closer conversation with border studies.

In Chapter 11, Gilbert examines the ways in which visas have been used as tools in the externalization of migration management in Canada. The author places this discussion in the context of the unresolved tension between migration and the globalized economy, and of the constitutive discourse of the 'illegality' attributed to Mexican migrants in North America. The author shows how Canada's visa policy has become as an 'effective' policy and discursive instrument to stop the flow of refugee claimant following a border externalization logic. It also analyses how the conjuncture of visa policy, public discourse and legislative reform has eroded the rights of refugee claimants in Canada. The chapter also looks at how the current narcoviolence in Mexico poses a particular challenge to the definition of 'refugee' and Canada's designation of Mexico as a 'safe country.'

The next contribution further expands on the role of Mexico in the externalization of migration management in North America. In Chapter 12 Dominguez and Iniguez examine the main transformations that have characterized Mexican border policies over the past decade. The authors argue that while a differentiated attention devoted to its northern and southern borders remains a defining feature of Mexico's policies, since the millennium the country's response to migration and security issues has evolved. Dominguez and Iniguez show how the externalization of US borders has influenced the design and implementation of border policies in Mexico. At the same time, Mexico has also attempted to externalize its southern borders toward Central America in order to control networks of transit migrants. The authors argue that the combination of the US and Mexican externalizations of border controls has produced an incipient process of informal regionalization of border management. The chapter also suggests that Mexico has belatedly taken some steps toward improving Central American migrants' human rights and combating human smuggling.

In Chapter 13 Dupeyron turns the attention to an international actor (the International Organization for Migration) and its role in the externalization of migration management in North America. The author demonstrates that through activities ranging from the designing of national migration policies to the deportation of undocumented migrants, the IOM is actively promoting its state members' migration policy objectives of 'secluding' unwanted migrants in their own home country and/or in precarious foreign guest worker jobs. Dupeyron traces how the IOM has become a central recipient of western states' externalized migration policies, and how it has presented itself as facilitator in the provision of "humane and orderly migration." The author highlights the problematic nature of the IOM's externalizing activities, and especially the ambiguity of the human rights-based narrative that the organization uses to justify its actions. To illustrate these dynamics, Dupeyron looks at the IOM's involvement in a guest worker contract between Guatemalans and Canadian farm employers.

In the concluding chapter, Geiger weaves together the main themes and questions raised in the volume and explore new avenues for future research on the externalization of migration policy in Europe, North America and beyond. The author identifies the emergence of 'migration management' and advanced technologies in the 1990s as crucial factors in enabling states to outsource and territorially shift their policies. New specialized actors have emerged that are today supporting national governments in these tasks by developing and offering certain tailored services and 'tools' for border and migration management. These actors form part of a new 'migration industry' with its own vested economic and political interests; together with states and their policy-makers these actors are complicit in a problematic and questionable depoliticization, pragmatization and commercialization of border and migration policies. According to Geiger, this state of affairs calls for continued critical scholarship and reflection on how cross-border flows are currently 'managed' and might be approached and regulated in the future.

Notes

1 This is particularly the case in Europe. Agencies involved in the externalization of migration policy include the EU Satellite Centre, EU Maritime Safety Agency, the European Asylum Support Office and the EU Agency for Large-Scale IT systems.
2 Despite the litigious nature of the American legal system, US courts have not been as active as their European counterparts in the realm of extraterritorial controls. This is a reflection of the official stance and jurisprudence on this subject, namely that on matters of entry and expulsion the federal government has a substantial degree of discretion (Taylor 2011, 4).
3 This rebordering practice is well encapsulated by a US Department of Homeland Security senior official when he claimed that "(t)he Guatemalan border with Chiapas is now our (i.e. US) southern border." Quoted in Miller (2014).
4 Lord Bingham of Cornhill's comment can be found in *Regina V. Immigration Officer at Prague Airport and another (Respondents) ex parte European Roma Rights Centre and others* (Appellants) [2004] United Kingdom House of Lords 55, par.26.
5 The PNR system was developed by the airlines for commercial purposes; nevertheless, due to security reasons (e.g. the fight against terrorism) some countries require airlines to provide them with some data prior to the passenger's entry into the country (e.g. USA, Canada, Australia and New Zealand). Advanced Passenger Information (API) system is also carried out by airlines but in contrast to PNR, it is on behalf of governments.
6 Council Directive 2001/51/EC of 28 June 2001, supplementing the provisions of Article 26 of the Convention implementing the Schengen Agreement of 14 June 1985 (Carrier Sanctions Directive).
7 Given their affiliation with the country that could impose fines, overseas immigration officers can, for instance, put considerable pressure on local agents when performing their duties. On this point, see McNamara (2013, 330).
8 Todd Miller, Border Patrol International "The American Homeland Is the Planet." www.tomdispatch.com/blog/175774/tomgram%3A_todd_miller,_the_border-industrial_complex_goes_abroad/.
9 In May 2013, the Council of the European Union gave the green light for EUBAM Libya, a civilian mission under the Common Security and Defence Policy (CSDP), to support the Libyan authorities in improving and developing the security of the country's borders. The deteriorating situation in the country has meant that the operation had to be put on hold.
10 The first example of border preclearance dates back to the early 1950s, when a US airline established this scheme at the Toronto Airport in Canada.
11 One of the most controversial cases is the preclearance arrangement introduced by the UK government at Prague Airport in 2001. In this scheme British immigration officials had the authority to determine whether a traveler could legally enter the UK before boarding the plane. The scheme was challenged in court (see note 4). In its ruling, the House of Lords argued that although the scheme was clearly discriminatory (the main target of these checks were individuals of Roma descent), it could still operate because of the lack of international legal provisions allowing entry of would-be asylum seekers before they have filed a claim.
12 In the western hemisphere, the US Coast Guard patrolled the coast of Ecuador to stop boats on their way to Mexico and Central America for "humanitarian reasons" (Taylor 2011, 10). The Coast Guard also flew surveillance planes out of a US military base in Ecuador.
13 'Seahorse Atlantic' was established in 2006 with the participation of Mauritania, Morocco, Senegal, the Gambia, Guinea Bissau and Cap Verde.

14 In this initiative, which took over from an Italian-led operation (Operation Mare Nostrum), Frontex coordinates the deployment of patrol vessels and aircrafts around Malta and Italy's territorial waters.
15 The *Hirsi* case deals with a challenge by migrants intercepted at sea by the Italian Navy and returned to Libya.
16 An interesting example is that of the category of the 'offshore excised person,' namely a detainee held in an airport zone where national legislation does not apply.
17 Turkey has been a key benefactor. Under a 2007 Twinning project – titled "Support to Turkey's Capacity in Combating Illegal Migration and Establishment of Removal Centres for Illegal Migrants" – the EU agreed to finance the establishment of two removal centers in the country (EC 2007, 4–5).

Bibliography

Adey, P. 2004. Secured and Sorted Mobilities: Examples from the Airport. *Surveillance and Society* Vol.1, No. 4, pp. 500–519.

Ajana, B. 2013. Asylum, Identity Management and Biometric Control. *Journal of Refugee Studies*. Vol. 26, No. 4, pp. 576–595.

Andreas, P. and Biersteker, T. (eds.) 2003. *The Rebordering of North America*. New York: Routledge.

Andreas, P. and Snyder, T. (eds.) 2000. *The Wall Around the West: State Borders and Immigration Controls in North America and Europe*. Lanham, MD: Rowman and Littlefield.

Arbel, E. 2013. Shifting Borders and the Boundaries of Rights: Examining the Safe Third Country Agreement between Canada and the United States. *International Journal of Refugee Law*. Vol. 25, No.1, pp. 65–86.

Baly, S., Burn, J. and Keyzer, P. 2007. Interception and offshore processing of asylum seekers: The International Law dimensions. *UTS Law Review*, Vol. 9, pp. 7–25.

Bauman, Z. 1998. *Globalization: The Human Consequences*. New York: Columbia University Press.

BCP (Border and Customs Protection) 2014. 2014 Statistics. Available at www.cbp.gov/newsroom/stats.

Bigo, D. 2002. Security and Immigration: Toward a Critique of the Governmentality of Unease. *Alternatives*. Vol. 27, No. 1, pp. 63–92.

Billet, C. 2010. EC Readmission Agreements: A Prime Instrument of the External Dimension of the EU's Fight Against Irregular Immigration. An Assessment After Ten Years of Practice. *European Journal of Migration and Law*, Vol. 12, No.1, pp. 45–79.

Bloom, T. 2015. The Business of Migration Control: Delegating Migration Control Functions to Private Actors. *Global Policy*. Vol. 6, No. 2, pp. 151–157.

Boswell, C. 2003. The 'External Dimension' of EU Immigration and Asylum Policy. *International Affairs*. Vol. 79, No. 3, pp. 619–638.

Broeders, D. 2007. The New Digital Borders of Europe: EU Databases and the Surveillance of Irregular Migrants. *International Sociology*. Vol. 22, No. 1, pp. 71–92.

Buonanno, L., Nugent, N. and Cuglasan, N. 2015. Transatantic Governance. In Buonanno, L., Cuglesan, N. and Henderson, K. (eds.) *The New and Changing Transatlanticism: Politics and Policy Perspectives*. Abingdon: Routledge.

Casas-Cortes, M., Cobarrubias S. and Pickles, J. 2014. 'Good Neighbours Make Good Fences': Seahorse Operations, Border Externalization and Extra-Territoriality. *European Urban and Regional Studies*. DOI: 10.1177/0969776414541136.

Castles, S. 2004. Why Migration Policies Fail. *Ethnic and Racial Studies.* Vol. 27, No. 2, pp. 205–227.

Clayton, G. 2010. The UK and Extraterritorial Immigration Control: Entry Clearance and Juxtaposed Control. In Ryan B. and Mitsilegas, V. (eds.) *Extraterritorial Immigration Control.* Leiden: Martinus Nijhoff, pp. 391–423.

Collinson, S. 1996. Visa Requirements, Carrier Sanctions, 'Safe Third Countries' and 'Readmission': The Development of an Asylum 'Buffer Zone' in Europe. *Transactions of the Institute of British Geographers.* Vol. 21, pp. 76–90.

Coutin, S. B. 2010. Confined Within: National Territories as Zones of Confinement. *Political Geography.* Vol. 29, No. 4, pp. 200–208.

Donohue, L.K. 2013. Technological Leap, Statutory Gap, and Constitutional Abyss: Remote Biometric Identification Comes of Age. *Minnesota Law Review.* Vol. 97, pp. 407–559.

Ellermann, A. 2008. The Limits of Unilateral Migration Control: Deportation and Interstate Cooperation. *Government and Opposition.* Vol. 43, No. 2, pp. 168–189.

Flynn, M. 2014. How and Why Immigration Detention Crossed the Globe. *Global Detention Project Working Paper No. 8.* April.

Frenzen, N. 2010. US Migrant Interdiction Practices in International and Territorial Waters. In Ryan, B. and Mitsilegas, V. (eds) *Extraterritorial Immigration Control: Legal Challenges.* Leiden: Martinus Nijhoff Publishers.

Gammeltoft-Hansen, T. 2011 *Access to Asylum: International Refugee Law and the Globalisation of Migration Control.* Cambridge: Cambridge University Press.

Geiger, M. and Pécoud A. 2014. International Organisations and the Politics of Migration. *Journal of Ethnic and Migration Studies. Vol.* 40, No.6, pp. 865–887.

Gibney M. J. 2005. Beyond the Bounds of Responsibility: Western States and Measures to Prevent the Arrival of Refugees. *Global Migration Perspectives.* No. 22.

Gilboy, J. 1997. Implications of "Third Party" Involvement in Enforcement: The INS, Illegal Travelers, and International Airlines. *Law and Society Review.* Vol. 3, No. 3, pp. 505–530.

Gold, S. 2013. The Biometric Passport Imperative. *Biometric Technology Today.* No. 2, pp. 5–6.

Guild, E. 2009. *Security and Migration in the 21st Century.* Cambridge: Polity.

Guild, E. and Bigo, D. 2008.*Terror, Insecurity and Liberty. Illiberal practices of liberal regimes after 9/11.* New York: Routledge. Available at www.ceps.eu/ publications/%E2%80%98joint-operation-rabit-2010%E2%80%99-%E2%80%93-frontex-assistance-greece%E2%80%99s-border-turkey-revealing.

Guild, E. and Carrera, S. 2010. 'Joint Operation RABIT 2010' – FRONTEX Assistance to Greece's Border with Turkey: Revealing the Deficiencies of Europe's Dublin Asylum System. *Center for European Policy Studies.*

Guiraudon, V. 2003. Before the EU Border: Remote Control of the "Huddled Masses." In Groenendijk, V., Guild E. and Minderhoud, P. (eds.) *In Search of Europe's Borders.* Leiden: Brill, pp. 41–68.

Hollifield, J. F. 1994. The Migration Challenge: Europe's Crisis in Historical Perspective. *Harvard International Review.* Vol. 16, No. 2, pp. 26–69.

Hurwitz, A. 1999. The 1990 Dublin Convention: A Comprehensive Assessment. *International Journal of Refugee Law.* Vol. 11, No. 4, pp. 646–677.

Huysmans, J. 2006. *The Politics of Insecurity, Fear, Migration and Asylum in the EU.* Abingdon, Oxon: Routledge.

Huysmans, J. 1995. Migrants as a Security Problem: Dangers of "Securitizing" Societal Issues. In Miles R. and Thränhardt D. (eds.) *Migration and European Integration: Dynamics of Inclusion and Exclusion*. London: Pinter.

Kanstroom, D. 2012. *Aftermath: Deportation Law and the New American Diaspora*. Oxford: Oxford University Press.

Kenk, V.S., Križaj, J., Štruc, J. and Dobrišek S. 2013. Smart Surveillance Technologies in Border Control. *European Journal of Law and Technology*, Vol. 4, No. 2. Available at http://ejlt.org/article/view/230/378.

Khursheed, W. and Lazaridis G. (eds.) 2015. *The Securitisation of Migration in the EU: Debates since 9/11*. Palgrave Macmillan: London.

Kimball, A. 2007. The Transit State: A Comparative Analysis of Mexican and Moroccan Immigration Policies. *CCIS Working Paper 150*. University of San Diego

Kneebone, S. 2008. The Legal and Ethical Implications of Extraterritorial Processing of Asylum Seekers: The 'Safe Third Country' Concept. In McAdam, J. (ed.) *Forced Migration, Human Rights and Security*. Oxford and Portland: Hart Publishing, pp.129–154.

Laffont, J.J. 2008. Externalities. In Durlauf, S.N. and Blume, L.E. (eds.). *The New Palgrave Dictionary of Economics*, second edition, London: Palgrave.

Lahav, G. 2010. Immigration Policy as Counterterrorism: The Effects of Security on Migration and Border Control in the European Union. In Crenshaw, M. (ed.) *The Consequences of Counterterrorism*. New York: Russell Sage Foundation.

Lavenex, S. 2006. Shifting Up and Out: The Foreign Policy of European Immigration Control. *West European Politics*. Vol. 29, No. 2, pp. 329–350.

Lavenex, S. and Uçarer, E. (eds). 2002. *Migration and the Externalities of European Integration*. Lanham, MA: Lexington Books.

Lavenex, S. and Uçarer, E. 2004. The External Dimension of Europeanization: The Case of Immigration Control. *Cooperation and Conflict*. Vol. 39, No. 4, pp. 417–443.

Lavenex, S. and Wichmann, N. 2009. The External Governance of EU Internal Security. *Journal of European Integration*, Vol. 31, No.1., pp. 83–102.

Loescher, G. 1993. *Beyond Charity: International Cooperation and the Global Refugee Crisis*. New York: Oxford University Press.

Lutterbeck, D. 2006. Policing Migration in the Mediterranean. *Mediterranean Politics* Vol. 11, No. 1, pp. 58–82.

Makaremi, C. 2009. Governing Borders in France: From Extra-Territorial to Humanitarian Confinement. *Canadian Journal of Law and Society*. Vol. 24, No. 3, pp. 411–432.

Magner, T. 2004. A Less Than 'Pacific' Solution for Asylum Seekers in Australia. *International Journal of Refugee Law*. Vol. 16, No. 1, pp. 53–90.

Maguire, M. 2015. Migrants in the Realm of Experts: The Migration-Crime-Terrorist Nexus After 9-11. In Wadia, K. and Lazaridis, G. (eds.) *The Securitisation of Migration in the EU: Debates since 9/11*. London: Palgrave Macmillan.

Marin, L. 2011. Is Europe Turning Into a 'Technological Fortress'? Innovation and Technology for the Management of EU's External Borders. Reflections on FRONTEX and EUROSUR. In Heldeweg, M. A. and Kica, E. (eds.) *Regulating Technological Innovation: Legal and Economic Regulation of Technological Innovation*. London: Palgrave MacMillan, pp. 131–151.

Marquis, C. 2004.'France Seeks U.N. Force in Haiti', *New York Times*, 26 February.

McNamara, F. 2013. Member State Responsibility for Migration Control Within Third States: Externalisation Revisited. *European Journal of Migration and Law*. Vol. 15, No. 3, pp. 319–335.

Miller, T. 2015. Border Patrol International "The American Homeland Is the Planet". Blog Entry. Available at www.tomdispatch.com/blog/175774/tomgram%3A_todd_miller,_the_border-industrial_complex_goes_abroad.

Mountz, A. 2010. *Seeking Asylum: Human Smuggling and Bureaucracy at the Border*. Minneapolis: University of Minnesota Press.

Noll, G. 2003. Visions of the Exceptional: Legal and Theoretical Issues Raised by Transit Processing Centres and Protection Zones. *European Journal of Migration and Law*. Vol. 5, No. 3, pp. 303–41.

Ryan, B. and Valsamis M. 2010. *Extraterritorial Immigration Control: Legal Challenges*. Leiden; Boston: Martinus Nijhoff Publishers.

Salter, M. 2004. Passports, Mobility, and Security: How Smart Can the Border Be? *International Studies Perspectives*. Vol. 5, pp. 71–91.

Solimano, A. 2010. *International Migration in the Age of Crisis and Globalization: Historical and Recent Experiences*. Cambridge: Cambridge University Press.

Taylor Nicholson, E. 2011. Cutting Off the Flow: Extraterritorial Controls to Prevent Migration. Brief presented at conference 'Cutting Off the Flow: Extraterritorial Controls to Prevent Migration', Berkeley Law, April 22.

Tondini, M. 2012. The Legality of Intercepting Boat People Under Search and Rescue and Border Control Operations with Reference to Recent Italian Interventions in the Mediterranean Sea and the ECHR Decision in the Hirsi Case. *The Journal of International Maritime Law*. Vol. 18, pp. 59–74.

Torpey, J. 2000. States and the Regulation of Migration in the Twentieth-Century North Atlantic World. In Andreas, P. and Snyder, T. (eds.) *The Wall Around the West: State Borders and Immigration Controls in North America and Europe*. Lanham, MD: Rowman and Littlefield.

Turton, D. 2002. Forced Displacement and the Nation State. In Robinson, J. (ed.) *Development and Displacement*. Oxford: Oxford University Press.

Weiner, M. 1993. *International migration and security*. Boulder, CA: Westview Press

Weiner, M. 1995. *The Global Migration Crisis*. New York: Harper Collins

Yale-Loer, S., Papademetriou, D. and Cooper, B. *Secure Borders, Open Doors: Visa Procedures in the Post-September 11 Era*. Washington, DC: Migration Policy Institute. Available at www.migrationpolicy.org/pubs/visa_report.pdf.

Zaiotti, R. 2014. International Organizations, Transatlantic Cooperation and the 'Globalization' of Homeland Security, *Studia Diplomatica*. Vol. 67, No. 3, pp. 53–67

Zolberg, A. 1999. Matters of State: Theorizing Immigration Policy. In Hirschman, C. and Kasinitz, P. (eds.) *The Handbook of International Migration: The American Experience*. New York: Russell Sage.

Zolberg, A. 2003. The Archaeology of 'Remote Control', in A. Fahrmeir, O. Faron, and P. Weil (eds.) *Migration Control in the North Atlantic World: The Evolution of State Practices in Europe and the United States from the French Revolution to the Inter-War Period*. New York: Berghan Books.

2 Rising tide

Analyzing the relationship between externalization and migrant deaths and boat losses

Keegan Williams and Alison Mountz

Recent increases in deaths of migrants crossing at sea have reached historical highs among those trying to land on the sovereign territory of Western states. Among the southern member states of the European Union (EU), for example, the reported number of losses reached a record 3,072 between January and August 2014 (IOM 2014). Most recognize that this is likely a gross underestimate. Increases in deaths have also been accompanied by significant increases in resources dedicated to enforcement operations in the annual budgets of enforcement activities, although little existing scholarship tracks this relationship. In this chapter, we address this gap using new empirical evidence and discuss our findings through the lens of externalization.

Our title, 'rising tide,' references the increase in enforcement associated with more dangerous journeys at sea, investments that have become so significant in 2015 as to involve what some are calling the militarization of the Mediterranean. The title also references a popular saying: "A rising tide lifts all boats." Yet this play on words ignores the danger that rising tides pose to ships, especially the smallest and weakest, which applies both literally and figuratively to migrants and their lack of agency in these journeys at sea.

We first offer a brief history of externalization in order to provide context to contemporary migration and enforcement activities concentrated around the EU, Australia, and the United States (US). Our history traces the roots of externalization back to US enforcement in the Caribbean. We then describe how we built a global database on migrant boat losses and state operations at sea since 1980. Using statistical analysis, we find evidence to support the long-asserted claim by social scientists that greater enforcement at sea intensifies precarity and risk-taking among migrants attempting unauthorized entry (Hiemstra 2012; Koser 2000; Nadig 2002). We also inconclusively analyze how variations in enforcement in specific areas relate to future losses in other areas. Finally, we consider the meaning of findings with respect to the externalization literature.

Externalization is part of the securitization of migration (Bigo 2000). While many scholars discuss securitization as the process wherein migration is securitized and migrants criminalized in discourse (e.g. Ibrahim 2005), securitization also involves material forms of exclusion that accompany this shift in discourse. Externalization is one example. By scripting migrants and would-be asylum seekers as criminal and security threats, the rationale is set forth discursively for their distancing through exclusionary measures or bureaucratic management offshore.

The modern period of externalization can be traced back to US interceptions of Haitian and Cuban migrants in the Caribbean in the early 1980s (Mountz and Loyd 2014). The late 1980s and early 1990s saw the development and subsequent thickening of immigration control networks offshore. The 2000s showed the diffusion of externalization as an effective and widely accepted practice to shut down routes to asylum. Within this period, post-9/11 brought the movement of externalization into the public geographic imagination, building on fear and premised on national security imperatives. US detention of 'foreign enemy combatants' captured during its war on terror catapulted Guantánamo Bay into public discourse as an iconic space of offshore enforcement where domestic and international law could effectively be evaded at the cost of human rights. The most recent period of externalization can be characterized by freneticism around these impulses to move ever farther offshore and deeper into zones of origin and transit.

Most existing literature on externalization, which we review in the next section, originates in the EU and addresses the EU context. We believe that this relates to the very visible efforts by the European Union to open internal borders while fortifying the external border as part of its process of regionalization. We also speculate that this abundance of literature relates to the robust participation of European scholars in the political science of migration and border enforcement. This bias is apparent in our review of the literature below.

Our chapter proceeds as follows. We begin by providing background and exploring existing literature on externalization (emphasizing the European context), including measures taken by states with respect to operations and associated losses at sea and previous work that has endeavored to track migrant boats and related operations. We then discuss the methodology that we designed to study enforcement and losses on the Mediterranean over the last several years. In our penultimate section, we present our findings: descriptive analysis and comparison of data on operations and losses. Finally, our concluding section offers insights gleaned from this analysis, a summary of key findings, and remaining questions.

Context: a brief history of externalization

While scholars have endeavored to define the notion of securitization of migration (Huysmans 2000, 2006; Ibrahim 2005), fewer have offered much in the way of defining externalization. This is likely because externalization of border enforcement takes many forms: interception, offshore networks of civil servants, increased transit visa requirements, development of detention facilities in transit regions and airport waiting zones (Makaremi 2009), bilateral arrangements for policing and repatriation, and so on. Additionally, externalization is often not an explicitly stated policy, although many proposals have been made over the years to externalize through different measures, particularly in the EU context (Schuster 2005). In fact, many mechanisms of externalization evolved over time from informal practice to more formal policy. But they were often housed quietly in the pages of bilateral agreements between states which were written behind closed doors and not made available to the public (see Zaiotti, this volume). When so much happens in the informal sphere or closed knowledge circuits, important questions about methodologies and ontologies of exclusion arise. How do we know what we do about what Aristide Zolberg (1999) called "remote control"? These are questions that we grappled with in this research project.

In order to begin to answer these questions, a brief sketch of the history of the term externalization in existing literature proves useful. The term first emerged in writing by political scientists working primarily in European context and on the case of externalization undertaken by EU member states (e.g. Boswell 2003; Brochmann 2002; Lavenex 1999; Noll 2003; Pastore 2001).

Whereas the use of the term externalization can be traced to the late 1990s when it came into being in association with EU border enforcement and asylum processing, earlier enforcement and security practices show earlier seeds planted – if not labeled externalization. The use of the term "external security", for example, preceded the term "externalization" (Bigo 2000; Grabbe 2000). Bigo (2000) tracked the convergence of internal and external measures of securitization, while also looking at the successful rhetoric of securitization used by politicians to stir fear and unease.

Shared understanding and usage of the term externalization arose in the EU context. Schuster (2005), for example, maps early plans set forth by EU member states to carry out offshore detention and asylum processing in North African states. Others study the geographical expansion of the EU through externalization (Betts 2004; Boswell 2003; Salter 2004; Walters 2004).

Some of the early US interception and detention of migrants in the Caribbean (Koh 1994; Noble 2011) can be understood as early practices of externalization before the term came into wider usage. Writing in the North American context emerged later, and dwelled less on the term externalization. Coleman (2005, 2007) and Mountz (2006) both explored how the United States and Canada, respectively, advanced offshore border enforcement. Whereas Coleman looked at the role of US policing programs in Mexico, Mountz examined Canadian interception at sea and informal enforcement of foreign airports. Kernerman (2008) examined Canada's use of visas to restrict access to sovereign territory and the ability to seek asylum among European Roma trying to reach Canada.

Australia was also seen as a leader in pushing its borders offshore. Scholars there have traced earlier histories of offshore border enforcement (e.g. Mares 2002; Hugo 2001). Others set about comparing Australia and the European Union (e.g. Hyndman and Mountz 2008).

Externalization intersected with the literature on the securitization of migration (Bigo 2002; Huysmans 2000, 2006). While the term could retroactively be applied to the first, large preventative offshore enforcement measures undertaken in the 1980s, the term really came into more common usage in the late 1990s in reference to EU measures to shore up its external borders while erasing internal borders between member states that were part of the Schengen Acquis. These changes to border enforcement in the EU coincided with broader anti-asylum seeker public opinions and discourses that intensified globally throughout the 1990s (Squire 2009). As asylum seekers were demonized, associated with criminality and security risks, their credibility undermined, paths to asylum shrank in the 1990s and again in the 2000s.

With some exceptions, much of the literature on externalization has been authored by political scientists. To geographers who conduct spatial analysis, externalization entails three geographical moves: movement of border enforcement offshore, movement of processing and detention to transit routes and regions, and creation of island spaces to carry out this processing, whether on actual islands or the creation of the island in forms such as refugee camps or detention centers en route (Moutz 2011). Essentially the border is dislodged as a way to thwart progress of those seeking asylum toward sovereign territory where they accrue rights (Durieux and MacAdam 2004).

As noted by Zaiotti (this volume), externalization relies to varying degrees on the contracting out of these functions and responsibilities to other places and people, whether foreign authorities or third party contractors who run processing and detention in these sites. Externalization is generally premised on the evasion of signatory responsibilities, with a belief that offshore enforcement will slow arrivals and reduce processing. Often, states require private entities or third parties to carry out these functions; they are able to step in at the limits of what states can legally do in their enforcement operations without violating international refugee law and, specifically, the principle of *non-refoulement* (Ashutosh and Mountz 2011).

There exists little work tracking migrant losses at sea and enforcement operations. This relates to the hidden nature of these operations and to the racialized dehumanization and devaluation of migrant lives and losses of life. There have, however, been important, recent efforts to document losses in different regional contexts, including Central American and Mexican journeys to the United States, losses at sea among people from various countries of origin trying to reach Australia, and losses in the Mediterranean Sea among those headed for the EU. Researchers and activists working with No More Deaths have documented losses along the Mexican border, often by working with death certificates available at county morgues in the border regions. More informal efforts to memorialize deaths have also been happening throughout Mexico and in the northern border region. Weber and Pickering (2011) estimated "border deaths" while also addressing the politiciztion, shortcomings and impossibility of this accounting. This work of documentation continues on through their Border Crossing Observatory based at Monash University

(Border Crossing Observatory 2013). Another website based in Australia tracks losses and is named in honor of a notorious ship that sank en route to Christmas Island, off the coast of Java, Indonesia called SIEV X (Suspected Illegal Entry Vehicle) (Hutton 2015). In the European context, Thomas Spijkerboer is running the largest, most contemporary project that we know of to study migrant losses (Spijkerboer *et al.*, 2015). His research team has contacted morgues in several hundred municipalities in southern EU border regions to assemble data. UNITED for Intercultural Action also maintains and updates a list of documented refugee deaths throughout the European Union since 1993, which it claimed stood at 18,759 people as of 12 September 2013 (UNITED 2014).

Methodology

We employed a mixed methodology designed to locate, record, and analyze migrant boat losses and maritime interdiction operations to fill the gaps identified in the externalization literature. This involved collecting data from media aggregators and additional sources using content analysis for descriptive and time-series statistical analysis. Although we placed no restrictions on the time or geography for losses and operations, the use of online databases tended to generate a recency bias focused around the European Union, Australia, and the United States. As we will show, these biases were partially reflective of the characteristics of the phenomena. Qualitative steps of the approach were actively recorded in analytical notes during the life of the project for review and later reproduction. This subsection describes these steps in addition to the quantitative methods employed.

The primary purpose of the project was to create data sets on migrant boat losses and border-related operations at sea. We first sought to explore and determine what methods and data were available. Previous scholarship had approached these subjects through an eclectic set of methodologies, with existing quantitative methods focused on state, NGO, and, where possible, media reports. These projects were typically localized in their time periods or geographies and did not make use of statistical analysis to describe or explain findings. Most of these works, additionally, studied marine migration in general and did not focus on losses. A secondary purpose of the project, therefore, was to give descriptive statistics of boat losses and related operations on a global scale and use time-series modelling to measure variations between enforcement and boat losses over time. This would provide novel empirical evidence related to the frequent claim in migration studies that tougher enforcement in the name of deterrence is ineffective in reducing migrant flows (Barnes 2004, 50; Fan 2008; Carling and Hernandez Carretero 2011, 56). We defined a "boat loss" as any incident involving the sinking or destruction of a boat facilitating unauthorized entry of migrants into a state's jurisdiction. We defined an "operation at sea" as any state-led mission intended to prevent entry of undocumented migrants through at-sea interdiction (Palmer 1997, 1565).

Previous work demonstrated that accounts of boat losses and operations might be found in media aggregators or other online sources. We set out to make use of targeted queries on two large media databases, LexisNexis and Factiva, and search engines to locate reports on these phenomena. The plan was to qualitatively determine unique reports of boat losses or operations using content analysis and store these files in rich-text (.rtf) and portable document (.pdf) formats for recording in data sets. We were interested in a number of variables, further described in the next subsection, including ship name, number of deaths, number of survivors, flag country of the ship, country of origin of those onboard, desired destination, location of the loss, date of the loss, and related enforcement activity (modelled from Weber and Pickering 2011). These variables would be stored using different data types, such as numeric or string, to record information. For example, instead of categorizing enforcement activity, we opted to write textual summaries; this would allow us to return to code data as necessary. The resulting data sets were saved in multiple formats (e.g. .csv; .xls) for analysis in Stata 13.

As the project continued, we began to identify better keywords or phrases, decided to code additional variables, and used multiple reports for each loss or operation. The term 'migrant boat incident,' for instance, was biased, since many reports did not actively call boat losses 'incidents'; searching 'migrant boat AND (sank OR lost),' instead, used verbs common in media reports and thus resulted in more accurate searches. In attempts to exhaust reports from media aggregators, we employed multiple search terms and compared the output to our current data sets. As we analyzed the documents, we decided to code additional variables not previously considered. For example, while many reports had estimates of passenger survivals or deaths, it became apparent that they also recorded the number of passengers missing from a loss. Some variables we had sought to record, like ship name, were nearly universally absent from available reports, but were included as missing observations where appropriate. We also found that most incidents featured more than one report, some of which recorded different but important details for the project. This was especially problematic for studying enforcement-related activities. It was also relevant for finding the most accurate details on some variables, such as location of the loss, particularly where reports were in a foreign language (e.g. Turkish). Owing to these factors, we sometimes used and stored multiple reports of the same boat loss or operation for the data sets to enhance the completeness and reliability of the data.

By the end of 2014, we had collected 191 media reports on 175 boat loss incidents and 110 reports on 61 maritime interdiction operations and stored them per best practices in data management. We also catalogued 30 photographs related to boat losses that appeared in these reports. Information on desired variables was recorded on a rolling basis, along with relevant metadata (e.g. date of record; case identification number; identification of the recorder). Final case and report counts were obtained after quality assurance of all data. With respect to boat losses, some cases were removed after careful review revealed they were duplicate entries. Criteria for cleaning operational data were more complex, since determining which cases fit the project's definition was difficult. In the end, we took a conservative route by only retaining cases that we were highly confident met our definition.

The statistical analysis was conducted in two steps using a priori assumptions. In the first stage, we described the counts, central tendencies, and measures of variation within the data. The second stage used time-series modelling to analyze correlations between enforcement and boat losses over time. Due to limitations in the operational data, as discussed later, we focused on the European Union from 2006 to 2013. In our models, we assumed statistical significance beyond a 95 percent confidence interval using Huber-White standard errors. Substantive significance was measured using adjusted R-squared values, model F-scores, root mean-squared errors, Bayesian information scores, and the size of coefficient estimates. Statistical analysis and additional figures were produced using Stata 13. Passenger country of origin and enforcement activity word frequencies and concordances were analyzed in TextSTAT 2.9.

Data

This subsection summarizes and briefly discusses the data collected from the project. Excluding metadata, we coded 13 variables in the losses at sea data set. Twelve variables were coded using information directly from media reports and one variable (Migrant Route of Loss) was constructed by clustering boats along major known migrant routes from the location data. Table 2.1 summarizes each variable with a count, a brief description, and, where applicable, a measure of central tendency.

Most variables were complete for the 175 boat loss cases, with "Origin of Ship," "Passenger Origins," "Destination of Ship," and "Enforcement Activity" being mostly complete, and "Name of Ship" and "Flag of Ship" mostly or completely missing. The boat loss cases in the data set ranged from 1990 to 2014 and were globally dispersed.

We recorded ten variables for the operations at sea data set. All variables were coded directly from reports on operations. Variable names, counts, and descriptions are presented in Table 2.2. Like losses at sea, most variables were complete for operations at sea, with three exceptions ("Stated Purpose," "Stated Outcomes," and "Budget"). "End Date" could only be recorded for 58 operations since at least three operations (Australia's *Sovereign Borders* and *Relex* and Frontex's *Triton*) were ongoing at the time of this book. Operations ranged from 1980 to 2014 and had a limited geography compared to losses at sea, with all cases occurring in Australia, the US, or the EU.

Table 2.1 Summary of variables in losses at sea data set

Variable	Count	Description	Mean/Mode
Date (mm/dd/year)	175	Records date of loss	2009 (Mean)
Location of loss	175	Records location of loss	N/A (String variable)
Migrant route of loss*	175	Assigns migrant route to location	Central Mediterranean (Mode)
Flag of ship	0	Records country flag of ship	N/A (Missing)
Name of ship	16	Records name of ship	N/A (String variable)
Estimated passengers	175	Records total passengers	109.6 (Mean)
Passengers survived	175	Records total surviving passengers	50.9 (Mean)
Passengers lost	175	Records total passengers lost	16.6 (Mean)
Passengers missing	175	Records total missing passengers	42 (Mean)
Origin of ship	174	Records country of origin of boat	Turkey (Mode)
Passenger origins	138	Records country of origin of passengers	Afghanistan (Mode)
Destination of ship	164	Records country of destination of boat	Italy; Australia (Mode)
Enforcement activity	154**	Describes enforcement activities related to loss	N/A (String variable)

Notes: * Constructed variable. ** Only includes reports where enforcement activities are listed.
Source: Various

Findings

Descriptive analysis

Most reported boat incidents were concentrated around the EU, Australia, and the US, respectively, between 2007 and 2014. Correspondingly, reported maritime interdiction operations predominantly occurred around these places and times. The average boat loss had 110 passengers on board, 54 percent among them missing or dead after the incident. Based on limited reports, the most common enforcement activity in response to a loss was to rescue some people aboard. As explored in the next subsection, there were statistically significant relationships between the level of enforcement, as measured in operational budgets and days active, and the levels of boat incidents and passengers involved.

Table 2.2 Summary of variables in operations at sea data set

Variable	Count	Description	Mean/mode
Operation name	61	Records name of the operation	N/A (String variable)
Start date	61	Records start date of operation	2008 (Mean)
End date	58	Records end date of operation	2008 (Mean)
Migrant route	61	Assigns migrant route based on geography of the operation	Western Mediterranean (Mode)
Participating states	61	Lists states involved in the operation	Spain (Mode)
Operation target	61	Records targets of operation from available reports	N/A (String variable)
Stated purpose	59	Records purpose of operation from available state reports	N/A (String variable)
Stated outcomes	9	Records outcomes of operation from available state reports	N/A (String variables)
Budget (2014 Euros)*	48	Records real budget of operation from state reports	4.61 million Euros (Mean)
Days active	48	Records number of days in which the operation was active	166 days (Mean)

Note: * Adjusts nominal budgets for inflation.

Source: Various

More than two-thirds of reported boat losses occurred between 2007 and 2014, with an especially high frequency (96) of losses reported from 2011 to 2014 (Figure 2.1). This growth in incidents was completely located in three transit routes: the Central Mediterranean (North Africa-Italy/Malta), Eastern Mediterranean (Turkey-Greece), and Australia (Indonesia-Australia). Including all time periods, the most common migrant routes involved in boat loss incidents were the Central Mediterranean Sea (52 incidents), Australia (44 incidents), Eastern Mediterranean Sea (34), Caribbean Sea (22), Western Mediterranean Sea (8), Canary Islands and Eastern Atlantic Ocean (7), Gulf of Aden (6), Eastern Pacific Ocean (2), and Western Indian Ocean (1). Adding these categories together revealed 96 percent of all reported incidents occurred around the European Union (58 percent of the total), Australia (25 percent), and the US (13 percent). The most frequent countries of origin for boats affected by a loss incident were Turkey (32), Libya (30), Indonesia (28), Haiti and Morocco (10 each), and Tunisia (7).

The losses at sea data contained an estimated 19,173 migrants on board the 175 boats. This meant that the average number of migrants per boat was 110,

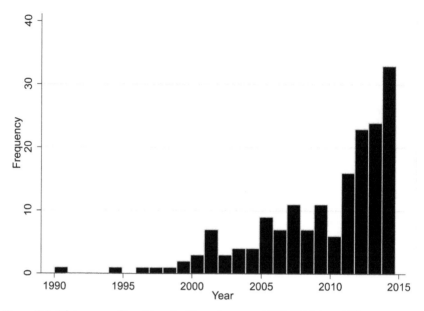

Figure 2.1 Histogram of boat loss incidents by year, 1990–2014 (n = 175)

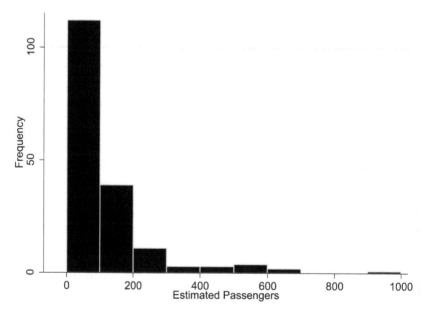

Figure 2.2 Histogram of estimated passengers onboard boats involved in incidents, 1990–2014 (n = 175)

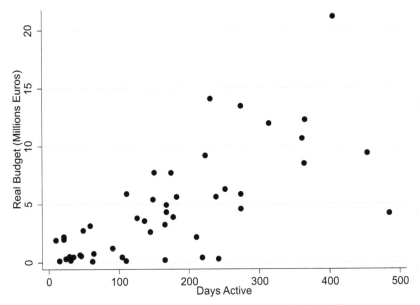

Figure 2.3 Distribution of operational budgets by days active (n = 48)

Source: data from Frontex (2014)

though there was a significant variation in this number. For example, the range of passengers varied from 4 to 920, and the highest frequencies of passenger estimates were between 0 and 200 persons (Figure 2.2). With respect to the fate of passengers, a minority (46 percent or 8,906) were reported to have survived, with a majority (54 percent or 10,267) reported missing or dead. Reports of persons missing were over 2.5 times more common than deaths, with a total of 7,355 passengers missing (likely lost at sea) and 2,902 confirmed dead. The average person, then, associated with a reported boat loss incident was missing or confirmed dead. These findings also varied by region, with the highest rates of passenger survival occurring in the Caribbean (51 percent), Australia (50 percent), and Central Mediterranean (47 percent). The highest rates of passengers confirmed dead occurred in the Gulf of Aden (42 percent), the Canary Islands and Atlantic (35 percent), and the Eastern Mediterranean (28 percent). Missing passenger rates were reported highest in the Gulf of Aden (48 percent), the Central Mediterranean (42 percent), and Australia (38 percent).

The most commonly reported country of origin for passengers involved with a lost ship were Afghanistan (27 cases), Somalia (17), Iraq (16), Syria (16), Palestine (14), Iran (12), Pakistan (11), and Haiti (10). Thirty-eight (or 76 percent) of 50 countries of origin reported were featured in five cases or less. By continent, Asia (126 cases) was the most frequent point of origin, followed by Africa (97), and Latin America and the Caribbean (22). A count of country of origin of passengers could be greater than the total number of cases, since it was possible for multiple countries of origin to be reported for each case. Boats involved in an incident were most commonly reported to be en route to Australia or Italy (40 cases each), Greece (30), the US (18), Spain (15), and Yemen (6).

The greatest heterogeneity found in the data came from enforcement activities. Authorities reportedly undertook search and rescue in no fewer than 82 (or 47 percent) of boat loss cases. Other commonly reported activities included boat interception (13 cases) and arrest of at least one person onboard (11). Coast guards were most likely to take part in an enforcement response (69 cases), followed by navies (15), border guards (14), and state police (13). A variety of reasons for losses were listed, including but not limited to capsize, fire, severe weather conditions, boat collision, explosion, violence onboard, and being fired upon by authorities.

The average operation in our data set started and ended in the same year, with 16 operations being reported as occurring over multiple years. Over 90 percent of recorded operations began between 2006 and 2013. The most common locations of operations by route were the Western and Central Mediterranean (15 cases each), Canary Islands and Atlantic (10), Eastern Mediterranean (8), Caribbean or other European Union (5 each), and Australia (3). Spain (44 cases) and Italy (42) were the most common of the 39 countries reported participating in operations, followed by Portugal (38), Germany (29), France and Romania (28 each), and a large number of European (including non-EU) states. The US participated in six operations and Australia in three. European operations and participation were more frequent, in part, because they operate annually and are reported more comprehensively. If Australian or American operations, for instance, were reported like their European counterparts, then Australia's number of operations would rise from 3 to 17 and the USA's from 5 to 10. Budget and days active were only available for EU operations. Adjusted for inflation, the average operation cost at least 4.5 million Euros during the 166 days of its activity (or about 27,268 Euros per day). As expected, there was a high and positive correlation ($r = 0.71$) between real budgets and days of operation, meaning that longer operations typically received larger budgets (Figure 2.3). Each additional day of an operation was associated with an additional 26,280 Euros in funding.

A comparison of enforcement levels and losses in Europe, 2006–2013

To test hypotheses and determine correlations between enforcement and boat losses, we compared and analyzed boat loss and operations data from the European Union between 2006 and 2013. We mapped the data to related operations in the same migration route during the same time period, resulting in 32 observations (four regions by eight time periods). The four regions analyzed were the Canary Islands and the Atlantic, Western Mediterranean Sea, Central Mediterranean Sea, and Eastern Mediterranean Sea. There were a total of 55

Table 2.3 Pearson's r scores between intensity of enforcement and migrant/boat
         losses (n = 32)

Correlated variables	Real budget	Days active
Boat loss incidents	0.24	0.47
Estimated passengers	0.65	0.67
Passengers survived	0.58	0.69
Passengers lost	0.33	0.42
Passengers missing	0.72	0.60
% passengers missing or lost	0.08	−0.11

boat loss incidents with 7,097 passengers in the EU during this time period and
44 maritime interdiction operations costing approximately 217 million Euros.

There were strong, positive correlations (Pearson's r) between the intensity
of operations, as measured by their real budgets and days active, and most
measures of migrant/boat losses (Table 2.3). These correlations demonstrated
that allocating more time and resources to an operation was associated with
more boat loss incidents as well as estimated passengers onboard. A notable
exception was the rate of passengers missing or lost, which increased slightly
with higher operational budgets, but decreased with number of days active.
These mixed results likely imply a weak relationship between rates of passengers missing or lost and operational intensity, particularly given the high
correlation between real budgets and days active.

We used ordinary least squares regression controlling for time to test two
hypotheses regarding the relationship between enforcement and migrant/
boat losses:

H1: Increased enforcement activity at sea decreases or has no relation-
    ship with migrant/boat losses.
H2: Increased enforcement activity in one region has no relationship
    with migrant/boat losses in other regions.

These hypotheses reflect claims made in migration studies that increased
enforcement leads to more losses and that increased enforcement does not
deter migrants but may lead them to take new routes. Table 2.4 displays the
results from the one-tailed tests of the relationship between increased enforcement activity and migrant/boat losses. Ten models were tested (five measures
of migrant/boat losses by two measures of operational intensity). In eight of the
ten models, a statistically significant relationship was found between increased
enforcement activities and migrant/boat losses. The relationship was strongest
(5/5 models significant) for days active: each 100 additional days an operation continued was approximately associated with one incident carrying 210
passengers. Of those 210 passengers, 100 would survive, 20 would die, and 90
would be missing. The relationship was weaker (3/5 models significant) for

Table 2.4 Results from regressions between enforcement and migrant/boat losses (n = 32)

Dependent variable	Coefficients (Adjusted R-Squared)	
	Real budget (millions Euros)	Days active
Boat loss incidents	0.09 (0)	0.01*** (0.17)
Estimated passengers	54.3*** (0.40)	2.1*** (0.42)
Passengers survived	23.6*** (0.32)	1.0*** (0.44)
Passengers lost	4.2* (0.07)	0.2** (0.12)
Passengers missing	26.5*** (0.48)	0.9*** (0.35)

Statistical significance beyond * 90% confidence interval; ** 95% confidence interval; *** 99% confidence interval.

operational budget: each 11.11 million additional Euros in budget was approximately associated with one incident carrying 603 passengers, 262 of whom would survive, 47 die, and 294 be missing. In general, models using operation days active had stronger measures of substantive significance.

Based on these results, we rejected the null hypothesis that increased enforcement activity at sea decreases or has no relationship with migrant/boat losses.

To test the hypothesis that increased enforcement activity in one region has no relationship with losses in other regions, each area had to be considered separately. This reduced the number of observations for each test to eight, then further reduced to seven by using lagged independent variables, weakening their statistical significance. The models were generated using the number of boat incidents and estimated passengers in each region as a function of the previous year's level of enforcement, by days of activity, in each other region. While this approach limited our ability to make claims regarding statistical significance, it was possible to look at each model's coefficients to gauge the substantive relationships between variables (Table 2.5). The substantive results were largely mixed, with particular pairs of migration routes affected differently by variations in level of enforcement. Increased enforcement in the previous year elsewhere was generally associated with fewer boat incidents and passengers for the Canary Islands and Atlantic route. In contrast, increased enforcement on all other routes was generally associated with higher numbers of boat incidents and passengers in the Central Mediterranean. Higher enforcement in the previous year in the Canary Islands and Atlantic or Eastern Mediterranean was associated with more boat incidents and passengers in the Western Mediterranean. Losses were higher in the Eastern Mediterranean when enforcement was increased in the Central or Western Mediterranean. Although these findings suggest that increased enforcement in specific areas may alter future losses elsewhere, it was not possible to reject the null hypothesis due to lack of data.

Table 2.5 Changes in expected future migrant/boat losses by increased enforcement elsewhere

Increase in previous year's enforcement	Current year's migrant/boat losses			
	Canary Islands and Atlantic	Central Mediterranean	Eastern Mediterranean	Western Mediterranean
Canary Islands and Atlantic	N/A	+	–	+
Central Mediterranean	–	N/A	+	–
Eastern Mediterranean	–	+	N/A	+
Western Mediterranean	–	+	+	N/A

Notes: + increased migrant/boat losses; – decreased migrant/boat losses.

Discussion and conclusions

Our findings showed that reported migrant/boat losses were concentrated since 2007 near the EU, Australia, and the US. Additionally, these areas were the regions of all reported operations. The findings also established that most migrants involved with incidents were from Asia and Africa, and determined estimates for their numbers as well as survival, death, and missing rates per boat. The most common enforcement activities and participating authorities related to losses were identified. Although a large number of reports lacked substantial detail on these activities, they demonstrated that a wide range of actors and responses took part. The findings, furthermore, identified a variety of commonly reported causes of migrant/boat losses. With respect to operations, we were able to determine their times, geographies, and participating actors. In the EU context, we were also able to establish operational budgets.

We found a strong and positive relationship between enforcement and migrant/boat losses during times and place where data were available. This relationship was relatively robust for different measures of operational intensity and losses. While we cannot prove causality in this case, we propose a limited number of explanations for these results:

1. There is no causal or correlational relationship between enforcement and losses, and the European case, 2006–2013, was an outlier.
2. There is no causal relationship between enforcement and losses. There is a positive correlation, but it is caused by a confounding or intervening factor.
3. There is a causal relationship between enforcement and losses.

Based on current limitations in the data, we cannot rule out the possibility that the European case was a statistical outlier to the global case. It is unlikely, however, that the results are an outlier for the EU case because they cover its entire geography during a significant time period. Assuming that they are not outliers, we suspect a limited number of potential confounding factors accounting for the relationship. One possibility is that a substantial recency bias in reporting the data collected exists, one that happens to correlate highly with increasing intensity of enforcement activities. While not feasible to fully gauge the impact of recency bias on our data, numerous reports document migrants taking riskier journeys as enforcement becomes tougher (e.g. Bialasiewicz 2012; Carling 2007). This means that increased incident reports might also be accounted for by increased enforcement. The incorporation of time into our regression models also usually found it to not have a statistically significant effect on migrant or boat losses, supporting the limited impact of recency bias as a consistent effect.

Another possibility is that, while there is a causal relationship between enforcement and boat losses, increased enforcement leads to higher levels of search and rescue. In this case, enforcement itself does not cause boat losses, but, intervened by search and rescue, follows it. We indirectly tested this explanation through the hypothesis that higher levels of enforcement activity decrease rates of passengers lost or missing in reported incidents; no statistically significant relationship was found. Although this finding could also be explained by the ineffectiveness of search and rescue, we propose that the evidence, instead, suggests that higher enforcement activity did not correspond to statistically significant increases in search and rescue.

Given these conclusions, the results support the explanation of a causal relationship between enforcement and losses, as suggested by the externalization literature. If more detailed data were available, it would be possible to check the direction of this causality using Granger causality tests or instrumental variable models.

A number of limitations limited our ability to collect data or conduct analysis. As noted, the use of electronic media restricted access to reports not made available online. This likely excluded some migrant/boat losses and operations preceding 2000. The structure of reporting typically excluded many important details, particularly regarding enforcement activities or specific details of boats. We were limited, therefore, in making claims about state activities related to loss incidents since what they did, in many cases, remained unclear. Inconsistencies between reports about the same incident required additional work and decision-making about which sources were most detailed or reliable, and how such reports could be put together. For example, more recent reports featured updated survival and deaths numbers, which we chose to use over earlier estimates. Reports on operations were substantially more limited than those on losses. As we found, the use of public, electronic media may be restricted in details on boat operations due to their politically sensitive nature. These sources can be supplemented using additional methods, like freedom of information requests. The EU, as it turned out, had an unusual wealth of information on its operations, which greatly helped us in analysis. One extension to our empirical basis for tracking the relationship between enforcement and losses would be to expand the numbers and descriptions of reports using additional methods. This would allow the use of additional tools of analysis over more time periods or geographies.

Overall, our analysis fills an important gap in the literature on externalization where a correlation between marine enforcement activity and migrant losses at sea tends to be asserted rather than explored empirically. As we have shown, however, our analysis at once affirms the correlation between enforcement and loss that is so often asserted by social scientists, while also opening up new questions ripe for further exploration with these and additional data, as they become available.

Bibliography

Ashutosh, I. and Mountz, A. 2011. Managing Migration for the Benefit of Whom? Interrogating the Work of the International Organization for Migration. *Citizenship Studies*, Vol. 15, No. 1, pp. 21–38.

Balibar, E. 2002. Three Concepts of Politics: Emancipation, Transformation, Civility. In Balibar, E. *Politics and the Other Scene*, Verso: London, pp. 1–39.

Barnes, R. 2004. Refugee Law at Sea. *International and Comparative Law Quarterly*, Vol. 53, No. 1, pp. 47–77.

Betts, A. 2004. The International Relations of the 'New' Extraterritorial Approaches to Refugee Protection: Explaining the Policy Initiatives of the UK Government and UNHCR. *Refuge*. Vol. 22, No. 1, pp. 58–70.

Bialasiewicz, L. 2012. Off-shoring and Out-sourcing the Border of Europe: Libya and EU Border Work in the Mediterranean. *Geopolitics*.Vol. 17, No. 4, pp. 843–866.

Bigo, D. 2000. When Two Become One: Internal and External Securitizations in Europe. In Kelstrup, M. and Williams, M. (eds.) *International Relations Theory and the Politics of European Integration*. Routledge: New York, pp. 171–204.

Bigo, D. 2002. Security and Immigration: Toward a Critique of the Governmentality of Unease. *Alternatives*. Vol. 27, No. 1, pp. 63–92.

Border Crossing Observatory. 2013. Australian Border Deaths Database. Available from: http://artsonline.monash.edu.au/thebordercrossingobservatory/.

Boswell, C. 2003. The 'External Dimension' of EU Immigration and Asylum Policy. *International Affairs*. Vol. 79. No. 3, pp. 619–638.

Brochmann, G. 2002. Citizenship and Inclusion in European Welfare States: The EU Dimension. In Lavenex, S. and Ucarer, E. (eds.) *Migration and the Externalities of European Integration*, Lexington Books: Oxford.

Carling, J. 2007. Control and Migrant Fatalities at the Spanish–African Borders. *International Migration Review*. Vol. 42, pp. 316–343.

Carling, J. and Hernandez-Carretero, M. 2011. Protecting Europe and Protecting Migrant? Strategies for Managing Unauthorised Migration from Africa. *The British Journal of Politics and International Relations*, Vol. 13, pp. 42–58.

Coleman, M. 2005. US Statecraft and the US–Mexico Border as Security/Economy Nexus. *Political Geography*. Vol. 24, No. 2, pp. 185–209.

Coleman, M. 2007. Immigration Geopolitics beyond the US–Mexico Border. *Geopolitics.* Vol. 39, No. 1, pp. 54–76.

Durieux, J. and MacAdam, J. 2004. Non-Refoulement through Time: The Case for a Derogation Clause to the Refugee Convention in Mass Influx Emergencies. *International Journal of Refugee Law.* Vol. 16, No. 1, pp. 4–24.

European Agency for the Management of Operational Cooperation at the External Borders of the Member States of the European Union (Frontex) 2014. Archive of Operations. Available from: http://frontex.europa.eu/operations/archive-of-operations/.

Fan, M. 2008. When Deterrence and Death Mitigation Fall Short: Fantasy and Fetishes as Gap-Fillers in Border Regulation. *Law & Society Review.* Vol. 42, No. 4, pp. 701–734.

Grabbe, H. 2000. The Sharp Edges of Europe: Extending Schengen Eastwards. *International Affairs.* Vol. 76, No. 3, pp. 519–536.

Guiraudon, V. and Lahav, G. 2000. A Reappraisal of the State Sovereignty Debate: The Case of Migration Control. *Comparative Political Studies.* Vol. 33, No. 2, pp. 163–195.

Hiemstra, N. 2012. *The View from Ecuador: Security, Insecurity, and Chaotic Geographies of U.S. Migrant Detention and Deportation.* Doctoral dissertation, Syracuse University.

Hugo, G. 2001. From Compassion to Compliance? Trends in Refugee and Humanitarian Migration in Australia. *Geoforum.* Vol. 55, pp. 27–37.

Hutton, M. 2015. SIEVX.com. Available from: http://sievx.com/.

Huysmans, J. 2000. The European Union and the Securitization of Migration. *Journal of Common Market Studies,* Vol. 38, No. 5, pp. 751–77.

Huysman, J. 2006. *The Politics of Insecurity: Fear, Migration and Asylum in the EU.* Routledge: New York.

Hyndman, J. and Mountz, A. 2008. Another Brick in the Wall? 'Neo-refoulement' and the Externalisation of Asylum in Australia and Europe. *Government and Opposition.* Vol. 43, No. 2, pp. 249–269.

Ibrahim, M. 2005. The Securitization of Migration: A Racial Discourse. *International Migration.* Vol. 43, No. 5, pp. 163–187.

International Organization for Migration (IOM). 2014. *Fatal Journeys: Tracking Lives Lost during Migration.* Available from: www.iom.int/files/live/sites/iom/files/pbn/docs/ Fatal-Journeys-Tracking-Lives-Lost-during-Migration-2014.pdf.

Kernerman, G. 2008. Refugee Interdiction before Heaven's Gate. *Government and Opposition.* Vol. 43, No. 2, pp. 230–248.

Koh, H. 1994. America's Offshore Refugee Camps. *University of Richmond Law Review.* Vol. 29, pp. 139–173.

Koser, K. 2000. Asylum Policies, Trafficking and Vulnerability. *International Migration,* Vol. 38, No. 3, pp. 91–111.

Lavenex, S. 1999. *Safe Third Countries: Extending the EU Asylum and Immigration Policies to Central and Eastern Europe.* Central European University Press: Budapest.

Makaremi, C. 2009. Governing Borders in France: From Extraterritorial to Humanitarian Confinement. *Canadian Journal of Law and Society.* Vol. 24, No. 3, pp. 411–432.

Mares, P. 2002. *Borderline.* UNSW Press: Sydney.

Mountz, A. 2006 Human Smuggling and the Canadian State. *Canadian Foreign Policy*. Vol. 13, No 1, pp. 59–80.

Mountz, A. 2011. The enforcement archipelago: Detention, haunting, and asylum on islands. *Political Geography*. Vol. 30(3), pp. 118–128.

Mountz, A. and Loyd, J. 2014. Transnational Productions of Remoteness: Building Onshore and Offshore Carceral Regimes Across Borders. *Geographica and Helvetica*. Vol. 69, pp. 389–398.

Nadig, A. 2002. Human Smuggling, National Security, and Refugee Protection. *Journal of Refugee Studies*. Vol. 15, No. 1, pp. 1–25.

Noble, D. 2011. *The U.S. Coast Guard's War on Human Smuggling*. University Press of Florida: Gainesville.

Noll, G. 2003. Visions of the Exceptional: Legal and Theoretical Issues Raised by Transit Processing Centres and Protection Zones. *European Journal of Migration and Law*. Vol. 5, No. 3, pp. 303–341.

Palmer, G. 1997. Guarding the Coast: Alien Migrant Interdiction Operations at Sea. *Connecticut Law Review*. Vol. 29, pp. 1565–1585.

Pastore, F. 2001. *Reconciling the Prince's 'Two Arms.' Internal-External Security Policy Coordination*. Paper for the Institute for Security Studies – Western European Union.

Salter, M. 2004. Passports, Mobility, and Security: How Smart Can the Border Be? *International Studies Perspectives*. Vol. 5, No. 1, pp. 71–91.

Schuster, L. 2005. *The Realities of a New Asylum Paradigm*. COMPAS Working Paper 20, University of Oxford Policy. Available from: www.compas.ox.ac.uk/publications/Working%20papers/Liza%20Schuster%20wp0520.pdf.

Squire, V. 2009. *The Exclusionary Politics of Asylum: Migration, Minorities and Citizenship*. Palgrave Macmillan: Chippenham.

UNITED for Intercultural Action (UNITED). 2014. *List of 18759 Documented Refugee Deaths Through Fortress Europe*. Available from: www.unitedagainstracism.org/wp-content/uploads/2014/10/List-of-Deaths-091213.xls%E2%80%9D.pdf.

Walters, W. 2004. Secure Borders, Safe Haven, Domopolitics. *Citizenship Studies*. Vol. 8, No. 3, pp. 237–260.

Weber, L. and Pickering, S. 2011. *Globalization and Borders: Deaths at the Global Frontier*. Palgrave Macmillan: London.

Zolberg, A. 1999. Matters of State: Theorizing Immigration Policy. In Hirschan, C., Kasinitz, P., and DeWind, J. (eds.) *The Handbook of International Migration: The American Experience*, Russell Sage: New York, pp. 71–93.

Part II

Externalizing migration management in Europe

3 Frontex and the shifting approaches to boat migration in the European Union

A legal analysis

Maarten den Heijer

Introduction

On 25 November 2014, Pope Francis addressed the European Parliament. Arguably more provocative than his call to not "allow the Mediterranean to become a vast cemetery," was his depiction of Europe having become "a grandmother, no longer fertile and vibrant."[1] The Pope was not referring to the European Union's enlargement fatigue, but to the perception that "the great ideas which once inspired Europe seem to have lost their attraction, only to be replaced by the bureaucratic technicalities of its institutions." Specifically in the context of immigration, the Pope lamented that courageous policies ensuring mutual support within the European Union (EU) and assisting countries of origin were missing, and that the EU was focusing instead on policies motivated by self-interest, which address only the effects and not the root causes of migration.

When it comes to the EU's policies on the plight of migrants undertaking the journey across the Mediterranean, it cannot be said that vision or inspiration is totally absent. It is obvious that the growing number of migrants attempting to enter the EU by sea in the last two decades poses all sorts of challenges in the sphere of controlling the external border, protecting refugees, combatting human smuggling and saving lives at sea. In all key policy documents, the three EU legislative institutions – the European Parliament, the European Commission and the Council of Ministers – stress that EU action should meet the goals of preserving life at sea, global responsibility-sharing, combatting human smuggling and trafficking, and respecting refugee rights.[2] The issue seems not to be that the requisite aims and values are lacking, but that they are so difficult to translate into one coherent whole. In its attempts to formulate concrete proposals for action or legislation, the EU tends to become mired in the potentially contradictory nature of the multitude of policy aims, but also in the controversy of how to distribute burdens and responsibilities amongst the EU, its member states and third countries.

That the EU strategies in respect of boat migrants are becoming embroiled in what Pope Francis termed "bureaucratic technicalities" is very much epitomized by the most recent effort to ensure common action in this area: the Frontex Sea Borders Regulation (Regulation 656/2014), adopted in May 2014. This contribution closely examines that Regulation and argues that it consolidates and even bolsters the controversies inherent in the EU's external sea borders policies, rather than solving them. The Regulation lays down binding rules for member states and the EU external border agency Frontex on joint operations of maritime border control, with the aim of establishing

> clear rules of engagement for joint operations at sea, with due regard to ensuring protection for those in need who travel in mixed flows, in accordance with international law as well as increased operational cooperation between and countries of origin and of transit.[3]

But the Regulation fails to deliver concrete answers on essential issues such as the place of disembarkation of rescued or intercepted migrants, the exact scope of duties to engage in search and rescue activities and the guaranteeing of refugee rights. Moreover, the Regulation disregards a number of international law obligations on the treatment of migrants found at sea. Although it formulates a number of useful procedural standards and upgrades human rights standards in comparison to the Council Decision it replaces, the Regulation may well be branded an instrument which first and foremost masquerades unresolved tensions between member states. It must, therefore, not be expected to bring about a fundamental paradigm shift in Europe's approach to the thousands of migrants who risk their lives each year in search for a better future.

After explaining the Regulation's coming into being, this chapter critically engages with the manner in which the Regulation addresses the potential tension between immigration deterrence and safeguarding refugee rights and between immigration deterrence and the saving of lives at sea. It concludes by formulating three main reasons why this most recent EU attempt to bring about clarity and consistency in the EU's approach to boat migrants is fundamentally flawed.

Background

The Frontex Sea Borders Regulation is the result of a rather laborious series of negotiations and institutional conflict within the EU. It replaces Council Decision 2010/252/EU which was annulled by the Court of Justice in September 2012, on the grounds of the Council having exceeded the scope of its implementing powers.[4]

When Frontex was established in 2004, the core of its mandate was described rather broadly as rendering "more effective the application of existing and future Community measures relating to the management of external borders."[5] Frontex was founded in lieu of a supranational European corps of border guards (for which political support was lacking) and leaves primary responsibility for border control with the member states. It assists national border guard services by providing technical assistance and by facilitating cooperation among national border guards. It is mainly known for coordinating joint maritime controls in the Mediterranean and the Atlantic. Frontex is also tasked with maintaining and coordinating the Eurosur framework which was set up in 2013 and which went 'live' in December of that year.[6] Eurosur links intelligence systems of member states for the purpose of increasing reaction capability at the external borders.

The original Frontex Regulation did not refer to other EU instruments on migration, such as the EU asylum directives, and only contained a general affirmation in the recital that the Regulation respects fundamental rights.[7] The lack of a concrete human rights framework on how to deal with refugees or migrants who are in distress at sea contributed to a perception that Frontex was preoccupied with border security considerations.[8] These concerns were in part addressed in the revision of the Frontex Regulation of October 2011, which introduced a reference to the 1951 Refugee Convention and the prohibition of *refoulement*, and an obligation for Frontex to draw up a fundamental rights strategy.[9] Likewise, the Eurosur Regulation mentions that, in addition to preventing unauthorized border crossings and to counter cross-border criminality, the shared intelligence has the purpose of "contributing to ensuring the protection and saving the lives of migrants."[10] However, these instruments do not as such resolve such contested issues as the manner in which screening for refugees should occur, where persons claiming asylum should be disembarked or how to respond to situations of distress. These were issues that, together with a description of powers of migrant interdiction, were developed in Council Decision 2010/252/EU, which was replaced by the Frontex Sea Borders Regulation.

Council Decision 2010/252/EU came about after lengthy negotiations. After the Commission had published a study on the international law instruments in relation to illegal immigration by sea in May 2007,[11] it commissioned an informal working group consisting of representatives of member states, Frontex, the United Nations High Commissioner for Refugees (UNHCR) and the International Organisation for Migration (IOM) to produce guidelines for Frontex's maritime operations. The group failed to agree however on the implications of refugee law, the role of Frontex and the question of where to disembark intercepted migrants. The Commission then decided to prepare a draft decision on the basis of Article 12(5) of the Schengen Borders Code,[12] which allows for the adoption of additional, implementing, rules governing surveillance in accordance with the regulatory procedure with scrutiny (in which Parliament does not have co-legislative but only veto power). The Commission's proposal aimed to make explicit the duties to respect fundamental rights and the rights of refugees in Frontex operations, created a legal basis in EU law in accordance with international maritime law for searching and intercepting vessels, and provided modalities for disembarking intercepted or rescued persons.

The draft failed, however, to acquire the requisite support in the Schengen Borders Code Committee, but was nevertheless forwarded to Parliament and the Council in November 2009.[13] Although Parliament voted in majority against the draft decision, it did not muster the absolute majority required for vetoing it.[14] The Council, as the body charged with implementing the Schengen Border Code, subsequently adopted the proposal in April 2010 with some amendments. The most pertinent one was that it divided the decision into binding rules on interception and non-binding guidelines on search and rescue situations and disembarkation.

Upon unanimous request of Parliament's Committee on Civil Liberties, Justice and Home Affairs (LIBE), however, the president of Parliament decided to bring an action for annulment of the decision before the Court of Justice. The Court annulled the decision, because it contained decisions affecting fundamental rights in an essential manner which go beyond the scope of 'additional measures', and only the EU legislature is entitled to adopt such a decision.[15]

The Court's ruling initiated renewed negotiations, this time under the ordinary legislative procedure with the full involvement of the European Parliament. The adopted text lays down detailed rules on the detection of vessels carrying irregular migrants, interception in the territorial sea, and interception on the high seas, interception in the contiguous zone, search and rescue situations and disembarkation. The Regulation addresses a number of gaps that were present in the annulled Council Decision in the sphere of human rights, saving lives at sea and ensuring accountability. It makes duties of participating member states in search and rescue situations binding, it contains a detailed provision on ensuring refugee rights and observing the principle of non-refoulement and it circumscribes obligations in the sphere of incident reporting and the manner in which Frontex draws up operational plans together with the member state hosting the operation. It further lays down a solidarity mechanism in case a member state faces a situation of "urgent and exceptional pressure at its external border." Compared to the annulled Council Decision, the Regulation is certainly more balanced and encompassing. This can be attributed, firstly, to the co-legislative role of the European Parliament, which strived especially to make some human rights guarantees explicit; secondly, to the *Hirsi* judgment of the European Court of Human Rights of 23 February 2012,[16] which clarified the human rights framework applicable to interceptions of migrants at sea; and thirdly, to the Lampedusa migrant shipwreck of October 2013,[17] which stirred widespread outcry and calls for deeper EU involvement and enhanced solidarity between member states. Yet, it is also evident from its text as well as from the negotiations both in the Council and between the Council and the European Parliament, that the Regulation avoids the political hot potatoes as much as possible.

Refugee rights and immigration deterrence

One such hot potato is the reconciliation of the aims of respecting refugee rights and immigration deterrence. How does one deter the entrance of irregular migrants while guaranteeing that refugees are not returned to a country where their life or freedom is threatened? Should one allow an intercepted (or rescued) migrant access to a refugee status determination procedure? Can one make an agreement with a third country which allows for the summary return of intercepted or rescued migrants?

The Regulation establishes a somewhat uneasy compromise between fundamental rights and the liberty of states to treat boat migrants outside ordinarily applicable statutory guarantees. This resembles strategies in Australia and the US of creating parallel legal frameworks for migrants arriving by boat. Before discussing the Regulation itself, this section first makes some comparative observations on US, Australian and European practices concerning boat immigration.

It has been observed that one reason why externalized maritime border controls, i.e. controls away from ports or coastal areas (which may extend so far as the territorial waters of another country), are a popular instrument of immigration deterrence, is that that they would allow for circumventing ordinarily applicable obligations in the sphere of protecting refugees.[18] Under the United States "Wet foot, Dry foot" policy, for example, a Cuban caught on the seas between the two countries is liable to being sent summarily home, while Cubans who make it to shore can invoke the provisions of the US Immigration and Nationality Act and are hence subject to the ordinary immigration procedures (Wasem 2009). The US Supreme Court confirmed, in the well-known *Sale/Haitian Centers Council* case, that neither the Immigration and Nationality Act nor the 1951 Refugee Convention, entertains obligations on the part of the US authorities outside US territory.[19] This legal construction allows for stripping migrants of fundamental rights. In Presidential Order 12807 of 24 May 1992, which was adopted in response to the exodus of Haitians after the coup that ousted President Aristide, the special position of refugees was noted, but only to the extent that "the Attorney General, in his unreviewable discretion, may decide that a person who is a refugee will not be returned without his consent."[20]

Disconnecting the ordinary immigration regime from the one applying at sea also underpins and legitimizes the programme of offshore deterrence of present-day Australia. That programme is meant to live up to the current government's election promise to turn back every boat that enters Australian waters illegally and which consists of either diverting any boat on its way to Australia to Indonesia or picking up the migrants and bringing them to offshore processing centers in Papua New Guinea and Nauru, with which agreements to that effect have been concluded. The interdictions are accompanied by the government broadcasting videos telling would-be immigrants that if they try to make the trip by boat illegally, "there is no way you will ever make Australia home."[21] Legally, the turnbacks are possible due to a series of amendments to Australia's Migration Act which marked the beginning of Australia's 'Pacific Solution' in 2001.[22] The amendments purported to excise certain places and installations along migration routes, such as Christmas Island and the Cocos Islands, from the Australian migration zone with the effect of making it impossible for "offshore entry persons" to make valid visa applications (which is required for making a formal asylum application) in these locations. To rule out the eventuality that asylum seekers would reach Australia's mainland by boat, the Australian Parliament passed a law in May 2013 to excise the mainland itself from the migration zone (ABC News 2013). The new policies of the Abbott government resulted in a sharp fall of the number of boat arrivals – less than 200 arrivals in 2014 compared to more than 20,000 in 2013. This resembles the decrease in boat arrivals after the first version of Australia's Pacific Solution was installed in response to the Tampa incident in 2001, also involving offshore processing in Nauru. Back then, 43 boats with a total number of 5,516 migrants arrived in the year 2001, and only one boat carrying one migrant in 2002 (Philips 2014). The numbers had started rising again when the newly elected Labor government fulfilled its government promise to close the center at Nauru in 2008.

The most clear European example of an interdiction programme involving summary returns of the kind employed in Australia and the US are the Italian push-backs of 2009. These were made possible after the Italian government had concluded an agreement with the Qaddafi regime in which the latter pledged to immediately accept all migrants found at sea who had embarked in Libya.[24] The push-backs, which were ill-received by civil society and a number of intergovernmental bodies, contributed to a decline of migrants intercepted in the Central Mediterranean route from almost 40,000 in 2008 to 4,500 in 2010 (Frontex 2011). The official position of the Italian government at the time was that migrants still at sea could not rely on Italian, EU or international law to effectuate an entry.[25] Civil unrest erupting in Libya (and Tunisia) made the continued implementation of the Italian–Libyan agreement impossible and caused a massive spike of migrants along this route in 2011.

It is not difficult to grasp the common logic behind the United States', Australian and Italian policies: if the goal is to minimize the number of irregular entries, one should i) make an agreement with another country to the effect of making summary returns to that country possible (Cuba, Libya, Papua New Guinea, Nauru); ii) ensure that intercepted migrants cannot invoke statutory guarantees on access to an asylum procedure or to courts; and iii) proceed with actually intercepting and summarily returning migrants at sea.

This logic is however challenged by fundamental rights. The judgment in *Sale* has been heavily criticized and in the European legal order at least, a different human rights paradigm in respect of boat migrants has come to prevail.[26] In the 2012 *Hirsi* judgment, in a case brought by 24 migrants who were part of a larger group of migrants who had had been intercepted in May

2009 in one of the first Italian push-back operations and were returned to the port of Tripoli, the European Court of Human Rights found violations of the prohibition of inhuman treatment, the right to an effective remedy and the prohibition of collective expulsion.[27]

The judgment made it clear that any interdiction policy which does not allow migrants the opportunity to claim asylum or to mount a legal challenge against return before the return is enforced will violate the European Convention on Human Rights. The European Court of Human Rights (ECtHR) formulated a set of rather clear conditions which must be met if a state wishes to return migrants found at sea. These can be summarized as follows:

i)   The returning country must verify, even if no individual asylum claim is made, whether the country of return fulfils its international obligations in relation to the protection of refugees.
ii) The migrants must have access to a procedure to identify them and to assess their personal circumstances.
iii) The personnel conducting the interviews must be trained to do so and the migrants must be assisted by interpreters or legal advisors.
iv) The migrants must be informed about their destination and be able to challenge their transfer before an independent tribunal, before the transfer is enforced.

It is obvious that these conditions rule out any policy of summarily returning migrants, even to countries which can generally be considered safe. Although the ECtHR did not say that refugee screening cannot be done on board ships or in some offshore facility, the Court stresses that proper procedures ought to be in place and that migrants should always be granted the possibility to bring legal challenges with suspensive effect. In the current European context, this can only imply that all intercepted migrants are brought to European shores for processing.

This may indeed be a hard pill for some European states to swallow. Although no European country has since the 2009 Italian push-backs been officially engaged in a policy of summarily returning migrants, there are strong indications that such practices continue to occur. Non-government organizations (NGOs) have reported large numbers of push-back cases involving Bulgaria and Greece along the land and sea border with Turkey.[28] There are also reports – and even videos – of the Spanish border guard immediately returning migrants to Morocco who had managed to climb the fences around the Spanish enclaves Ceuta and Melilla in Morocco.[29] In a move that resembles Australia's excision of certain territories from its migration zone, the Spanish government proposed an amendment to its immigration law in October 2014 which would allow rejecting migrants who present themselves at the border of Ceuta and Melilla and returning them to Morocco, presumably legalizing a practice that is already ongoing – but hard to reconcile with the standards developed in the *Hirsi* judgment.[30] The *Hirsi* standards were confirmed by the ECtHR in its later judgment in *Sharifi*, a case concerning a group of migrants from Afghanistan, Sudan and Eritrea, who had travelled by boat from the Greek town of Patras to Bari, Italy in 2008 and 2009, where the Italian border police intercepted them and immediately deported them back to Greece.[31]

The reluctance on the part of at least some European states to allow migrants entry into a proper refugee status determination procedure or return procedure makes it easier to understand why the Frontex Sea Borders Regulation, which was adopted two years after the *Hirsi* judgment, only half-heartedly seeks to incorporate the standards of that judgment. In proposing the Regulation, the European Commission expressly sought to take account of the *Hirsi* standards.[32] Instead of merely alluding to the prohibition of returning asylum seekers to an unsafe country (non-refoulement) without setting forth how that prohibition should be guaranteed in practice, as was done in Decision 2010/252/EU, the Regulation sets forth a number of substantive and procedural guarantees. Thus, it is stipulated that "when considering the possibility of disembarkation in a third country," the participating member states "shall take into account the general situation in that country."[33] Furthermore, it is laid down that "all means are used to identify the intercepted or rescued persons" before being returned to a third country.[34]

But a closer reading of the Regulation text reveals that in three areas relating to procedural standards, the drafters have left room for ambiguity. The Regulation provides as basic rule that disembarkation shall occur in the coastal member state in the case of interception in the territorial sea or the contiguous zone, and holds that in the case of interception on the high seas, disembarkation may take place in a third country. If that is not possible, disembarkation shall take place in the host member state.[35]

The rule of disembarkation in the coastal member state in the situation of interception in the territorial waters is in line with the clarification in the recast Asylum Procedures Directive that asylum applications made in the territorial waters fall within the directive's scope and are therefore to be examined by the coastal member state.[36] The Regulation does not, however, address the possibility of asylum claims made by migrants intercepted in the territorial waters entirely satisfactory. First, by way of exception, Art. 6(2)(b) of the Frontex Sea Borders Regulation allows participating units to order a vessel found in the territorial sea "to alter its course towards a destination other than the territorial sea or the contiguous zone." This compromises the guarantees of the Asylum Procedures Directive and is difficult to reconcile with the *Hirsi* ruling. Second, the Regulation does not address the situation where a coastal member state experiences systemic flaws in the asylum procedure and in the reception conditions for asylum seekers. Both the ECtHR and the Court of Justice of the European Union (CJEU) have declared that in such situations, member states may not transfer asylum applicants to the member state in question pursuant to the Dublin Regulation.[37] This is especially relevant in the context of Frontex operations, as Greece and Italy, two likely candidate countries to host Frontex missions, have been declared unsafe for (particular categories of) asylum seekers.[38] Although this case law specifically deals with Dublin transfers, it would not be unreasonable to extend its implications to interceptions at sea where one member state wishes to transfer intercepted migrants who may wish to claim asylum to the coastal member state. The Frontex Sea Borders Regulation avoids the issue, and hence fails to clarify for participating member states how to address the issue of disembarkation in a member state where the asylum system is flawed.

Also problematic are the Regulation's standards on disembarkation of intercepted or rescued persons in a third country. First, the Regulation does not require that the identification is done by trained personnel, and neither requires the presence of interpreters and legal advisors on board participating units, but merely holds that such 'details' are to be contained in the operational plans to be drafted for each Frontex mission, and only 'when necessary'.[39] The operational plans, however, are secret. It is highly questionable whether they are appropriate instruments for guaranteeing fundamental rights. The adjective 'when necessary' further begs the question when proper communication and legal advice is not deemed essential for refugee status determination.

Second, the Regulation says that migrants have to be informed of their destination "in a way," instead of a language, "that those persons understand or may reasonably be presumed to understand". One MEP involved in the negotiations with the Council explained that this seemingly minor detail may have profound consequences:

> When insisting on informing persons on board of the place of disembarkation in a way rather than in a language they understand, the Council indicated that this would allow member states to point just on a map to show people where they would be disembarked. Fine so far. However, if a common language does not exist, how could refuges then 'express any reasons for believing that disembarkation in the proposed place would be in violation of the principle of non-refoulement' in a way Frontex would understand? They could wave or shout. But probably every person on board would wave or shout no matter if he or she is in need of protection or rather looking for a better life in Europe. And how, if everybody waves or shouts, can they make clear that an interpreter would be needed in their case? As mentioned above, interpreters will be provided only if necessary, possibly via a radio interpreter's help line.
>
> (Keller 2014)

This reminds of the so-called 'shout test,' which was employed by the US Coast Guard intercepting Haitian migrants in 2004. The test purportedly allowed a person to be interviewed for pre-establishing his refugee status, but only if he, in the words of one US Coast Guard Commander, aggressively insists he fears for his life (*Duluth News Tribune* 2006). A commentator observed that "only those who wave their hands, jump up and down, and shout the loudest – and are recognized as having done such – are even afforded, in theory, a shipboard refugee pre-screening interview" (Frelick 2004–2005). It was reported that only 9 of 1,850 interdicted Haitians were transferred to the migrant reception center on Guantánamo Bay for status determination on the basis of this form of pre-screening and that only one of those was found to be a refugee (Wasem 2010).

Third, and arguably most problematic in view of the *Hirsi* judgment, is that the Regulation does not say anything about legal remedies with suspensive effect. There is no mention of a right of appeal against return to a third country. Given that a right of appeal and making that right effective by providing information and access to legal assistance is well-nigh impossible to reconcile with a desire to disembark intercepted or rescued migrants in a third country, the omission of such guarantees in the Regulation suggests rather clearly that some member states simply insisted during the negotiations to maintain that possibility.

Overall, the Regulation establishes a very specific regime for dealing with migrants who may claim asylum who are found at sea, outside the ordinary framework of EU asylum law. The EU directives on asylum, which regulate such matters as the reception conditions, the standards of the asylum procedure and the eligibility criteria for asylum, are, according to the definitions on the scope of the directives, applicable "to all third country nationals and stateless persons who make an application for international protection on the territory, including at the border, in the territorial waters or in the transit zones of a member state." Applications made at the high seas or beyond are hence excluded from their scope.[40] If the latter applications are made in the context of a Frontex mission, they are governed by the Regulation, which provides a set of substandard guarantees, omitting such key elements as the right of access to an asylum procedure, the right to legal assistance, the right that a decision is given in writing, the right of appeal, etcetera. Further, the Regulation has its own standards on returns and apprehension, which are not at all aligned with the norms set out in the Returns Directive (2008/115/EC) setting out the common procedures to be applied in member states for returning illegally staying third country nationals. The conclusion is accordingly that very much akin to the United States and Australian precedents, the Regulation excludes boat arrivals from ordinary statutory guarantees.

A further omission is that the Regulation is remarkably silent on interceptions in the territorial sea of third countries. In earlier policy documents, the wish was expressed to clarify the scope of member state powers also in respect of operations in the territorial sea of third countries.[41] The issue is topical, as for example Spain has concluded detailed arrangements with a range of countries in Northwest Africa, including Morocco, Senegal, Mauritania and Cape Verde, which foresee in joint sea patrols of the Spanish *Guardia Civil* and the border authorities of the partner countries, and which aim to prevent migrant vessels from exiting the territorial sea of the latter countries.[42] These patrols have been highly instrumental in closing the Atlantic migration route to the Canary Islands and continental Spain. Frontex has been involved in these operations, as it coordinated the various Hera operations in the seas adjacent to Mauritania, Senegal and Cape Verde. Spain's bilateral arrangements have raised questions about their implications for the migrants' fundamental right to leave a country, the safeguards applicable during interception and the division of responsibilities between the various states involved (ibid.).

Saving lives at sea and immigration deterrence

Another hot potato concerns the question of how to execute border surveillance "while contributing to saving lives."[43] Compared to 'ordinary' situations of distress at sea, the combination of 'irregular migrants' and 'distress' tends to generate questionable dynamics of burden avoidance and burden shifting.

Obviously, if a state elects to render assistance to a migrant vessel at sea it makes itself the most plausible candidate for having to carry the migrant burden – i.e. to bring the migrants to shore and thus allowing them to effectuate entry. Private shipmasters may for similar reasons be discouraged to engage in rescue operations. For them, there are not only potential delays and commercial losses involved, they may also be simply refused entry into any nearby port.[44] Furthermore, if allowed entry into port, they must seriously take into account the possibility of being investigated or even charged for involvement in migrant smuggling (*Statewatch News* 2007).

This gives rise to all sorts of avoidance behavior which is regularly reported on by the press and NGOs in various European countries. Recall, for example, the incident of May 2007, when the shipmaster of a tuna pen flying the Maltese flag had rendered assistance to a group of 28 irregular migrants whose ship had sunk and who, for both financial and security reasons, refused to allow the migrants on board. Because Malta, Italy nor Libya were under a clear obligation to allow the migrants to disembark in one of their ports, the migrants were compelled to desperately cling to the buoys of a tuna fishing net for three days before finally being brought to Lampedusa.[45] In another incident in July 2006, the Maltese government had refused to allow the disembarkation of 51 migrants rescued by the Spanish fishing trawler *Francisco y Catalina* by affirming that it had been Libya's responsibility to rescue them and, given that a Spanish vessel ended up doing so, the migrants had become Spain's responsibility. After a stand-off lasting eight days, it was agreed that Malta, Libya, Italy, Spain and Andorra would all take in a share of the migrants.[46] A further peculiar example of how shipmasters are not all too keen on rendering assistance is an incident in August 2014, when the Rescue Centre in Rome called on 76 ships which were in the area of rescue of a vessel in distress, but where within one minute there were only six ships left on the radar screen. All the others had switched off their radar signal.[47] Well-documented as well is the fate of the so-called left-to-die boat, of whom all but 11 of the 72 passengers died, and which gave rise to an investigation of the Parliamentary Assembly of the Council of Europe. The report revealed, amongst other things, how a military helicopter had dropped water and biscuits in the adrift vessel but never returned, and how a large military vessel allegedly ignored the boat's distress calls after half of the passengers had already died (Strik, 2012).

Avoidance behavior lies at the heart of disagreements between EU member states, especially in Malta and Italy, on the interpretation of relevant international law standards on search and rescue at sea. One such dispute concerns the definition of rescue in the International Convention on Maritime Search and Rescue (SAR Convention) and in particular the duty to deliver rescued persons to a 'place of safety.'[48] Malta maintains that disembarkation should always occur at the nearest safe port to the site of the rescue, which is often a port in Italy. Italy, on the other hand, is of the opinion that rescued migrants should be brought to a port of the country in whose search and rescue (SAR) region the rescue took place.[49] Malta, strategically located in the middle of the Mediterranean, has a large SAR zone relative to its size (which encloses the Italian island of Lampedusa) and has for that reason rejected amendments to the SAR Convention of 2004 which bestow 'primary responsibility' for ensuring that survivors are disembarked from the assisting ship and brought to a place of safety on the government responsible for the SAR region.[50] A majority of other Contracting States, including Italy, did accept the amendments. Malta feared that the new arrangement would be an open invitation to migrant smuggling, since vessels carrying migrants would simply have to enter the closest neighboring SAR area, launch a distress call and the government of that SAR zone would effectively be obliged to come to their assistance and guarantee a place of safety.

Malta and Italy also disagree when a vessel is in distress, which under the Law of the Sea triggers duties of search and rescue. Klepp describes how the Armed Forces of Malta adhere to the definition that distress is the imminent danger of loss of lives. From that perspective: if a ship is sinking it is in distress. If the boat is not sinking, it is not in distress. The Italian navy, on the other hand, would rescue every heavily overloaded boat.[51]

Avoidance behavior is also discernible on the part of the smugglers and migrants themselves. There are quite a few incident reports where migrants aim to be rescued by one member state in particular – often Italy. Thus, after a group of 220 Syrian migrants had been taken on board a Maersk cargo ship some 285 nautical miles southeast of Malta in October 2014, they refused to be transferred on Maltese patrol boats.[52] After having exhausted further attempts to transfer the migrants onto a boat to Malta, the Armed Forces of Malta made contact with the Italian government which agreed to take charge of the group. In a similar incident one month earlier, some 300 mostly Syrian migrants who had been rescued by the cruise liner Salamis Filexonia refused to disembark in Cyprus, insisting that they be taken to Italy instead.[53] After a stand-off that lasted a day, Cypriot police negotiators coaxed the migrants off the ship. The Armed Forces of Malta have sometimes been accused of providing migrants in distress with water, food or fuel and driving them to Italian waters, but have themselves always denied such allegations, saying instead that it often happens that migrants refuse to be transferred onto Maltese patrol vessels, in which case they remain in close vicinity in order to render assistance should the need arise.[54]

The Frontex Sea Borders Regulation is much more explicit on search and rescue situations than Council Decision 2010/252/EU. It provides binding rules on search and rescue (as opposed to guidelines in the Council Decision) in the form of quite detailed instructions for participating units in case they are confronted with distress situations. These rules were adopted despite opposition of six Mediterranean countries (Italy, Cyprus, France, Greece, Malta and Spain).[55] These countries called the proposed provisions on search and rescue "unacceptable for practical and legal reasons" and argued that international maritime law already dealt 'amply' with such situations.

These countries do have a point, as international law already circumscribes search and rescue obligations extensively. Art. 98 of the United Nations Convention on the Law of the Sea (UNCLOS) contains the primordial duties on the part of every shipmaster "to render assistance to any person found at sea in danger of being lost" and "to proceed with all possible speed to the rescue of persons in distress." It further obliges coastal states to "maintain adequate and effective search and rescue services." These general duties are worked out in detail in the SAR Convention. The Frontex Sea Borders Regulation confirms that it does not affect these obligations and for the most part reiterates them.[56] From the outset, it appeared that the key added value of the Regulation would lie in harmonizing divergent interpretations of member states on such issues as what distress amounts to and what a place of safety means. However, this turned out to be a bridge too far.

The Regulation 'solves' the issue of what distress exactly is by inventing a sliding scale of three phases of urgency, namely a "phase of uncertainty," a "phase of alert" and a "phase of distress."[57] All three phases are defined in an open manner – by merely mentioning examples of situations belonging to each phase. Especially remarkable is that the Regulation does not differentiate between the types of action to be undertaken in each of the phases of urgency. It merely holds that in all three phases, participating units are obliged to transmit all relevant information to the Rescue Coordination Centre and to place themselves at the disposal of that Rescue Coordination Centre.[58] Also, "while awaiting instructions from the Rescue Coordination Centre," they must "take all appropriate measures to ensure the safety of the persons concerned." Although this may guarantee that participating units remain alert also in situations not requiring immediate action, the Regulation fails to specify when rescue operations must be initiated. It assigns primary responsibility for such a decision to the Rescue Coordination Centre, and puts participating units under the vaguely formulated residual obligation to take appropriate measures to ensure safety. Hence, the Regulation leaves it to each Rescue Coordination Centre and participating unit to decide whether it is appropriate to remove every overloaded and unseaworthy vessel from the sea as soon as it leaves port, or to come into action only when a boat is actually sinking.

Neither does the Regulation provide clarity on what 'a place of safety' is. It holds that the member state hosting the Frontex mission and the other participating member states shall cooperate with the responsible Rescue Coordination Centre to identify a place of safety after a rescue operation.[59] The Rescue Coordination Centre shall then designate a place of safety. Importantly, however, in case a place of safety cannot be agreed on "as soon as reasonably practicable," the Regulation provides as residual rule that disembarkation should occur in the host member state.[60] That rule goes some way in addressing the frequent controversies between member states over allowing rescued migrants entry into a port. However, Malta has always firmly resisted such a rule (which already existed in the previous non-binding guidelines) and for that reason has refused to host any Frontex mission since 2010.[61]

The freedom of Malta to simply ignore the common EU rules in this manner lays bare the more fundamental issue of how to ensure common action and member state solidarity in the area of external border control. The Frontex pool of assets and border guards is made up of voluntary contributions of member states. The voluntary character of member state participation has obvious consequences for Frontex's effectiveness. According to the former director of Frontex, the dependence of Frontex on voluntary contributions of member states compromises its ability to fulfil its mandate.[62] Moreover, it fails to guarantee that a member state facing particular pressure at its external borders is appropriately assisted. Apart from Malta's refusal to host Frontex missions, the United Kingdom (UK) has also announced that it would no longer participate in search and rescue operations in the Mediterranean (see Follis, this volume).[63] The UK, despite being outside Schengen, did participate in most Frontex operations, but is principally opposed to Operation Triton, the EU's successor to Operation Mare Nostrum which is said to have saved the lives of around 150,000 migrants.[64] It is precisely the success of Mare Nostrum in terms of lives saved which explains the UK's reluctance to participate. Just as Malta, the UK government is concerned that search and rescue operations in the Mediterranean act as a pull factor for illegal migration, only encouraging more people to make the dangerous crossing in the expectation of being rescued. The discussion on Mare Nostrum and Triton shows that there is not only hardly a common vision on how to approach the very issue of boat migration, but also that any member state with a different view can simply choose to withdraw itself from any common action.

Conclusion

No one should have expected the Frontex Sea Borders Regulation to be the panacea for all the tragedies in the Mediterranean and the difficulties in finding a common European answer. The Regulation is in fact a stronger instrument than the Council Decision it replaces, both in terms of fundamental rights and search and rescue obligations. Yet, the Regulation is inherently flawed for three main reasons. First, because it avoids hard choices on difficult issues of how to share the migrant burden and how to cooperate with third countries, it tends to legitimize questionable unilateral practices of member states. This is most evident from its substandard guarantees against removals to third countries, but also transpires from some of the compromises in the sphere of search and rescue. It is important to recall, therefore, the duty under EU law to implement secondary EU legislation in conformity with the EU Charter on Fundamental Rights. But it would have been better if these guarantees were respected fully in the text itself.

Second, similar to its predecessor, the Regulation proceeds from the questionable presumption that boat migrants may be treated in a fundamentally different manner than migrants who present themselves in or at the border of a state. The Regulation does not refer in any way to the guarantees for persons subjected to 'ordinary' border controls laid down in the Schengen Borders Code, and refers neither to guarantees for asylum seekers laid down in EU asylum law. It thus creates a parallel regime for migrants found at sea, opening up procedural possibilities for expedient and summary returns, but also for apprehending migrants and the taking of other coercive measures, outside a comprehensive human rights framework. There are notable parallels here with the manner in which Australia has created a special immigration regime for 'unauthorized boat arrivals' and the United States policy of legally separating migrants based on whether their feet are dry or wet.

Third – and one might be tempted to welcome this in view of the two preceding points – the Frontex Sea Borders Regulation provides only a limited set of answers to a limited set of situations. It only enters into play if it is decided that a Frontex mission should be started and only applies to member states who on a voluntary basis decide to participate in a Frontex mission. It does not apply to unilateral maritime controls of member states, which are still far more common than those undertaken under the aegis of Frontex. It follows that there are no common EU rules on individual member state maritime controls. Given the prevailing disputes between member state arising from such operations and the fact that all external border controls are in the scope of Union law, the choice of the EU legislator to adopt rules solely for Frontex operations, which it justifies on the basis of the principles of subsidiarity and proportionality, can be seriously questioned.[65] The result is that, although the EU has fully harmonized the manner in which border controls 'at' the external border must be conducted by virtue of the Schengen Borders

Code, it allows for altogether divergent control practices of member states at sea. This challenges the very integrity and coherence of the EU's external border controls regime.

Notes

1 Address of Pope Francis to the European Parliament, Strasbourg, 25 November 2014.
2 See e.g. Commission documents COM(2011) 743 final (The Global Approach to Migration and Mobility) and COM(2008) 69 final.
3 Reg. 656/2014, recital 1. Also see Stockholm programme [2010] OJ C115/01, par 5.1.
4 CJEU 5 September 2012, Case C-355/10 (*Parliament v Council*).
5 Reg. 2007/2004, Art. 1(2).
6 Reg. 1052/2013, Art. 6
7 Reg. 2007/2004, Recital 22.
8 See e.g. S. Carrera (2007).
9 Reg. 1168/2011, new Articles 1(2) and 26a.
10 Reg. 1052/2013, Art. 1.
11 SEC(2007) 691.
12 Reg. 562/2006.
13 COM(2009) 658 final.
14 Motion for a resolution on the Draft Council Decision supplementing the Schengen Borders Code as regards the surveillance of the sea external borders in the context of operational coordination coordinated by the European Agency for the Management of Operational Cooperation at the External Borders, B7-0227/2010, 17 March 2010. See further procedural file 2009/2755(RPS).
15 CJEU 5 September 2012, Case C-355/10 (*Parliament v Council*).
16 ECtHR [Grand Chamber] 23 Feb. 2012, *Hirsi Jamaa a.o. v Italy*, no. 27765/09.
17 See e.g. Statement by President Barroso following the European Council meeting of 25 October 2013 and European Council Conclusions of 24/25 October 2013, Council doc. ST 169 2013 INIT.
18 Gammeltoft-Hansen (2011, 41), referring in this connection to states playing "sovereignty games."
19 US Supreme Court 21 June 1993, *Sale v Haitian Centers Council*, [1993] 509 US 155.
20 Executive Order 12807, 'Interdiction of Illegal Aliens', 24 May 1992.
21 YouTube video 'No Way. You will not make Australia home', published 15 April 2014. www.youtube.com/watch?v=rT12WH4a92w.
22 Migration Amendment (Excision from Migration Zone) Act 2001; Migration Amendment (Excision from Migration Zone) (Consequential Provisions) Act 2001.
24 See extensively Pascale (2010).
25 See *Hirsi Jamaa a.o. v Italy*, par. 64-66.
26 For criticism see e.g. Koh ok(1994, 2417); Hathaway (2005, 318).
27 *Supra* note 16.
28 Human Rights Watch 29 April 2014, 'Containment Plan: Bulgaria's Pushbacks and Detention of Syrian and other Asylum Seekers and Migrants' (Report), www.hrw.org/sites/default/files/reports/bulgaria0414_ForUpload_0.pdf; Amnesty International 31 March 2014, 'Bulgaria: suspension of returns of asylum-seekers to Bulgaria must continue' (Report), www.amnesty.org/en/documents/EUR15/002/2014/en/; ProAsyl 7 November 2013, 'Pushed Back: Systematic human rights violations against refugees in the Aegean Sea and at

the Greek Turkish land border' (Report), www.proasyl.de/en/press/press/news/
pro_asyl_releases_new_report_pushed_back/; Amnesty International 29 April
2014, 'Greece: Frontier of hope and fear migrants and refugees pushed back at
Europe's border' (Report), www.amnesty.org/en/documents/EUR25/004/2014/
en/; Amnesty International 9 July 2013, 'Greece: Frontier Europe: Human
rights abuses on Greece's border with Turkey (Report), www.amnesty.org/en/
documents/EUR25/008/2013/en.
29 Eldiario 16 October 2014, Vídeos: La Guardia Civil deja inconsciente a palos a
un inmigrante y lo devuelve atado de pies y manos, www.eldiario.es/desalambre/
VIDEOS-Guardiua-Civil-ilegalmente-inmigrante_0_314268729.html. For a com-
pilation of accounts see UNHCR, 'Syrian Refugees in Europe. What Europe Can
Do to Ensure Protection and Solidarity', July 2014, UN High Commissioner
for Refugees (UNHCR), Syrian Refugees in Europe: What Europe Can Do to
Ensure Protection and Solidarity, 11 July 2014, available at: http://www.refworld.
org/docid/53b69f574.html.
30 Additional Provision 10 to the Organic Law 4/2000, of 11 January, Regarding
the Rights and Freedoms of Foreign Nationals Living in Spain and Their Social
Integration (Spanish Alien Law), adopted by Spanish Congress on 11 December
2014. Amnesty International Press Release 31 October 2014, 'Spain: Immigration
Law Amendment Will Violate Rights', Amnesty International Press Release
31 October 2014, 'Spain: Immigration Law Amendment Will Violate Rights',
www.amnesty.org/en/documents/EUR41/006/2014/en.
31 ECtHR 21 October 2014, *Sharifi a.o. v Italy and Greece*, no. 16643/09.
32 COM(2013) 197 final.
33 Reg. 656/2014, Art. 4(2).
38 Ibid, Art. 4(3).
34 Ibid, Art. 10(1)(a) and 10(1)(b).
35 Directive 2013/32/EU, Art. 3(1).
36 ECtHR 21 January 2011 [Grand Chamber], *M.S.S. v Belgium and Greece*, no.
30696/09, CJEU 21 December 2011, Case C-355/10 (*N.S. and M.E.*); ECtHR 4
November 2014 [Grand Chamber], *Tarakhel v Switzerland*, no. 29217/12.
37 See *M.S.S. v Belgium and Greece* and *Tarakhel v Switzerland*.
38 Reg. 656/2014, Art. 4(3)
39 S. Keller (MEP, Greens), 'LIBE Special April 2014: New rules on Frontex operations at sea', 16 April 2014, www.ska-keller.de/en/topics/migration/borders/
libe-special-april-2014-new-rules-on-frontex-operations-at-sea.
40 See Directive 2013/33/EU, Art. 3; and Directive 2013/32/EU, Art. 3(1).
41 COM(2006), 733 final, par. 34.
42 The agreements themselves are confidential. See for an extensive analysis Garcia
Andrade (2010).
43 Reg. 656/2014, recital 1.
44 For examples see den Heijer (2012, 231).
45 The incident has been widely reported. For a summary see Migration Policy
Group (2007, 11).
46 Migration Policy Group, *Migration News Sheet*, Brussels, August 2006, pp. 16–18.
47 BBC News 12 November 2014, 'Mediterranean migrants: EU rescue policy
criticized'.
48 International Convention on Maritime Search and Rescue 1979, Annex par.
1.3.2.
49 See Klepp (2011, 539 at 548–551) and Mallia (2013, 11).
50 Malia (2013, 11). See also *Times of Malta* 26 April 2009, 'Shrinking Malta's search
and rescue area is 'not an option''.
51 See Klepp (2011, 539 at 554).
52 *Times of Malta* 26 October 2014, 'Rescued migrants refused to come to Malta'.

53 *International Business Times* 26 September 2014, 'Nearly 300 Migrants, Mostly Syrians, Refuse To Disembark In Cyprus after Rescue'.
54 Human Rights Watch Sept. 2009, 'Pushed Back, Pushed Around. Italy's Forced return of Boat Migrants and Asylum-Seekers'; *Malta Today* 19 August 2014, 'AFM denies Italian media report claiming it pushed migrants away'.
55 *ANSA* 15 October 2013, 'Immigration: EU Council divided over Frontex guidelines'. See also Council docs. 10349/13 of 14 June 2013 and 14612/13 of 10 Oct. 2013.
56 Reg. 656/2014, recital 14 and 15.
57 Ibid, Art. 9.
58 Ibid. Art. 9(2)(a).
59 Ibid, Art. 10 (1)(c).
60 Ibid.
61 *Times of Malta* 2 July 2011, 'Army can honour 'all international obligations', AFM commander insists'. Briefing European Parliamentary Research Service 13/12/2013, 'Irregular immigration in the EU. Some national perspectives on arrival of immigrants'.
62 I. Laitinen (Frontex director), 'An inside View', EIPASCOPE 2008/3.
63 *BBC News* 28 October 2014, 'UK opposes future migrant rescues in Mediterranean'.
64 UNHCR 17 October 2014, 'UNHCR concerned over ending rescue operation in the Mediterranean'.
65 COM(2013) 197 final, Paragraph 5

Bibliography

ABC News 2013. Parliament excises mainland from migration zone. 16 May.

Carrera, S. 2007. *The EU Border Management Strategy. Frontex and the Challenges of Irregular Immigration in the Canary Islands.* CEPS Working Document No. 261/ March.

Duluth News Tribune. 2006. Coast Guard told not to encourage asylum claims by Haitian migrants, 23 January.

Di Pascale, A. 2010. Migration Control at Sea: The Italian Case. In: Ryan B. and Mitsilegas V. (eds.). *Extraterritorial Immigration Control. Legal Challenges.* Leiden/Boston: Martinus Nijhoff.

Financial Times. 2014. 'Australia's 'stop the boats' asylum policy divides opinion', 6 November.

Frelick, B. 2004. "Abundantly Clear": Refoulement. *Georgetown. Immigration Law Journal* Vol. 19, pp. 245–275.

Frelick, B. (2004–2005). 'Abundantly Clear': Refoulement, 19 *Georgetown. Immigration Law Journal*, 245 at 246.

Frontex. 2011. Risk Analysis Quarterly. Issue 1, Jan–March. Available online: http://frontex.europa.eu/assets/Publications/Risk_Analysis/FRAN_Q1_2011.pdf

Gammeltoft-Hansen, T. 2011. *Access to Asylum. International Refugee Law and the Globalisation of Migration Control.* Cambridge: Cambridge University Press.

Garcia Andrade, P. 2010. Extraterritorial Strategies to Tackle Irregular Immigration by Sea: A Spanish Perspective. In Ryan B. and Mitsilegas V. (eds). *Extraterritorial Immigration Control. Legal Challenges,* Leiden/Boston: Martinus Nijhoff.

Hathaway, J.C. 2005. *The Rights of Refugees under International Law.* Cambridge: Cambridge University Press.

den Heijer, M. 2012. *Europe and Extraterritorial Asylum,* Oxford: Hart Publishing.

Keller, S. 2014. LIBE Special April 2014: New Rules on Frontex Operations at Sea', 16 April. www.ska-keller.de/en/topics/migration/borders/libe-special-april-2014-new-rules-on-frontex-operations-at-sea.

Klepp, S. 2011. A Double Bind: Malta and the Rescue of Unwanted Migrants at Sea, a Legal Anthropological Perspective on the Humanitarian Law of the Sea. *International Journal of Refugee Law*. Vol. 23, pp. 538–557.

Koh, H.H. 1994. The "Haiti Paradigm" in United States Human Rights Policy. *Yale Law Journal*, Vol. 103, pp. 2391–2435.

Mallia, P. 2013. The Challenges of Irregular Maritime Migration. *Jean Monnet Occasional Paper No. 4/2013.*

Migration Policy Group. 2007. *Migration News Sheet*, Brussels, July, p. 11.

Philips, J. 2014. Boat Arrivals in Australia: A Quick Guide to the Statistics. *Parliament of Australia Research Paper Series* 2013–14.

Statewatch News 2007. 'Italy/Tunisia: Fishermen on trial for rescuing migrants', 7 September.

Strik, T. 2012. Parliamentary Assembly of the Council of Europe, 'Lives lost in the Mediterranean Sea: who is responsible?', 5 April, doc. 12895.

Wasem, R.E. 2009. Cuban Migration to the United States: Policy and Trends. *CRS Report for Congress*. Washington: US Congressional Research Service.

Wasem, R.E. 2010. US Immigration Policy on Haitian Migrants. *CRS Report for Congress*. Washington: US Congressional Research Service.

4   Ethnography up the stream

The UK 'let them drown' policy and the politics of bordering Europe

Karolina Follis

On October 15, 2014, Lord Hylton, a crossbench peer (i.e. an independent member) in the United Kingdom (UK) House of Lords, issued a formal query to the Minister of State at the Foreign and Commonwealth Office, Baroness Anelay of St Johns. He asked what naval or air-sea rescue contribution the UK government would make to prevent refugees and migrants from drowning in the Mediterranean. The Tory politician submitted a written answer stating that,

> We do not support planned search and rescue operations in the Mediterranean. We believe that they create an unintended "pull factor", encouraging more migrants to attempt the dangerous sea crossing and thereby leading to more tragic and unnecessary deaths. The Government believes the most effective way to prevent refugees and migrants attempting this dangerous crossing is to focus our attention on countries of origin and transit, as well as taking steps to fight the people smugglers who willfully put lives at risk by packing migrants into unseaworthy boats.
>
> (UK Parliament 2014)

The parliamentary record for that or the following days does not reflect any debate on the issue. However, just under a fortnight later, as the year-long sweeping Italian navy Mediterranean rescue operation 'Mare Nostrum' was winding down without an equivalent successor, after having saved an estimated 150,000 people, it emerged that Baroness Anelay's answer was indeed the official policy of the UK government.

On October 27, the Guardian publicized the decision to refrain from supporting further rescue missions under the headline "UK axes support for Mediterranean rescue operation" (Travis 2014a). The paper followed up a day later with an article citing a statement in support of the policy issued by the Home Secretary Theresa May, which sought to reframe it as a matter of a European consensus rather than a singular decision reached in a London government department. The statement said

Ministers across Europe have expressed concerns that search-and-rescue operations in the Mediterranean … [are] encouraging people to make dangerous crossings in the expectation of rescue. This has led to more deaths as traffickers have exploited the situation using boats that are unfit to make the crossing.

(Travis 2014b)

Later, it was reported that "May is believed to have been a leading figure among the European interior ministers" (Travis 2014c) who in early October had decided to replace rescue with a European Union (EU) border protection operation 'Triton' run by Frontex "with a third of resources and a limited remit to patrol within 30 miles of the Italian coast" (ibid.).

As the story became front page news in late October 2014, prompting outraged editorials and condemnation by migrant advocates and human rights NGOs, James Brokenshire, the immigration minister serving in the Home Office under May, was dispatched to the House of Commons to defend the policy against its parliamentary critics, recruiting not only from the Labour opposition, but also from the ranks of the Liberal Democrats (at the time the minority partner in the UK coalition government) and even the Conservative party itself. By then dubbed the "let them drown policy," it was described by the MPs who opposed it as "discreditable" and "a barbaric abandonment of British values" (Travis 2014c). In response, Brokenshire insisted that the unintended consequence of rescue is more deaths. He called for stopping the emergency measures "at the earliest opportunity," and for widely publicizing their discontinuation in North Africa, ostensibly in the interest of preventing further loss of life (Travis 2014c).

The immorality and twisted logic of this proposition has been discussed elsewhere (Andersson 2014b; Hodges 2014; Amnesty International 2014; OHCHR 2014).[1] I concur with Ruben Andersson's observation that "the idea of stopping rescue operations […] to deter others from embarking on the journey [across the Mediterranean is] a proposition as absurd as removing seatbelts in cars to make drivers more risk-averse" (Andersson 2014b). But in this chapter I recall the unveiling of the 'let them drown policy' with a different intention. I want to pause on the openly anti-humanitarian aspect of the UK government's position which, striking as it may be, has not thus far been fully examined.

Students of the European border regime have noted on many occasions the paradox that migration and border control practitioners typically claim to protect the 'area of freedom, security and justice' in a humane manner and in accordance with the EU fundamental rights framework. They do so even as they engage in such inherently violent practices as immigration raids, detention and deportation procedures (Hall 2010; Andersson 2014; Feldman 2012; Dzenovska 2014; Cuttitta 2014). The UK government's tough-minded argument for strict border controls *in place of* rescuing migrants called for a withdrawal of this humanitarian element from EU bordering practice. As such it stood out against the mainstream European rhetoric on boat migration, where humanitarian claims have typically gone hand in hand with those of security and law enforcement. I want to suggest that the moment when James Brokenshire called for the swift ending of emergency rescue measures as a way to counter the migratory crisis is significant because it was a moment when the humanitarian mask came off. It exposed with stark clarity the deadly nature of European border politics which is of course apparent to anyone who cares to pay attention, but which in official discourse is usually obscured by human rights rhetoric and the technocratic jargon of "migration management" (Maguire and Murphy 2009).

Drawing on the discussions concerning the 'let them drown' policy and engaging the literature on the EU border regime, in this chapter I want to make a two-fold argument, consisting of a conceptual observation and a methodological proposition. Conceptually, I want to build on the recent influential accounts of the EU border regime as a decentered apparatus (Feldman 2012), assemblage or industry (Andersson 2014; Gammeltoft-Hansen and Sørensen 2013). These accounts, rooted in a wealth of empirical evidence, are compelling in their engagement with the multilayered nature of the contemporary border politics. But I also note what they leave out. In observing that the governance of borders and human mobility is decentred, enacted in a variety of sites by a multiplicity of differently empowered actors, these anthropologists and political sociologists have little patience for national governments, legislative bodies and their politics. As Balibar insisted (2002), borders are everywhere and they proliferate. To understand how this process occurs it does indeed make little sense to attribute too much agency and significance to any one actor or site, central or otherwise.[2] However, the expert knowledge that underpins the dynamics of the EU border regime travels along particular paths and those paths are thoroughly politicized. They lead downstream, from experts to functionaries, and upstream, from practitioners, via experts to the political centers. I argue that we could benefit from revisiting the relationship between the sites where conventional national and EU politics is exercised and the design and functioning of the EU border regime. One way to do that is to track how knowledge travels. In this chapter I am interested in the flow upstream.

From that follows my methodological point, which is that central nodes of EU and national politics (parliamentary chambers, governmental offices and the discourses they produce) are unjustifiably neglected as ethnographic field sites for the study of the contemporary European border regime. In the case of the 'let them drown' policy, the shedding of the humanitarian mask by the UK government is an ethnographic opportunity, an entry point into the study of shifts in border politics. In what follows I want to recover that opportunity and to do so I will proceed in three steps. First, I briefly consider the theoretical question of where political power is located within the EU border regime. Second, I discuss my own fieldwork with Council of Europe parliamentary rapporteurs to highlight how bordering knowledge and expertise travels upstream to inform political decisions such as the 'let them drown' policy. Third, I conclude with the argument that while the practical implementation, the consequences of and the resistances to this and related policies (cf. Mountz 2012) can only be grasped through ethnographic attention to situated detail, we must also understand how such decisions are reached in the first place, or else we might misunderstand the distribution of power within the border regime and thus miss the opportunities to challenge it.

Political power in the EU border regime

In *The Migration Apparatus*, the anthropologist Gregory Feldman describes the world of EU migration policymaking as acephalous, that is headless, "a decentralized apparatus of migration management composed of disparate migration policy agendas, generic regulatory mechanisms and unconnected policy actors and policy 'targets'" (Feldman 2012, 5). Feldman's ethnographic focus is on the technocrats responsible for the day-to-day perpetuation of the apparatus and on the "*mediated* practice of policymaking" which, as he argues following Foucault, renders the migrant "an object of information, never a subject of communication" (Foucault 1977, cited in Feldman 2012, 6). In spite of tensions within the apparatus, migration policymaking in the EU is converging within a political context that Feldman describes as a contest of "right vs. right" (2012, 25). The traditional left has eroded and what remains is a political scene dominated by the right bifurcated into neoliberalism and neo-nationalism. Between them,

> [They] share a strong desire to crack down on illegal migration ... [and] a concern for a strong security establishment, even though neoliberals want it to clarify the circuits through which discrete and mobile individuals move and neo-nationalists want it to protect a more rooted and well defined national (or local) collective. It is thus no coincidence that the EU has had more success in reaching agreements on the negative aspects of migration management, for example border security and migrant return than in other migration policy domains.
>
> (Feldman 2012, 9)

In the course of pursuing his, as he calls it, "nonlocal" ethnography, Feldman met with many of the public officials, academics and non-government organization (NGO) and inter-governmental organization (IGO) workers for whom the "negative aspects of migration management" constitute their day-to-day trade. Consequently, they are represented as flesh and blood people in his account, committed "to law, order and sovereignty" but also experiencing and expressing conflicts of conscience (Feldman 2013, 137).[3] Politics is reduced to context, to the interplay of abstract ideological orientations (neoliberal and neo-nationalist) which frame and restrict the unfolding of

EU migration policy. In a Foucauldian fashion, power resides in dynamic systems and processes rather than in any particular institution or agent, be it at the EU or the national level.

I do not propose to dismantle this account, nor am I suggesting that the apparatus does have a (single) head after all. Insofar as we are looking for an insight into the complex system where Frontex and Europol risk analysts mix and mingle with International Organization for Migration (IOM) experts, United Nations High Commission for Refugees (UNHCR) refugee advocates and representatives of national public administrations and law enforcement agencies, where each actor enters with his or her own personal and political agenda, needs, ambitions and demands of their superiors, to that extent Feldman's account goes a long way. But it has comparatively less to tell us about how forces of political leadership shape the EU border regime, or about the processes whereby consequential decisions are taken and business as usual gets temporarily interrupted to resume in altered conditions. This is the Weberian dimension of the border regime grounded in the understanding of politics as leadership in action, "striving to share power or striving to influence the distribution of power" (Gerth and Mills 1968 [1946], 78). According to Weber,

> [w]hen a question is said to be a political question, when a cabinet minister or official is said to be a political official, or when a decision is said to be politically determined, what is always meant *is that interests in the distribution, maintenance or transfer of power are decisive for answering the question and determining the decision in the sphere of activity.*
>
> (Gerth and Mills 1968, 78, emphasis added)

From this perspective we must distinguish between those actors within the border regime who are bureaucrats and those who are political officials in the above sense. Bureaucrats are servants of the state endowed with 'bureaucratic authority' (ibid., 196) that is the ability to discharge their duties according to predetermined rules. We may substitute the term 'technocrat' to reflect the complexity of technical and technological expertise required of the contemporary equivalent of the Weberian file master (ibid., 197). We may consider the differences between the nature of bureaucratic and technocratic authority, which are considerable (although beyond the scope of this chapter). We should also observe that no rules are ever absolute and that complex political demands and organizational tensions require administrative discretion which comes in a variety of context-specific forms (Bouchard and Carroll 2002). In the end however, whether they preside over files or databases, these officials are not the ones who bear the *political* responsibility for the work that they do.[4] There are other kinds of responsibility, to be sure. Moral and legal come to mind, but *political* responsibility which is inherently public in nature does not accrue directly to servants of the state.

Similar constrains frame the authority of law enforcement, which is about enforcing the law such as it is, not about shaping it. This is why border guards and police officers, no matter the rank, always claim to be apolitical or, to quote my own informants from my study of the Polish–Ukrainian border, "concerned with the practice" and not with the principles of the policing of borders and mobility (Follis 2012, 91).[5] Likewise, from an emic perspective, complex endeavors like capacity building and such EU policy blueprints as the Integrated Border Management and Common European Asylum Policy are described as "not [being] about ideology" but rather amounting to "a technical ... matter" (ibid., 159). We may counter in a Foucauldian vein, that these things are in fact the very embodiment of a specific border politics. However, that doesn't change the fact that to the insiders, the principles are a given and *not* disputing them is part of the ethos of their service.

Political officials on the other hand have an essential interest in power understood as political rule; that is the ability to impose their will, ideology and interests onto the collective agenda. The capability to do so comes, in theory, with accountability to the public for the consequences of their actions (Keohane 2002). The principles of policy and the laws to be enforced are sometimes forged in the heat of the battles politicians fight to assert their agendas. More often they emerge from acts of political conformity. Political decisions can also be deeply influenced by powerful industries and their economic interests (Andersson 2014; Carmel 2013). Such dynamics may radically narrow the room for any kind of challenge to the status quo, as in Feldman's discussion of "right vs. right" (Feldman 2012, 25). They can also produce stalemates making any change or reform impossible, as is the case with the United States (US) immigration policy (Zatz and Rodriguez 2014). However, if, as Chris Shore and Sue Wright argued, "[p]olicy is the ghost in the machine – the force which breathes life and purpose into the machinery of government and animates the otherwise dead hand of bureaucracy" (Shore and Wright 1997, 4), the ghost cannot be conjured without political agency and approval.[6] This is the case even if the tense and halting nature of modern politics makes the roles of different actors often difficult to tease apart.

This discussion relies of course on ideal types. In the contemporary reality of EU border politics boundaries between bureaucrats, law enforcement and political officials can be blurry and not always apparent at first sight (cf. Walters and Haahr 2000). Nevertheless we must not confuse bureaucrats, technocrats, civil servants and law enforcement officers with active career politicians. Whether as those who aspire to, or run for office, or as law-makers and members of the executive, politicians are key actors who "define the terms of political contestation" (Holmes 2000, 63). They have created the frames for the EU border regime to assume the shape it has now. It follows that they, and not the bureau- or technocrats, are the ones who could *create the conditions* for it to change. This could happen only if the publics politicians purport to represent so demanded.

Ethnography up the political stream

In April 2014, a few months before the 'let them drown' policy made news in the UK, I had the opportunity to speak to Christopher Chope, a backbench Tory Member of Parliament (MP) in the UK House of Commons. I spoke to him about his work as a parliamentary rapporteur for the Committee on Refugees, Migrants and Displaced Persons of the Parliamentary Assembly of the Council of Europe (PACE). In the UK House of Commons, Chope represents a wealthy constituency on the southern coast of England. In the May 2015 General Election his seat was targeted by the UK Independence Party (UKIP), the Eurosceptic and anti-immigrant party led by the charismatic Nigel Farage (Daily Echo 2014). UKIP did very well in the European Parliament election in 2014 and was projected to seriously upset the status quo in the general election that followed. In the end the Tories won a decisive victory, Chope retained his seat and some of UKIP's positions are represented in the new Parliament by the Conservative Party's backbench. In PACE Chope was one of the eight representatives of the UK Conservative Party. PACE is the parliamentary body of the Council of Europe (CoE), the Strasbourg-based international organization of 47 member states whose goal is to advance peace, democracy and human rights in Europe (separate from the EU, not to be confused with the European Council). PACE members are currently serving MPs representing national parliaments who work in thematic committees and meet in plenary sessions to discuss and develop recommendations on issues within the mandate of the CoE. The Assembly's discussions do not translate directly into any policy decisions, but they are distributed to the EU and national parliaments and frequently cited in policy documents. They contribute to the development of soft law and CoE conventions, which become binding when national parliaments ratify them. The informal interview I conducted with Chope took place at the CoE headquarters in Strasbourg and was part of a research project on responsibility for migrant boat disasters, where I wanted to understand how European human rights bodies account for such catastrophes and who is thought to be responsible when migrants die at sea (Follis 2015). However, as is usually the case, the fieldwork yielded other insights as well.

I noted for example the sharp distinction that the national MPs whom I interviewed drew between civil servants, and themselves as politicians. They saw themselves as key actors supported by those whose role is to run the machinery of governments and international institutions. At the CoE and in the national parliamentary offices this understanding was mutual. The staff, that is the often highly skilled and knowledgeable individuals who prepare and run meetings, compile documents and transmit communiqués, were quite secure in their identity as civil servants. They distinguished their own role not just from that of the politicians, but also from the roles of "experts" (usually representatives of EU or UN agencies) and NGO workers.[7] At the time of my fieldwork, Christopher Chope was responsible for drafting a report on the arrival of seaborne migrants in Italy (PACE 2014). The topic was an urgent one for the committee, particularly since the much publicized PACE inquiry into the 2011 case of the 'left-to-die' boat (PACE 2012). In that case, 63 of 72 passengers of a boat en route from Libya to Lampedusa died when the vessel was left to drift on the Mediterranean for two weeks, in spite of sending distress signals which were received and recorded by relevant authorities (Shenker 2011). Tineke Strik, the Committee member who drafted that report and pushed it through the approval process in the Assembly, is a senator from The Netherlands representing the Greens. On the several occasions I spoke to her and heard her speak in public she made no excuses for her left-wing politics. However, PACE committees strive for balance in their reporting and it was in keeping with this principle that another report on a related issue would be assigned to an MP from the opposite end of the political spectrum.

Christopher Chope volunteered to report back to the Migration Committee (and in due course to the Assembly) on how Italian authorities are coping with the arrival of "mixed migratory flows" to their coastal areas. He was approved for the role in 2011 and it took him over two years to compile the report. The idea that PACE was in need of insight into that particular problem emerged from the debates in the Migration Committee. But as is the case with all reports, the rapporteur, once selected, pursues the work as he or she sees fit. Constraints stem from limited resources, but not any from any sort of prescription regarding what can and cannot be the subject of inquiry. The rapporteurs decide what issues they take on, and how they approach them. Indeed at the very start of our conversation Chope affirmed that, "I am a politician. My role is to call things as I see them and to propose solutions."

Political work on a PACE report begins with the framing of the subject. It has to be done subtly, so as not to alienate fellow Assembly members who represent different political orientations, but who will ultimately vote on the final text (when a report passes the vote in a plenary session, it becomes an official PACE document). It is advisable to allude to the basic principles underpinning the work of the Council of Europe, which are national sovereignty, rule of law, human rights and democracy. Likewise, it makes sense to establish legitimacy by referencing respectable authorities who guard standards hailed as universal. In the case of most migration related issues, the UNHCR and the Fundamental Rights Agency of the EU (FRA) usually count as such authorities. The title of Chope's report was "The large-scale arrival of mixed migratory flows on Italian shores," where "*mixed* migratory flows" was the operative concept. 'Mixed migration' is a key term in current migration policy jargon, validated by the fact the UNHCR and other authorities use it in contexts where in the past they might have invoked 'refugee flows.' Unlike 'refugee flows' however, which implies the need for and the entitlement to protection, the concept of 'mixed migration' makes no promises. It reflects the presumption that boat migrants are differential legal subjects. Some of them are indeed people who would be recognized as refugees under most

European countries' refugee status determination procedures, but the vast majority are economic migrants or asylum seekers who would fail to secure legal status, and who therefore ought to be returned to their home countries or granted only temporary permission to stay on humanitarian grounds. Italy, according to Chope, is dealing with mixed flows, not with refugees. Therefore the issue that Italy is facing and that Chope set out to investigate is not one of how to process and accommodate the new arrivals, but rather how to sift the wheat of the few 'legitimate refugees' from the chaff of everyone else, and how quickly and efficiently to send home the majority who are not entitled to international protection and who ostensibly are in no need of it in the first place. He explained:

> You probably think that the people on the boats are poor people, desperate people and that we should feel sorry for them and that we should help them. But that is not true. They are actually people who are quite well off. They have the money, a lot of money, to pay the smugglers and to get on the boats, and to get to Europe. Why do they do that? Because they know that when they arrive, all they need to do is ask for asylum and they will be on their way. They know how the EU asylum system works and they play it. They are middle class people who are effectively jumping the queue to come to our shores. They could apply for visas. They could look for work in the UK for example and be sponsored by an employer. That is the fair way, the legal way. But they chose not to do that and the EU makes it easy for them not to do that … It really is time to rethink the whole concept of asylum.[8]

As I probed whether the "rethinking" extended to the concept of non-refoulement, Chope turned the conversation to "pull factors" and noted that the fact that European states generally do not send people back to certain countries is one of them. He then asserted that rescuing migrants at sea, even though a noble humanitarian goal in itself, has become a pull factor too. He described how the human smugglers turned the Italian operation *Mare Nostrum* into a convenient system for delivering migrants to European shores. He explained that rather than packing human cargo on rickety dinghies in Libya, the smugglers now (that is when *Mare Nostrum* was in effect) use larger vessels, which he called "mother ships." They then offload the passengers onto lifeboats and send them off into Italian territorial waters, knowing that they will be rescued. "It is a huge business," he said, "of organized gangs which need to be targeted." When migrants die, smugglers are to blame, not European states. European states are straining under the burden which, he implied, they brought upon themselves by being far too generous in their treatment of migrants.

The report Chope ultimately assembled is diplomatic, but nonetheless it reflects the views of its author. For example, in a section titled "The Counter-effects of Mare Nostrum," after acknowledging the number of lives saved, the report says,

[Mare Nostrum] is also thought to have contributed to a much increased flow of seaborne migrants setting out from North Africa and heading for Italy. The tenfold increase in January 2014 compared with January 2013 is considered to be related to the greater certainty on the part of migrants that they will reach their chosen destination alive.

(PACE 2014, par. 17)

The paragraph then goes on to recount the "mother ship" story, which is sourced to the Italian navy. It ends by noting that "many migrant boats are equipped with a mobile phone which has the direct number for the Italian Coastguard" (PACE 2014, par. 17).

The recommendations appended to Chope's findings are unusually sparse and they stop short of advocating for the end of rescue. They do however contain a suggestion "to introduce a new international crime, whether or not defined as a crime against humanity, when a person receives a financial benefit directly or indirectly for transporting people in a vessel which is unsafe for the purpose" (par. 4.1). They also suggest that "the authorities of the countries concerned … open negotiations on the modalities and conditions of return to countries of embarkation of migrants intercepted in international waters" (par. 4.3). One does not need particularly refined between-the-lines reading skills to understand that this means finding ways around the principle of non-refoulement. The humanitarian mask is still on, but it covers a scowl.

How can we explain the shedding of the mask by the UK government in October 2014? I am not suggesting that one MP's findings about the situation on the Mediterranean directly influenced the Foreign and Home Offices' decision to refrain from lifesaving operations. Rather, I highlight the need to explore ethnographically sites such as the Council of Europe reporting process alongside other international venues where European border politics are articulated and where they crystallize. I call this approach 'ethnography up the stream' because it focuses on the process whereby politicians come to develop an understating of their subjects of interest and how that translates into political positions, which eventually come to inform policy decisions. Participating MPs interact with their peers from across Europe debating matters safely under the "gloss of harmony" (Müller 2013), but simultaneously they feed back to their political bases back home the knowledge and perspectives that otherwise may have not come on their radar. The language of the documents they produce may be muddled and strategically ambiguous, but the ethnographer's immersion in the institutional context helps peel back the layers of obfuscation.

Chope's work passed muster at the Assembly and made its rounds through the offices of parliamentarians and government officials in various European states. Although I did not have the chance to ask him about this in person, I imagine that he made sure that it was received by his fellow members in the House of Commons. As the discussions around Mare Nostrum unfolded informed by the views of rescue-sceptics like Chope, the electoral threat from UKIP to the Tories continued to mount. By declaring an end to UK support for rescue operations, we can imagine Theresa May thinking that she killed two birds with one stone. The 'let them drown' policy strikes a UKIPian Eurosceptic chord in that it refuses to engage in any Brussels-led nonsense other than a basic EU border enforcement operation. It also makes the anti-immigrant voter happy by suggesting that the rules of access to European territory are not there to be gamed, even at the price of death. Yet judging by the reaction in the House of Commons and by the commentaries even the ordinarily Tory-friendly press, this might have been a step too far.[9]

Conclusion

In this chapter I have argued that the anthropology of the EU border regime would benefit from a renewed analysis of the relationship between the sites where conventional national and EU politics is exercised and the design and functioning of the EU border regime. I reflected on the locations of power within the management of borders and human mobility, ultimately to contend that European politicians and not the bureau- or technocrats created the conditions for the EU border regime to assume the shape it has now. Politicians therefore are the ones who, if their constituents so demanded, could *create the conditions* for it to change. Drawing on my project of ethnography up the stream, I have shown how bordering knowledge and expertise travels not just downstream, from experts to practitioners, but also upstream to inform political decisions such as the UK 'let them drown' policy.

The October 2014 decision to refrain from preemptive rescue represents the first official admission in Europe of that which up to that point we had only read between the lines, namely that that the UK, and possibly the EU as a whole, is in the process disentangling its border policy and practice from the humanitarian imperative of assisting first and asking questions later to an a priori orientation of not taking migrant survivors in. Whatever we think of the insufficiency, hypocrisy and other flaws of ordinary humanitarian responses (Barnett and Weiss 2008; Fassin 2012), this decision demanded a challenge both at the highest level and at the myriad openings that exist as political knowledge travels up the stream.

Epilogue

The first effective challenge to the 'let them drown' policy came six months after it was first made public, that is in the spring of 2015 after the disaster of April 19, when approximately 800 people died in a single incident of a boat carrying migrants that capsized 120 miles south of Lampedusa. This event set a new record for a single disaster death toll at a time when the numbers of deaths at sea were already increasing exponentially compared to the same time of year in 2014.[10] The tragedy, and the subsequent special meeting of the European Council on April 23 to tackle the migrant crisis on the Mediterranean, unfolded during the UK's general election campaign, prompting the leaders of the contending parties to speak out, and causing a rift in the Prime Minister David Cameron's cabinet. Ed Miliband, at the time still the leader of the Labour Party vying for electoral victory, called for 'Mare Nostrum' to be reinstated and declared that if he were prime minster going to the European Council special meeting, he would call for renewed search and rescue efforts. His was a moral appeal as he argued "there is no trade-off between controlling immigration, showing basic humanity and living up to our moral responsibilities as a country. That reflects the values of the British people" (Travis 2015a). Meanwhile, UKIP's Nigel Farage took the opportunity to reiterate his familiar points about the need to oppose "unlimited immigration" and any joint EU action on the Mediterranean, which according to Farage would only lead to further extension of the powers of Brussels (Mason 2015).

As to the argument that search and rescue constituted a pull factor for migrants and should therefore be withheld, David Cameron appeared to perform an about face, as "the extensive media coverage convinced Downing Street and Tory election strategists that voters see the tragedy in the Mediterranean as a humanitarian crisis rather than an immigration issue" (Travis 2015a). In Luxembourg, at the special summit on April 23, Cameron agreed to deploy the Royal Navy flagship HMS Bulwark, which happened already to be near the epicenter of the migrant crisis, having just led the parade of ships at the centenary celebration of the Gallipoli campaign in the Dardanelles. Under the new agreement, the British assault vessel would assist with search and rescue between the Libyan and Italian coast, albeit under the strict condition that there would be no expectation that any of the saved migrants would be admitted into the UK. The plan for this move, according to a story published on April 22 by *The Guardian's* Home Affairs editor, faced strong opposition from the Home Secretary Theresa May and the foreign secretary Philip Hammond. May and Hammond "were … said not to be budging from their belief that such rescue operations actually create a 'pull factor' and lead to more deaths by encouraging more migrants to risk the dangerous sea crossing" (Travis 2015a). At the same time they insisted that the principal priorities in the area were shutting down the migrant smuggling routes and establishing a rapid returns program (ibid.). The home secretary declared also that the UK would not participate in the EU plan to relocate 40,000 Syrian and Eritrean migrants arriving in Italy and Greece to other EU Member States (European Commission 2015). In the meantime, in the first month of its deployment HMS Bulwark rescued over 2,700 migrants who were subsequently disembarked in Italian ports. The vessel's mission expired on July 5, 2015 (Travis 2015b). A ship one fifth of its size, the HMS Enterprise, was sent to the area with the principal aim of searching, identifying and disposing of smugglers' boats, not of rescuing migrants (Travis 2015c). And thus the tension between tough-minded anti-humanitarianism and the imperative, apparently now embraced by Cameron, to adhere to the moral minimum of "saving lives," is set to continue playing out for some time at the highest levels of the UK government. It shall unfold in the shadows of preparations for the referendum on UK's continued membership in the EU (set to take place before the end of 2017) and against the wider background of EU debates over burden-sharing, migrant quotas, the boundaries of the planned military operation to sink smugglers' boats, and other ways to respond to the situation in the Mediterranean (Nielsen 2015). But regardless of the final outcome in any one of these tangled controversies, the argument that rescuing people only makes more of them arrive has already been made and broadcast. The suspicion that our humanitarian impulses only attract an undeserving mob of unwanted "economic migrants" has been firmly planted, in the words of François Crépeau, exactly like the enduring conviction "that the social safety net encourages idleness" (Jackson 2015). The view has been given credibility by a major European political force and will continue to attract supporters to the anti-humanitarian model of border control. Based on the antagonistic tone of the debate in the UK, and increasingly also elsewhere in the EU, it appears that opposing, resisting and challenging this anti-humanitarianism will require an unprecedented mobilization of all migrants' rights advocates, regardless of their specific ideological and activist orientations, at all stages of the political process.

Notes

1 In a scathing editorial in the right-leaning *Daily Telegraph*, Dan Hodges summarized the government's reasoning thus:

 We understand that by withdrawing this rescue cover we will be leaving innocent children, women and men to drown who we would otherwise have saved. But eventually word will get around the war-torn communities of Syria and Libya and the other unstable nations of the region that we are indeed leaving innocent children, women and men to drown. And when it does, they will think twice about making the journey. And so eventually, over time, more lives will be saved.

 (Hodges 2014)

2 See Rumford's comments on Balibar's idea of the "polysemic nature of borders" (Rumford 2008, 39–44).
3 In a related article Feldman shows that these functionaries are on occasion capable of stepping outside the frame of their prescribed role and engage in thinking (in the Arendtian sense) which is the precondition of political action (Feldman 2013).
4 Responsibility has two conventional senses: the "capability of fulfilling an obligation or duty" and the "state or fact of being accountable" (OED n.d.). I use it here in the latter sense. To be accountable means to be answerable, which entails a relationship: accountability is the responsibility of an agent to someone else, for example the electorate (Keohane 2002, 1124) or, as recently suggested by Goodhart (2011), to specific norms (see Follis 2015).
5 A body of literature in criminology addresses the issue of discretion in policing. See Bordner 1983; Lowe 2011.
6 For a related argument developed with reference to the political science literature on EU integration studies see Ripoll Servent and Busby 2013.

7 The role of the NGO workers can be thought of as implicitly political as they come to committee meetings as advocates, in the case of the migration committee on behalf of refugees and asylum seekers (Keck and Sikkink 1998).
8 Interview with Christopher Chope, April 8, 2014. Christopher Chope's comments were reconstructed from field notes. The interview was not recorded.
9 See the Galloway and Sheridan motion at www.parliament.uk/edm/2014-15/477 (accessed on March 2, 2015).
10 On 23 April 2015, IOM reported that between January and 21 April 2015, 1,727 people lost their lives in the Mediterranean. With these figures, "the 2015 death toll now is more than 30 times last year's total at this date (April 21) when just 56 deaths of migrants were reported in the Mediterranean" (IOM 2015).

Bibliography

Amnesty International 2014. *Amnesty Condemns UK Opt Out of Search and Rescue for Refugees and Migrants*. Available from: www.amnesty.org.uk/press-releases/amnesty-condemns-uk-opt-out-search-and-rescue-refugees-and-migrants [27 February 2015].

Andersson, R. 2014a. *Illegality, Inc. Clandestine Migration and the Bussiness of Bordering Europe*. University of California Press, Berkeley.

Andersson, R. 2014b. Mare Nostrum and migrant deaths: the humanitarian paradox at Europe's frontiers, *50.50 Inclusive Democracy,* 30 October. Available from: www.opendemocracy.net/5050/ruben-andersson/mare-nostrum-and-migrant-deaths-humanitarian-paradox-at-europe%E2%80%99s-frontiers-0 [27 February 2015].

Balibar, É. 2002. *Politics and the Other Scene*. Verso, London.

Barnett, M. and Weiss, T. M. (eds) 2008. *Humanitarianism in Questions. Power, Politics, Ethics*. Cornell University Press, Ithaca, New York.

Bordner, D. 1983. Routine policing, discretion and the definition of law, order and justice in society, *Criminology*, vol. 21, no. 22, pp. 294–304.

Bouchard, G. and Wake Carroll, B. 2002, Policy-making and administrative discretion: the case of immigration policy in Canada, *Canadian Public Administration/ Administration Publique Du Canada*, vol. 45, no. 2, pp. 239–57.

Carmel, E. 2013. Lampedusa and marketized surveillance in the Mediterranean: a political drama in two acts, *E-International Relations,* 23 October. Available from: www.e-ir.info/2013/10/25/lampedusa-and-marketized-surveillance-in-the-mediterranean-a-political-drama-in-two-acts/ [20 May 2015].

Cuttitta, P. 2014. 'Borderizing' the island. Setting and narratives of the Lampedusa 'Border Play', *ACME: An International E-Journal for Critical Geographies*, vol. 13, no. 2, pp. 196–220.

Daily Echo 2014. "I am going to win Christchurch for UKIP," says candidate as town is named in 100 seats to target, *Daily Echo,* 4 September. Available from: www.dailyecho.co.uk/news/11453691._I_am_going_to_win_Christchurch_for_UKIP___ says_candidate_as_town_is_named_in_100_seats_to_target/ [20 May 2015].

Dzenovska, D. 2014. Bordering encounters, sociality and distribution of the ability to live a 'normal life,' *Social Anthropology*, vol. 22, pp. 271–287.

European Commission 2015. European Commission makes progress on Agenda on Migration, Press Release, 27 May. Available from: http://europa.eu/rapid/press-release_IP-15-5039_en.htm [17 June 2015].

Fassin, D. 2012. *Humanitarian Reason: A Moral History of the Present*. University of California Press, Berkeley.

Feldman, G. 2012. *The Migration Apparatus: Security, Labor, and Policy Making in the European Union.* Stanford University Press, Palo Alto.

Feldman, G. 2013. The specific intellectual's pivotal position: action, compassion and thinking in administrative society, an Arendtian view, *Social Anthropology*, vol. 21, pp. 135–54.

Follis, K. 2012. *Building Fortress Europe. The Polish–Ukrainian Frontier.* University of Pennsylvania Press, Philadelphia.

Follis, K. 2015. Responsibility, emergency, blame: reporting on migrant deaths on the Mediterranean in the Council of Europe, *Journal of Human Rights*, vol. 14, no. 1, pp. 41–62.

Foucault, M. 1977. *Discipline and Punish*, Vintage, New York.

Gammeltoft-Hansen, T. and Sørensen, N. (eds) 2013. *The Migration Industry and the Commercialization of International Migration.* Routledge, London and New York.

Gerth, H. H. and Mills, C. W. (eds) 1968 [1946]. *From Max Weber. Essays in Sociology.* Oxford University Press, New York.

Goodhart, M. 2011. Democratic accountability in global politics: norms, not agents, *The Journal of Politics*, vol. 73, no. 1, pp. 45–60.

Hall, A. 2010, "These people could be anyone": fear, contempt (and empathy) in a British immigrant removal centre, *Journal of Ethnic and Migration Studies*, vol. 36, pp. 881–98.

Hodges, D. 2014. Drown an immigrant to save an immigrant: why is the Government borrowing policy from the BNP? *The Telegraph*, 28 October. Available from: www.telegraph.co.uk/news/politics/11192208/Drown-an-immigrant-tosave-an-immigrant-why-is-the-Government-borrowing-policy-from-the-BNP.html [27 February 2015].

Holmes, D. R. 2000. *Integral Europe: Fast-Capitalism, Multiculturalism, Neofascism.* Princeton University Press, Princeton.

IOM 2015. *Survivors of Mediterranean Tragedy Arrive in Sicily.* Available from: www.iom.int/news/survivors-mediterranean-tragedy-arrive-sicily [20 May 2015].

Jackson, G. 2015, UN's François Crépeau on the refugee crisis: "Instead of resisting migration, let's organize it", *The Guardian*, 22 April. Available from www.theguardian.com/world/2015/apr/22/uns-francois-crepeau-on-the-refugee-crisis-instead-of-resisting-migration-lets-organise-it, [20 May 2015].

Keck, M. E. and Sikkink, K. 1998. *Activists Beyond Borders: Advocacy Networks in International Politics.* Cornell University Press, Ithaca.

Keohane, R. O. 2002. The concept of accountability in world politics and the use of force, *Michigan Journal of International Law*, vol. 24, pp. 1121–40.

Lowe, D. 2011. The lack of discretion in high policing, *Policing and Society: An International Journal of Research and Policy*, vol. 21, no. 2, pp. 233–47.

Maguire, M. and Murphy, F. (eds) 2009. Managing migration? The politics of truth and life itself, a special issue', *Irish Journal of Anthropology*, vol. 12, pp. 1–80.

Mason, R. 2015. Nigel Farage opposes EU action to tackle migrant deaths in Mediterranean, *The Guardian*, 20 April. Available from: www.theguardian.com/politics/2015/apr/20/nigel-farage-opposes-eu-action-to-tackle-migrant-deaths-in-mediterranean [17 June 2015].

Mountz, A. 2012. Mapping remote detention. Dis/location through isolation, in *Beyond Walls and Cages: Prisons, Borders and Global Crisis*, eds J. M. Loyd, M. Mitchelson and A. Burridge, University of Georgia Press, Athens.

Müller, B. (ed.) 2013. *The Gloss of Harmony. The Politics of Policy-Making in Multilateral Organisations.* Pluto Press, London.

Nielsen, N. 2015. EU ministers discuss migrant scheme, *EU Observer*, 16 June. Available from: https://euobserver.com/justice/129134 [17 June 2015].

OHCHR 2012. "Let them die, this is a good deterrence" – UN human rights expert, press release 30 October. Available from: www.ohchr.org/EN/NewsEvents/Pages/DisplayNews.aspx?NewsID=15239&LangID=E#sthash.0UCs6nNm.dpuf [27 February 2015].

PACE 2012. *Lives Lost in the Mediterranean Sea: Who Is Responsible?* Available from: http://assembly.coe.int/ASP/XRef/X2H-DW-XSL.asp?fileid=18095&lang=en [30 August 2012].

PACE 2014. *The Large-Scale Arrival of Mixed Migratory Flows to Italian Coastal Areas.* Available from: http://assembly.coe.int/nw/xml/XRef/Xref-DocDetails-EN.asp?fileid=20941&lang=EN [27 February 2015].

Ripoll Servent, A. and Busby, A. 2013, Introduction: Agency and influence inside the EU institutions, *European Integration Online Papers*, 17(1), Article 3, available from: http://eiop.or.at/eiop/pdf/2013-003.pdf, [20 May 2015].

Rumford, C. 2008. *Cosmopolitan Spaces: Europe, Globalization, Theory.* Routledge, New York and London.

Shenker, J. 2011. Aircraft carrier left us to die, say migrants, *The Guardian,* May 8. Available from: www.theguardian.com/world/2011/may/08/nato-ship-libyan-migrants [17 June 2015].

Shore, C. and Wright, S. 1997. *Anthropology of Policy: Critical Perspectives on Governance and Power.* Routledge, London.

Travis, A. 2014a. UK axes support for Mediterranean migrant rescue operation, *The Guardian,* 27 October. Available from: www.theguardian.com/politics/2014/oct/27/uk-mediterranean-migrant-rescue-plan [27 February 2015].

Travis, A. 2014b. Home Office defends decision for UK to halt migrant rescues, *The Guardian,* 28 October. Available from:www.theguardian.com/world/2014/oct/28/home-office-defends-uk-migrant-pull-factor [27 February 2015].

Travis, A. 2014c. Migrant rescue operations must be stopped at earliest opportunity, says minister, *The Guardian,* 30 October. Available from: www.theguardian.com/uk-news/2014/oct/30/home-office-minister-rescue-migrants-must-be-stopped-mediterranean [27 February 2015].

Travis, A. 2015a. UK Cabinet split over EU plans to expand sea search and rescue of migrants, *The Guardian*, 22 April. Available from: www.theguardian.com/world/2015/apr/22/uk-cabinet-split-over-eu-plans-to-expand-sea-search-and-rescue-of-migrants [17 June 2015].

Travis, A. 2015b. UK could withdraw from migrant rescue missions in Mediterranean, *The Guardian*, 16 June. Available from: www.theguardian.com/uk-news/2015/jun/16/uk-set-to-withdraw-hms-bulwark-migrant-rescue-missions-mediterranean [17 June 2015].

Travis, A. 2015c, HMS Bulwark's replacement yet to rescue any migrants in the Mediterranean, *The Guardian*, 27 July. Available from: www.theguardian.com/world/2015/jul/27/hms-bulwark-replacement-has-yet-to-rescue-any-migrants-in-mediterranean, [29 October 2015].

UK Parliament 2014. *Daily Hansard,* Wednesday 15 October. Available from: www.publications.parliament.uk/pa/ld201415/ldhansrd/text/141015w0001.htm [27 February 2015].

Walters, W. and Haahr, J. H. 2000. *Governing Europe. Discourse, Governmentality and European Integration.* Routledge, London and New York.

Zatz, M. S. and Rodriguez, N. 2014. the limits of discretion: challenges and dilemmas of prosecutorial discretion in immigration enforcement', *Law & Social Inquiry*, vol. 39, no. 3, pp. 666–89.

5 The politics of negotiating EU Readmission Agreements

Insights from Morocco and Turkey

Sarah Wolff

Introduction

This chapter analyses the politics of instrumentation of European Union (EU) Readmission Agreement (EURA) from a third country perspective. It scrutinizes the role of EU incentives and third countries' preferences in the negotiations of this policy instrument, which has been originally designed as '[a] new policy instrument[s] in the construction of rational and orderly immigration regimes' (Baldwin-Edwards 1997: 511). It evidences the transformations affecting the governance of migration, through the study of this specific policy instrument used in the EU's remote control policy with third countries. Through the case studies of Morocco and Turkey, it demonstrates that policy instruments are inherently political and cannot be taken as neutral technical devices. EURA are used as venues of political negotiations both for the EU and third countries and have become a space of contestation and resilience for third countries. For ten years, both countries refused to sign an EURA and therefore share a similar position of 'hard bargainers'. In December 2013, though, a 'negotiation turn' took place, Turkey signed the EURA and Morocco committed to sign an EURA within the framework of a Mobility Partnership (MP) in June 2013. The surge of influx of Syrian refugees and of mixed migration flows towards Europe in 2015 has turned the issue of the cooperation of third countries and in particular of Turkey into a prominent issue on the EU agenda (Wolff, 2015a). This chapter argues that third countries are not passive actors when confronted to the externalization of border controls and are able to influence to some extent the EU. This thus enlightens the understanding of the instrumental dimension of remote control policies (see Chapter 1, this volume) and is to be contextualised within new strategies of the EU to gain support of its neighbours as exemplified by the EU-Turkey Joint Action Plan concluded in October 2015 by the European Commission.

Beyond the functional need of EURAs to co-opt third countries in the EU's fight against irregular migration, this chapter shows that a series of obstacles forced the EU to revise the design of EURA. New EU incentives were offered: a Mobility Partnership (MP) to Morocco and visa liberalization to Turkey. Yet, in spite of a fine-tuning of EU incentives over time, this chapter finds that third countries' political domestic and regional dynamics constrains the politics of EURA instrumentation. Concerned with the implications for EU external migration policy more broadly, the meanings and representations carried by EURAs in third countries, following Le Galès and Lascoumes, are also relevant to this study. For these authors a public policy instrument is

> a device that is both technical and social, that organizes specific social relations between the state and those it is addressed to, according to the representations and meanings it carries. It is a particular type of institution, a technical device with the generic purpose of carrying a concrete concept of the politics/society.
>
> (Lascoumes and Le Galès 2007)

This necessarily implies looking at the issues of power and appropriateness of EURAs as external migration policy instruments.

In nature, EURAs are both agreement and incentive-based policy instruments. As a bilateral agreement their purpose is to return irregular migrants. Third countries readmit their own nationals and third country nationals having transited through their territory. The Council opens EURAs negotiations on the basis of a recommendation from the Commission. After several rounds of negotiations, the Commission as lead negotiator issues a proposal to the Council to adopt the decision authorizing the signature of the EURA by qualified majority voting.[1] The European Parliament needs to give its assent. In the third country, EURA can be ratified by Parliament, depending on the domestic constitutional arrangements.

EURAs' negotiations take place in a multi-level governance setting both within the EU and with third countries, challenging the traditional state-centric approach to public policy instruments (Kassim and Le Galès 2010). Originally, readmission agreements at national level date back to the 19th century (Coleman 2009) and have been widely used after World War II. After the Amsterdam treaty, EU member states delegated this competence at EU level. It aroused a lot of attention in the literature given the turf wars it generated between the European Commission and EU member states (House Of Commons 2011). Compromising on a shared competence, the Commission 'has not withdrawn its claim on exclusivity' (Coleman 2009: 75). It is regularly in conflict with the Council, which since Lisbon has reasserted its ultimate political role over JHA issues such as visa liberalization (Wolff 2015b).

Negotiations and implementation of EURAs with third countries are also often undermined by EU member states' informal bilateral readmission mechanisms. Those 'non-standard agreements' take the form of memorandum of understanding and letters of exchange with third countries. Allowing for flexible and informal readmission, they fall outside parliamentary and/ or judicial scrutiny (Cassarino 2010) and undermine the credibility of EU readmission policy as well as human rights and international protection guarantees (European Commission 2011b).

EURAs are also incentive-based instruments coupled with migration, border management operational and financial support, visa facilitation/ liberalization or mobility partnerships (MP) (Trauner and Kruse 2008). The nature of incentives nonetheless varies depending on EU's geographical and strategic priorities. As of April 2013, out of 15 EURAs in force with third countries,[2] only seven were coupled with a visa facilitation agreement, mainly in the Western Balkans plus Georgia, Turkey and Ukraine. Until the Arab Spring, visa facilitation was never offered to Southern Mediterranean countries.

If EURAs' policy drift, inefficiency, security focus as well as fundamental rights deficiencies have been extensively researched, fewer studies have investigated the role of third countries. The concept of 'partnership' at the core of the *Global Approach to Mobility and Migration* (GAMM) has been criticized, especially towards Southern countries (Lavenex and Stucky 2011; Adepoju et al. 2010). Research on EU-Morocco migration governance stressed the importance of domestic organizational factors (Wunderlich 2012) while the AKP's government's adherence to the EURA would be the consequence of the European Commission's leadership (Bürgin 2012). The Europeanization literature has also looked at the impact of EURA on third countries' policy, polity and politics. Following its signature, Cabo Verde reformed its drug-trafficking and irregular migration policies (European Commission 2008). In Albania, institutional and procedural changes enabled to enforce the third country nationals' clause (Dedja 2012).

Building on the above-mentioned work, this chapter analyses the politics of instrumentation of EURA from a third country perspective, notably at the level of negotiations. Turkey and Morocco account for a most different systems research design as they display important differences in their relationship to the EU and in their political, social and economic systems. Yet they share a similar hard bargainer's position and a recent shift in EURA negotiations. Investigating Morocco's and Turkey's resilience to sign an EURA, this chapter finds that EU incentives had to be revised and fine-tuned over time by linking it to a MP and trade negotiations for Morocco and to a visa liberalization dialogue for Turkey. These incentives have nonetheless been constrained by domestic and regional factors. Finally, beyond their 'hard bargainer' discourses, this chapter finds that Morocco's and Turkey's border management and migration control practices fit the meanings and representation of EU migration governance carried over by EURA as a policy instrument. Document and content analysis of EU, Moroccan and Turkish officials' declarations and press releases are used to process-trace the negotiations. A number of targeted interviews were also held between April 2013 and July 2013 with EU officials from DG Enlargement, DG JHA, Turkish and Moroccan officials, NGOs, as well as member states. It was completed by interviews in Morocco in July 2014 and June 2015.

Explaining the politics of EURAs: EU incentives and third countries' preferences

This chapter researches why third countries negotiate EURA and whether the path of negotiations can lead to changes in the design of the policy instrument. In doing so, it analyses the 'power dynamics and social relations that underlie the selection of instruments' but also what EURA negotiations with hard bargainers' countries can reveal about the way 'instruments change over time and their (intended and unintended) consequences for politics and policy' (Kassim and Le Galès 2010: 7). This implies exploring the 'wider social and political context in which instruments are adopted and operationalized' (Kassim and Le Galès 2010: 11) by a third country.

I adopt a four-steps approach. First, I analyse EU incentives since 'there is no single third country that is happy to sign an EURA',[3] which are seen as 'EU monologues where little interest exists on the other side' (Roig and Huddleston 2007: 374). Negotiating an EURA is costly domestically, especially when readmitting third country nationals. Countries signing an EURA are usually motivated by (i) the perspective of enlargement and (ii) the perspective of visa facilitation/visa liberalization (Roig and Huddleston 2007). This is the case for EU's Eastern neighbours and the Western Balkans where EURAs combined to visa facilitation regimes help to 'mitigate the negative side effects of the eastern enlargement' (Trauner and Kruse 2008). This was the case for Albania where the prospect of pre-accession motivated high-level officials to comply with the EURA requirements, in spite of 'high domestic costs' (Dedja 2012: 131). The 'external incentive model' has theorized the successfulness and credibility of EU incentives, which rely on international and domestic factors as well as material gains and legitimacy issues (Sedelmeier 2011). I therefore expect that Turkey and Morocco should be inclined to sign an EURA where EU incentives are 'clear, credible and sizeable'. My first hypothesis is:

> The more the external incentive is clear, credible, sizeable and temporally close the more likely is the signature of an EURA.

Second, wider domestic political dynamics matter. Signing an EURA domestically depends on the costs it implies for adopting the new rule by the government. Costs can be material but also institutional and societal through veto players. Societal mobilization, supportive formal institutions as well as administrative capacities can foster or hamper EU conditionality, which will affect differently policy outcomes (Tsebelis 1999) depending on their preferences (Ganghoff 2003). EURA are usually high-level and informal involving Moroccan and Turkey veto players that can range from the heads of state and government, coalition governments, constitutional courts. They are empowered differently, depending on the nature of the regime (nondemocratic, presidential, parliamentary). My second hypothesis is therefore:

The higher the number of veto players domestically, the more difficult will be the EURA.

Third, due to the regional nature of migratory fluxes though (see editorial), it is also necessary to analyse the broader regional power dynamics of EURAs negotiations. The third hypothesis is:

The higher the costs for the regional position of the third country, the more difficult will be the EURA.

Fourth, the domestic appropriateness of EURAs in the third country is relevant. In the sociological tradition, actors engaged in a negotiation can be socialized to EU norms such as the concepts of 'circular migration' and the 'control of borders'. EURA negotiations can lead to 'persuasion' and 'socialization' strategies to politicize or depoliticize EURA as a policy instrument and impact its appropriateness at domestic level (Sedelmeier 2011). This requires looking at the public debate; the media and the discourse of policy-makers at home in order to understand which norms are considered appropriate. EURA negotiations can lead to 'intense (discursive) struggles and re-produce meanings, subjects and resistances' (Kunz and Maisenbacher 2013: 196). Through politicization it become 'contested amongst a widening circle of political actors' (Hooghe and Marks 2006: 205–222). Inversely de-politicization removes an instrument from this platform of debate and contestation, by putting forward its technical and output-oriented nature (Majone 1996). New EU migration policy instruments designed around the concept of 'circular migration', 'partnerships' and 'cooperation' since the mid-2000s helped the Commission to break away from the old coercive style (Kunz and Maisenbacher 2013: 197). Yet, EU incentives are at odds with the functional reality of migration fluxes, thereby reflecting highly political choices and power struggles, which contrast with the 'logic of appropriateness' of third countries.

Since 2011, migrants arriving from Libya are on the surge and the Central Mediterranean route identified by Frontex has seen a sharp increase in migrants trying to reach Italy and Malta. In 2014, the numbers reached their highest peak with 170,760 irregular migrants intercepted on that route (compared to 4500 in 2010) (Frontex 2015). And yet up until 2015 the discussions on visa facilitation had not progressed much between the EU and its partners. It only with the unprecedented influx of Syrian refugees and the October 2015 'deal' with Turkey that discussion on visa facilitation and even on Turkish membership made it to the agenda. This has definitely affected the 'logic of appropriateness' in third countries. My fourth hypothesis is therefore:

The more appropriate is EU external migration policy in the eyes of a third country, the higher are the chances of the EURA to be signed.

EURA negotiations with Morocco

Between the start of the negotiations with Morocco in 2000 and the political agreement reached in March 2013 on a Mobility Partnership (MP), EURA negotiations were stalled. While the draft text of the EURA was received by Morocco in April 2001, 'informal preparatory meetings, as well as discussion within the EU-Morocco Association Council, was necessary to convince Morocco to comments formal negotiations finally in April 2003' (Coleman 2009: 150). Between April 2003 and November 2005, eight rounds of unsuccessful formal negotiations took place (Coleman 2009: 151). In 2004 though, at the occasion of the negotiations on the ENP Action Plan, the Council declared that its key priority included the

> effective management of migration flows, including the signing of a RA with the European Community, and facilitating the movement of persons in accordance with the acquis, particularly by examining the possibilities for relaxing the formalities for certain jointly agreed categories of persons to obtain short-stay visas.
>
> (Council of the European Union 2006)

In total, as of 2013, there were 15 rounds of unsuccessful negotiations; the last round taking place on 10 May 2010 (Belguendouz 2013). A breakthrough happened nonetheless in March 2013 with a political agreement on a MP, which was then signed in June 2013.

EU incentives' evolution over time

Very early on, high-level Moroccan officials were concerned of being the 'Gendarme' on behalf of the EU and still continue to do so.[4] Unlike other Euro-Mediterranean Association Agreements, Morocco's does not have a readmission clause. In 2000, a 'permanent dialogue on immigration' was initiated between Morocco and the High-Level Working Group on Asylum and Migration (HLWG) (Council of the European Union 2000). The HLWG, created in 1998, prepared 'cross-pillar Action Plans for the countries of origin and transit of asylum seekers and migrants' including Afghanistan, Somalia, Sri Lanka, Iraq, Albania and Morocco. The Moroccan authorities considered the Action Plan towards 11 October 1999 as 'lack[ing] balance, particularly in its emphasis on the "security dimension"' (Council of the European Union 2000). Difficulties of implementation, the lack of EU member states' commitment and the lack of measures of 'effective implementation of existing readmission agreement' were raised. Morocco also had the 'impression of imbalance in the Action Plans and the countries at which the plans are directed feel that they are the target of unilateral policy by the Union focusing on repressive action' (Council of the European Union 2000). In 2002 the EU provided €70 million for the development of Northern Morocco to encourage EURA negotiations (GADEM 2010), along with several other financial incentives since then (Wolff 2012). They failed though to facilitate EURA negotiations until 2013.

The Commission realized early on that EURA negotiations would be extremely difficult unless EU member states would be ready to offer credible incentives such as visa facilitation. In the case of Morocco progress did not happen until 'it became a very comprehensive programme and the perspective of visa facilitation was integrated in a "package deal"'.[5] The later was negotiated in parallel to the ENP Action Plan for 2012–2016 negotiations, also known as the Advanced Status Action Plan, which involved negotiations on a Deep and Comprehensive Free Trade Agreement.[6] According to DG Home Affairs officials, EURA negotiations sensibly shifted from the moment the Commission was able to convince EU member states of the value of the MP as a substantial incentive. The Arab Spring played in favour of the Commission in its plea towards EU member states providing a 'momentum to change the approach'. Before 'visa facilitation was absolutely out of question for MS'.[7]

A joint political declaration on a Mobility Partnership (MP) between the EU and Morocco was agreed in June 2013, following a high-level commitment of President Barroso and Commissioner Malmström.[8] It was preceded by a joint document by Lady Ashton and President Barroso on *A Partnership for Shared Democracy and Prosperity* and a DG Home-led communication on 'A dialogue for migration, mobility and security with the southern Mediterranean countries' (European Commission 2011a). Content analysis reveals a shift towards more 'mobility' in the Commission's discourse. The word 'mobility' is used 40 times in the document, amongst which 15 times under the form of 'mobility partnership'. References to 'security' are less prominent (12 times), 'irregular migration' nine times, 'readmission' 7 times and 'control' only 3 times. Incentives include a financial package of capacity building measures, technical assistance on legal migration (i.e., to develop 'active labour market policy programs', avoiding brain drain, diminishing fees for remittances and diaspora investment). In line with the EU concept of circular migration, migrants will come temporarily to Europe and be supported through a series of measures, including 'voluntary return arrangements' to go back to their home country. Security support is also offered through the conclusion of 'working arrangements with Frontex', developing border management capacities, cooperation on the 'EUROSUR project', with the European Asylum Support Office (EASO) and with Europol (Council of the EU 2013).

The June 2013 Joint Declaration on the MP specifies further the partnership around four main objectives: (i) to manage the labour migration more effectively, (ii) to strengthen cooperation on migration and development 'in order to exploit the potential of migration and its positive effects of Morocco and European countries', (iii) combat illegal immigration, human being trafficking and smuggling, to promote an effective readmission policy respectful of fundamental rights and 'ensuring the dignity of the people concerned' and (iv) to comply with international instruments on the protection of refugees. The EURA negotiations should, in this context, be 'resumed' (Council of the EU 2013: 7). The new MP however does not yet challenge the long-term structuring role that EURA has had on EU-Moroccan migration cooperation (Halpern 2010: 45). First, this structuring has been exploited by Morocco. The stalling of the negotiations of the EURA has been paralleled by a satisfactory cooperation through bilateral 'non-standards agreements' between EU member states and Morocco. Morocco has concluded readmission agreements, although only for Moroccan nationals, with Germany (1998), France (1993, 2001), Portugal (1999), Italy (1998, 1999) and Spain (1992, 2003) (Di Bartolomeo et al. 2009: 45). This led to 'unintended consequences', drifting away from the original goal of EURAs (Halpern 2010: 45). Pursuing negotiations with the EU in parallel helped nonetheless Morocco to gain influence on the EU's agenda by forcing its way in and putting forward more 'comprehensive' migration demands on the table.

Second, structural differences between the Commission and EU member states remain. For the Commission the MP is an opportunity to provide a forum where to discuss visa facilitation issues with Morocco as a credible incentive. For EU member states though the MP constitutes a new opportunity to pursue national preferences (Renslow 2012) and to remain in control of cooperation with Morocco. In the Joint Declaration on the MP, it is thus worth noting that out of the 37 new projects listed, 28 tackle irregular migration, 14 legal migration, 7 migration-development issues and 6 international protection. By opting for issue-linkage with the MP, EU member states keep a firm hand on pilot projects where they take the lead. This also increases their bargaining power vis-à-vis the Commission and Morocco (Kunz and Maisenbacher 2013).

The MP reveals an inherently political choice behind issue-linkage of policy instruments, displaying power struggles through a technocratic language that occasionally covers it (Bache 2013). This is illustrated by the absence of visa liberalization in the MP. The objective is to establish a visa facilitation agreement for some categories of travellers such as students, researchers and businessmen (Council of the European Union, 2013).

The MP is also not a Common Agenda on Migration and Mobility (CAMM), envisaged by the revised 2011 GAMM and that would focus on the notion of 'partnership' (Belguendouz, 2013). The two main objectives are therefore to ease EU visa policy towards some categories of Moroccan citizens and to sign an EURA. In 2014, the text of the MP was considered 'far from ideal' for EU member states,[9] who prefer 'to sign the EURA first and that later on [to] discuss visa facilitation'.[10] This contrasts with the official position of Menouar Alem, the Moroccan Ambassador to the EU, who asked 'why should a country like Morocco, the last stop before "the European Eldorado", take all the responsibility?' (EPC 2012). Instead the Ambassador called for visa facilitation, blaming the 'double standard' discourse of the EU

(EPC 2012) and reiterating one of the constant Moroccan demands to facilitate channels of legal migration to the EU (Roig and Huddleston 2007).

Following the signature of the June 2013 Mobility Partnership, Morocco has also engaged in a new immigration and asylum policy launched in September 2013 by King Mohammed VI. This reform took everyone by surprise, including the EU Delegation in Rabat. Since then, a new Directorate in charge of Migration Affairs has been created within the Ministry of Moroccans Residing Abroad which now expands to Migration Affairs. A national strategy of immigration and asylum was launched and an exceptional operation led to the regularization of 18.694 people out of 27.643 demands (around 68%) at the end of 2014.[11]

In the meantime, visa facilitation agreement negotiations started in January 2015 on the basis of a mandate from the Council of 5 December 2013. Having refined their incentives, Moroccans officials have started in the meantime to link the visa facilitation dossier to the one on the negotiations on the Deep and Comprehensive Free Trade Agreement (Wolff, pers. comm., 2015).

Linking the EURA negotiations to the MP confirms our first hypothesis according to which EU incentives need to be clear, sizeable and credible. Our analysis reveals two additional dimensions. First, there are politicization dynamics at hand, the Commission having managed to push for visa facilitation while EU member states remain the gatekeepers of the incentive through the choice and design of the MP. Second, power interdependences are reflected in the wording of the June 2013 MP declaration, which is conditioned to the implementation of both 'visa and readmission facilitation agreements' (Council of the European Union 2013). Morocco managed to link the EURA to a broader agenda with the MP, which deals as much as with borders, than with migration and development.[12] Dynamics of (de)politicization are therefore key in understanding the impact of EU incentives.

Domestic and regional context: the politicization of EURA

Beyond EU incentives, domestic veto players and regional dynamics are also key in driving Morocco's preferences on the EURA negotiations.

First, in spite of Morocco being governed since 2011 by the Justice and Development Party (PJD), an Islamist party, the official line remains that Morocco is not the 'EU Gendarme'. The PJD has been co-opted by the Makhzen and the king remains the main arbitrator. The Benkirane I government was led by a coalition government that included both ex-communists and pro-royalists (Ottaway 2012). With the stepping down of the islamo-nationalist party Istiqal from the coalition, the PJD had to negotiate with the Rassemblement National Indépendant (RNI), historically set up by the monarchy in order to build a coalition for the Benkirane II government. The king's ultimate political role, including on the EURA, remains, as illustrated by the French Ministry of Interior Manuel Valls' visit in July 2012 to discuss visa facilitation arrangements with France (Duhem 2012). Even though the PJD won the elections on a social justice and antipoverty political programme, migration policy towards Sub-Saharan migrants is absent, perpetuating past policies of previous governments. The only measure on migration mentioned measure 156, which aims at combatting discrimination against Moroccans throughout the world, focusing on emigration rather than immigration policy, like in the old days (Political Programme 2011).

On the ground, Moroccan readmission is also not so much at odds with the structuring of the EURA. Security cooperation between Morocco and Spain has been reinforced since Benkirane is Prime Minister. A bilateral agreement to facilitate visa procedures for some categories of citizens was signed in 2011 (Durham 2012). The implementation of EU projects has for long been monopolized by the Moroccan Ministry of Interior reproducing therefore its repressive side (Wunderlich 2012: 1426). The PJD has therefore little say in a domestic setting where Moroccan migration policy veto players remain high-level and can easily influence EURA negotiations. The 2013 immigration reform is the result of a royal decision that the government is implementing. With respect to H2, what matters is not the number of veto players at the domestic level but rather the nature of the political system. Secondly, domestic practices of readmission confirm that EURA are not being negotiated in a complete political vacuum, on the contrary. Morocco has put forward the 'technical, legal and ethical difficulties' of the EURA. Technical issues include the length of detention, the proof of the nationality of the irregular migrant as well as the 'technical' issue that most of the irregular migrants come from the Southern border of Morocco, namely Algeria.[13]

Beyond this official discourse on domestic costs, there is a gap with the practice of readmission, which is structured by securitization. Irregular migration has been criminalized with law no. 02–03 passed unanimously by the Parliament after the 16 May 2003 attacks on Casablanca (Wolff 2012: 75). Later on, following the events of 2005 in Ceuta and Melilla, Morocco recruited 9000 supplementary agents into the army and adopted a new policy to improve its border control capacities (Lahlou 2007). Return operations of Sub-Saharan migrants are regularly taking place. In 2003 Nigerians were returned from the Oudja airport (416 people), Nador (on 3 November 2003, 207 people), Fes-Saïs (480 people on 20 December 2003), as well as from Tanger and Rabat (Belguendouz, 2010). Several other instances have been reported by the GADEM (GADEM 2010). Readmission with EU member states is also considered as quite advanced, and was successful in the Canary Islands with Spain.[14] The official discourse of 'ethical difficulties' also contrasts with the situation of Sub-Saharan migrants in Morocco, who, according to Doctors Without Borders are in 'precarious living conditions', 'forced to live in and the widespread institutional and criminal violence that they are exposed to' which influence their 'medical and psychological needs' (Doctors Without Borders 2013). In fact, the NGO maintains that the period since December 2011 has seen a sharp increase in abuse, degrading treatment and violence against Sub-Saharan migrants by Moroccan and Spanish security forces … [There are] shocking levels of sexual violence that migrants are exposed to throughout the migration process and demands better assistance and protection for those affected.

(Doctors Without Borders 2013)

Since February 2015 when 1200 migrants were arrested in the north of Morocco (FIDH 2015), it seems that for the time being arrestations and deportations of migrants have been suspended.

Rather, interviews reveal that two main regional concerns have driven Morocco's position in negotiating with the EU. First, a key concern is that the EU has been unsuccessful to secure EURAs with Cotonou countries, therefore fearing to become the country of return by proxy for African countries refusing to reaccept their nationals (Coleman 2009). Interviewees called for the EU to get involved with African countries, but also more specifically with Algeria. According to Moroccan officials, 95 per cent of Morocco's readmission with EU countries concerns migrants coming from the Algerian border.[15] At the same time, the EU has asked Morocco to revise its 90 days visa-free policy towards Algeria, Tunisia and Libya as well as towards Mali, Niger, Senegal, Guinea and Ivory Coast (Belguendouz 2010).

This evidences the importance of the perception and image of Morocco amongst its regional partners. As Interviewee C puts it 'Morocco wants to be the best student amongst ENP neighbours, but does not want to be the worst student vis-à-vis its African partners either'. Morocco wants to continue to have a good relationship with its African partners and avoid any kind of accusation by Algeria, which vetoes its accession to the African Union and with whom relations are poisoned by the Western Saharan conflict. This regional dimension pushed Morocco to influence the EU to adopt a more comprehensive and regional approach in its migration instrument. This regional consultative process, by focusing on intergovernmental operational cooperation and the exchange of best practices, is believed to favour trust and cooperation in an area 'characterised by great uncertainty in a high degree of policy interdependence' (Köhler 2011). The Rabat Process enabled Morocco, confronted to a high degree of uncertainty, to find more networking opportunities and to influence the *Global Approach to Migration* at the 2005 Hampton Court EU Summit (Wolff 2012). At the same time, it helped to forge its regional leadership role vis-à-vis African partners in the field of migration management. Morocco displayed a strong preference for information-based and operational support via a regional consultative process, which favours practical cooperation instead of the EURA. Concerns vis-à-vis other regional partners remained constant demands of Moroccan officials.[16] They are reflected in the June 2013 Political Declaration on the MP which specifies that the EURA negotiations should be accompanied by 'the promotion of active and efficient cooperation with all regional partners' (Council of the European Union 2013: 6).

Interim conclusion

The analysis of EURA–Morocco negotiations confirms our first hypothesis. Since financial incentives were not enough and in the absence of credible EU incentives under the form of visa facilitation/mobility discussions, the EU had to adapt its strategy and decided to link up EURA negotiations to the discussion on a MP. This however only happened after the Arab Spring, member states resisting such issue-linkage until then, through the leadership role of the Commission mainly in coupling EURA negotiations to the MP. Yet, this 'learning' process hides the politics of MP instrumentation whereby EU member states remain the gatekeepers of EU migration policy towards Morocco. Hypotheses 2, 3 and 4 are then inter-related. At domestic level, what matters is not the number of veto players but rather the fact that the ultimate decision-making power lies with the king (hypothesis 2 is disconfirmed). EURAs' appropriateness (hypothesis 4) is linked to the findings on broader regional political dynamics (hypothesis 3). The analysis finds that surprisingly, in spite of an official discourse resisting playing 'the Gendarme on behalf of the EU', there is a gap with the practice of Moroccan readmission with regional partners and its demands for 'ethical' concerns in EURA negotiations. Rather, the role and image of Morocco as a regional migration player play a bigger part in the politics of instrumentation. Hypothesis 3 is therefore confirmed and hypothesis 4 is disconfirmed.

EURA negotiations with Turkey

The opening of accession negotiations in October 2005 was one of the most controversial EU decisions. In spite of the initial opposition of Austria and Cyprus, the EU was confronted to a 'normative entrapment' to consider Turkish application, with no valid reason to oppose it (Schimmelfennig, cited from Thomas 2011: 113). The European Commission also 'certified that Turkey had made significant progress in complying with the EU's political norms' (Schimmelfennig, cited from Thomas 2011: 114). Since 2006 though accession negotiations have been blocked due to the Cyprus issue as well as by France on some chapters of accession. In its 2012 progress report on Turkey, the Commission raised its concerns on 'Turkey's lack of substantial progress towards fully meeting the political criteria', on the 'respect for fundamental rights' especially towards the Kurdish minority (European Commission 2012). It is in this political context that the first round of EURA negotiations started in May 2005. In June 2012, Turkey agreed to 'initial' the EURA but refused to sign it in the absence of a credible EU commitment on visa liberalization. This took the form of an EU roadmap that was subsequently negotiated in the Council and finalised in November 2012.

Visa liberalization dialogue: still not credible enough?

In 2002, even before the opening of Turkish accession negotiations, the Council mandated the Commission with an EURA negotiating directive. The Commission, and in particular DG Home, 'rallied support internally for the creation of a link [with the readmission agreement negotiations] with the start of accession talks' (Coleman 2009: 181). Accordingly, before the European Council agreed to the candidate status of Turkey in 2004, the Commission 'repeatedly called upon the Member States to use their bilateral relations and diplomatic contacts to push Turkey for a prompt start of negotiations for a Community readmission agreement' (Coleman 2009: 181). The first round of EURA negotiations took place between May 2005 and December 2006, but was put on hold until 2009. Turkish partners started to ask for equal treatment with the Western Balkans that had just been given visa liberalization. Interview H confirmed that surprisingly before that Turkey had not requested visa liberalization and focused until 2007 on financial demands for border management.[17] Turkey then aligned with biometric passports requirements and put in place an Integrated Border Management strategy.[18]

However, instead of opening a visa liberalization dialogue, EU ministers of interior committed to a very loose 'dialogue on visa, mobility and migration' during the JHA Council of 24 and 25 February 2011, which was 'the diplomatic equivalent of a slap in the face' (Stiglmayer 2012: 99–109). DG Home and DG Enlargement pushed MS to commit and to link the EURA to visa liberalization, the two Directorate- Generals working hand in hand.[19] After 2009, the negotiations made some progress on 19 articles but some conflicts occurred on five articles. One of the key concerns for Turkey was to secure some funding from the EU to support resettlement policies from the European Refugee Fund (Frenzen 2010). Also it feared that in the absence of a strong EURA, there would be an increase in the log of complains to the European Court of Human Rights in Strasbourg.[20]

Accordingly,

> three further formal negotiation rounds took place on 19 February (Ankara), 19 March (Ankara) and 17 May 2010 (Brussels). An additional meeting between the Chief Negotiators was held on 14 January 2011 in Ankara. Those meetings brought the negotiations to the end at the level of Chief Negotiators.
>
> (European Commission 2012)

The re-launching of the negotiations in January 2011 took place under the European Commission leadership that managed to overcome resistance from Germany and France to open up a visa liberalization dialogue (Burgin 2012). Endorsing a cost-benefit approach, Burgin argues that 'the political gain of the Commissions' offer to consider visa exemption for Turks outweighed the financial and social costs of readmitting irregular immigrants and the lack of credible EU membership perspective' (Burgin 2012: 108).

In June 2012, visa liberalization talks started as part of a broad political re-launch of the negotiations for EU accession. The conditions in which the roadmap was offered to Turkey highlight nonetheless the contradictions and the lack of credibility of this incentive. According to interviews, 'Turkey's' position was that we would initial the EURA only when the mandate in the Council on VL would be secured'.[21] Following the reverse logic, EU member states wanted first Turkey to sign the EURA and then to provide Turkey with a roadmap. A visa liberalization roadmap is technically a European Commission document. Member states are only officially consulted. Yet, the consultation became more of a political negotiation in the Council and took longer than usual, lasting until November 2012.[22] Transmitted to Turkey, the roadmap is at the time of writing under consideration by Ankara. This created a 'real problem of trust (though not with the Commission)',[23] with a concern that visa liberalization would never happen. Prior experience from the Balkans has shown that once the roadmap negotiations started, 'it is highly likely that it will be completed' (Knaus 2013). This was the case with five Western Balkans that opened negotiations with a roadmap in 2000, leading to visa liberalization three years later (Knaus 2013). Yet, it seems that Turkey feels different from other candidate countries, rightly so for having been discriminated in the past.

The roadmap specifies that 'progress in the visa liberalization process should be founded on the performance based approach and conditioned on effective and consistent implementation by Turkey of those requirements vis-à-vis the EU and its Member States' (European Union, 2013). The main elements include mobility of bona fide travellers; improving border management especially on the Greek–Turkish and Bulgarian–Turkish borders and with Frontex and Europol; improving migration management through cooperation with EU immigration liaison officers, information on countries of origin concerning illegal migration, promoting joint return flights and raising awareness about the risk of illegal migration in public information campaign; the provision of assistance and protection to asylum seekers; the fight against terrorism and the fight against transnational organized-crime. The roadmap identifies several legislative and administrative reforms that Turkey needs to embark upon in order to 'establishing a secure environment for visa-free travel'. These areas include document security, migration and border management, public order and security, as well as fundamental rights (Council of the European Union 2012: 13).

Like for the Kosovo roadmap, the concept of 'reinforced consultation' indicates a stronger involvement of the Council in the process (ESI 2012). This confirms the high-level political nature of the roadmap. The Commission needs to take 'into utmost consideration the political discussions in the Council' (Council of the European Union: 5). Like Kosovo, but unlike the Western Balkan countries, it is also expected that progress will be benchmarked against performance indicators, which include 'Commission's assessments of the expected migratory and security impacts of the liberalisation of the visa regime with Turkey' (Council of the European Union 2012: 28).

EU member states' different views on Turkish accession have also weakened the credibility of the EU's incentive. In 2012, during discussions in the Working Party for Enlargement and Countries Negotiating Accession to the EU, France 'maintain[ed] a reserve' and argued that 'preventing illegal immigration from third countries through Turkey would require an alignment of Turkish policy with the EU visa policy regarding these countries'. This was opposed by 'a number of other delegations' that thought it could only be asked at the moment of the visa liberalization (Council of the European Union 2012: 2). Divergent EU member states positions contribute to perpetuate distrust in the negotiations.

The lack of clarity, especially from EU member states, has weakened the power of the visa liberalization dialogue, in the form of a roadmap, as a credible EU incentive. Turkey is now hesitant in signing the EURA before getting satisfaction on the EURA. Hypothesis 1 is therefore confirmed but needs to be analysed in conjunction with domestic and regional factors.

Domestic and regional political dynamics

If Turkish migrant legislation can be defined as 'conservative', its visa policy towards Middle East and Caucasus neighbours has been rather liberal. In 2009, visa requirements were abolished mutually with Syria, Albania, Libya, Jordan, Tajikistan, Azerbaijan, Lebanon and Saudi Arabia, leading to the 'construction of a new Schengen area in the Middle East' (Paçacı Elitok and Straubhaar 2011: 125). It is also in line with the Turkish 'zero problem' foreign policy adopted by Erdogan to re-establish Turkish regional leadership. EU requirements in the roadmap are putting at risk this liberal model while requiring Turkey to embrace Schengen, a model that it has been criticizing for discriminating against Turkish citizens. Turkey has also agreed in 2009 to visa-free travel with Russia and Iran, enabling for the latter 'large numbers of regime opponents to flee the country and enjoy temporary protection in Turkey before settling elsewhere in the West' (Evin et al. 2010: 30). Reforming Turkish visa policy would be an economic challenge but could also undermine its regional position, which would suffer from EU requirements on visa, and 'would not be good for business'.[24]

Turkey, fearing to have to readmit non-Turkish nationals, has adopted a 'delaying tactic' vis-à-vis the EU by securing readmission obligations from other countries before agreeing on an EURA (Coleman 2009: 180). This regional readmission policy is viewed as 'a solution to this problem while distributing the responsibility for transit migration over the region, and creating a scope for return and readmission to countries of origin' (Coleman, 2009: 179). Hence the liberal visa-free policy is matched by the same conditionality that the EU is applying to Turkey through the EURA. Turkey has signed formal readmission agreements with Greece (2002), Syria (2003), Romania (2004) Kyrgyzstan (2004), Ukraine (2005), Russia (2011) and negotiations have been completed with Pakistan (2011). Discussions are on-going with Azerbaijan, Bangladesh, Belarus, Bosnia and Herzegovina, FYROM, Georgia, Lebanon, Libya, Moldova and Uzbekistan (House of Commons 2011).

A Council document analysing the bilateral practice of readmission with Turkey concludes that in spite of the absence for 'nearly all' EU member states of any bilateral readmission agreement with Turkey, paradoxically 'nearly all responding delegations are however able to carry out returns to Turkey' for Turkish nationals only. Only Romania, Greece and the UK have readmission arrangements with Turkey. In 2010, Greece returned almost 100,000 Turkish nationals, followed by 9035 returned by Germany and only 2500 by The Netherlands. Regarding the return of non-Turkish nationals, the UK is the only country to admit this possibility while for instance 'Norway added that transits of third country nationals in Turkey are not allowed even if the returnees are escorted by police' (Council of the EU, 2012).

The Turkish Ministries of Foreign Affairs and EU Affairs have been coordinating EURA negotiations. The Ministry of Interior was involved only from an expert perspective, accordingly because there is less trust of the AKP into the Ministry of Interior (*Financial Times* 2014). Opinions towards what should be done diverge within the Turkish government. Some officials consider that reforms such as the one on asylum law are in fact useful to advance Turkish legislation and to comply with international normative requirements. While the political system is also highly centralized with the final authority in signing the EURA residing in the prime ministers' decision, interviews revealed that there is a greater diversity of views. This has to do more generally with the diversity of views regarding Turkish accession to the EU.

Turkish migration and asylum stakeholders usually put forward three main critical arguments (Tolay, 2012: 54): (i) the costs of change, (ii) the unfairness towards Turkey, and this is often the case with the discourse on 'equal treatment' and (iii) the faultiness of certain EU migration policy, and some 'hypocrisy' on the Schengen visa policy. Distrust is also an official argument, the Turkish ambassador asking whether Turkey can really trust the Council in granting a visa-free regime to the EU. Yet other research shows that Turkish NGOs and elite levels frequently refer to European norms of fundamental rights and freedoms to support asylum and migration reforms in Turkey (Tolay 2012: 50). High-ranking officials and bureaucrats 'tend to appreciate the fact that, in the EU, there is an existing official framework, a clear and intentional immigration policy and allocated means that allows for a more comprehensive and consistent state policy towards migration' (Tolay 2012: 51).

Therefore, it is possible to identify in practice a phenomenon of 'Europeanization' of Turkish JHA domestic legislation. On border management, if initially, Turkey said it found it too costly to agree to an EURA and to reaccept also non-Turkish national, it has been cooperating intensively with Frontex to reduce those numbers. The Greek–Turkish border is one of the main points of entry for irregular migrant, one of the main 'hot spots of irregular migration' especially on the Evros river. Turkish borders need to be policed over around 3000 km of land border and 6500 km of sea borders together with migration source countries such as Syria, Iraq, Iran, notwithstanding Kurdistan (Lagrand 2010). Several security actors police the border. If the General Directorate of Security is in charge of border control of people, the Gendarmerie is in charge of the Iran and Iraq borders, the Land Forces of the rest of the land borders and the Coast Guards of sea borders (Lagrand 2010). The detections at the border crossing between Greece and Turkey have declined significantly since 2010 (Frontex 2012). Frontex conducted several operations on the Turkish/Greek border including the Poseidon Joint Operation at the sea border but also land operations to identify irregular border crossings. Several cases revealed that migrants are also using lorry transports to enter the EU, for instance from Turkey to Slovakia or to Bulgaria and Romania (Frontex 2012). The Eastern migratory route is also used for smuggling of cigarettes and for all sorts of organized-crime activities such as human being trafficking and smuggling of Middle East migrants into Europe (Europol 2013) and the smuggling of Iraqi and Kurdish migrants (Europol 2012).

Political difficulties between Greece and Turkey have traditionally hampered an effective readmission policy between the two countries. Recently though the readmission agreement signed between Greece and Turkey in 2001 was implemented in 2010 (Strik 2013). This relatively good working cooperation at the operational level, combined to a Syrian refugee crisis and changing migratory fluxes, highlights a gap between Turkish official position of 'hard bargainer' and the practice.

Finally, changes in Turkish migration and asylum law, requested in the roadmap, are not only linked to the EU incentive of visa liberalization but also to fluctuating regional migratory routes. For a long time Turkey's cooperation with European countries was driven by the presence of important diasporas in Germany and the conclusion of guest workers programmes in the 1960s on labour migration with Austria, Belgium, The Netherlands, France, as well as Sweden (Kirişçi 2007) which were framed in an 'emigration' narrative. In the 1970s, some Turkish refugees came to Europe to flee the military regime. Nowadays, in Turkey, the Kurdish minority mainly lodges asylum applications (Kirişçi 2007). With the war in Afghanistan, Turkey has become a key transit country for Afghan migrants, but also from Iraq, Iran or Pakistan. The current Syrian crisis is certainly impacting Turkish's strategy vis-à-vis the EU. Since the beginning of the Syrian crisis, 9 million people have fled the country going mostly to neighbouring countries, with 6.5 million in Turkey, Lebanon, Jordan and Iraq. 150,000 have claimed asylum in the EU, mostly in Germany and Sweden (UNHCR, 2015). For some it is likely that the presence of important numbers of Syrian refugees on Turkish territory should naturally push Turkey to continue to liberalize its refugee protection policies, as exemplified by the 2013 Law on Foreigners and International Protection (Icduygu, 2015). Yet this research has shown that this will need to take into account EU, regional and domestic dynamics.

Interim conclusion

In December 2013, the EURA with Turkey was signed and the visa liberalization dialogue launched. As this chapter goes to press, the influx of Syrian refugees affecting Turkey and the EU validates hypothesis 1 and the need for the EU to offer credible incentives. Like in the case of Morocco hypotheses 2, 3 and 4 are interrelated. If the number of domestic veto players (hypothesis 2) is not relevant, regional costs and perception of Turkey by regional partners matter (hypothesis 3). Finally, the official position of 'hard bargainer' in EURA is challenged by the discourse-practice gap identified and migration practices which tend to invalidate the inappropriateness of EU demands' thesis (hypothesis 4 is disconfirmed).

Conclusions

The study of Morocco and Turkey in EURA negotiations reveals that beyond EU incentives, broader domestic and regional political dynamics are key to the study of the politics of EU migration instrumentation and in particular of remote control. There is also a need to differentiate between an official discourse of 'hard bargainers' and the practice of readmission, which reveals that EURA negotiations have structured Morocco's and Turkey's migration cooperation with the EU. Our most dissimilar system research design evidences that beyond the relevance of EU incentives, the differential empowerment of domestic veto players combined to regional factors explain what drives third countries' preferences and negotiations strategy on the EURA. The appropriateness of EU policies in the eyes of third countries is however tactically played out in official discourse, but does not hold as a strong factor in practice.

Linking back to the debate on EU migration policy instruments, this chapter corroborates that EURAs are not functional instruments set in stone, which respond to EU's migration policy rational needs. Rather, further research might look into how EU migration instruments are being structured by a complex process of politicization and (de)politicization dynamics, involving not only EU actors but also third countries. Politicization dynamics include EU turf wars between the Commission and EU member states. Reluctance from EU governments to lift up visa requirements for Turkish citizens is driven by electoral concerns and a fear that asylum-seekers application would increase (Stiglmayer 2012). Yet, with the Western Balkans the EU was able to suspend visa-free regime. In April 2013, the European Parliament has indeed given its approval to provisionally suspend visa-free regime for countries like Serbia and Macedonia that are thought to abuse the asylum application system in Germany, the Netherlands, France, Luxembourg and Belgium (Agence Europe 2013). As explained by Sander Luijsterburg, from the Dutch Permanent Representation, 'readmission and return policy' are key to 'help to win public support for other parts of migration policy' (EPC

2013). The Commission strategy to depoliticize EURA negotiations by coupling it to a more comprehensive and innovative approach such as the MP in the case of Morocco cannot hide political turf wars over the implementation of the projects by EU member states who remain in control of migration cooperation with Morocco.

At domestic level, EURA negotiations have been the object of high-level politicization by Moroccan and Turkish officials who have refused to police the EU's borders. This politicization has been motivated by the meanings and perception it carries for the regional position of Morocco and Turkey. However on the ground, this contrasts with de-politicized domestic readmission and migration practices.

More generally, beyond the case of Turkey and Morocco, this analysis calls for a reflection on EU migration policy instruments over time as complex political and cognitive processes. As social and political institutions, they structure power relations both within the EU and in relation to third countries. They do not always respond to the original intended effect and can escape the objectives assigned to them. This is specifically reflected in most of our interviews, which revealed an emerging debate on the very relevance of EURA as migration policy instruments. While Commission officials raised the validity of third country clauses, member states officials' views included withdrawing some of the EURA mandates given to the European Commission.

Acknowledgements

This paper was first presented at the workshop on 'The Politics of Migration Policy Instruments: Insights from EU External Relations', jointly organized by the School of Politics and International Relations–Queen Mary University of London and the Institute for European Integration–University of Vienna, London, 2–3 October 2013. I am thankful to all the participants of the workshop as well as Florian Trauner, Paul Copeland and Tim Bale for their insightful feedback.

Notes

1 Articles 79.3 TFEU and 218 TFEU.
2 Hong Kong, Macao, Sri Lanka, Albania, Russia, Ukraine, Macedonia, Bosnia and Herzegovina, Montenegro, Serbia, Moldova, Pakistan, Georgia, Cabo Verde, Armenia. Mandates were given to the European Commission for China, Algeria and Belarus but the negotiations have not yet been formally launched. Morocco can be considered as having formally launched the negotiations.
3 Interviewee C.
4 Interviewee E.
5 Interviewee C.
6 Interviewee C.
7 Interviewee C.
8 Interviewee F.

9   Interviewee F.
10 Interviewee F.
11 The numbers are from the Moroccan Ministry of Moroccans Resident Abroad and of Migration Affairs.
12 Interviewee F.
13 Interviewee E.
14 Interviewee G.
15 Interviewee E.
16 Interviewee E.
17 Interviewee H.
18 Interview A & H.
19 Interviewee A & B.
20 Turkey has indeed a series of cases in front of the European Court of Human Right in Strasbourg. One of the most symbolic cases is the 2009 *Abdolkhani and Karimnia v. Turkey* (Application no. 30471/08) that condemned Turkey for breaching the European Human Rights Convention (EHRC) for willing to return two Iranian refugees (who had been granted this status by UNHCR during their stay in Iraq) back to Iran, contravening therefore to the principle of non-refoulement.
21 Interviewee H.
22 Interviewee B.
23 Interviewee H.
24 Interviewee H.
25 Interviewee G.

Bibliography

Adepoju, A., van Noorloos F. and Zoomers, A. 2010. Europe's Migration Agreements with Migrant-Sending Countries in the Global South: A Critical Review. *International Migration.* Vol. 48, No. 3, pp. 42–75.

Agence Europe 2013. *JHA: MEPs say yes to provisional suspension of visa-free regimes for the Balkans.* Available from: www.agenceurope.com/EN/index.html.

Bache, I. 2013. Partnership as an EU Policy Instrument: A Political History. *West European Politics.* Vol. 33, No. 1, pp. 58–74.

Baldwin-Edwards, M. 1997. The Emerging European Immigration Regime: Some Reflections on Implications for Southern Europe. *Journal of Common Market Studies.* Vol. 35, No. 4, pp. 497–519.

Belguendouz, A. 2010. Expansion et Sous-Traitance des logiques d'enforcement de l'exemple du Maroc. *Cultures & Conflicts,* pp. 155–219.

Belguendouz, A. 2013. *Reflexions sur le projet de partenariat euro-marocain pour la mobilité.* Available from: http://nancy.maglor.fr/index.php?option=com_content&view=   article&id=1773%3Areflexions-sur-le-projet-de-partenariat-euro-marocain-pour-la-mobilite&catid=139%3Aaccueil&Itemid=83 [13 April 2013].

Bürgin, A. 2012. European Commission's Agency Meets Ankara's Agenda: Why Turkey Is Ready For a Readmission Agreement. *Journal of European Public Policy.* Vol. 19, No. 6, pp. 883–899.

Cassarino, J.-P. 2010. *Readmission Policy in the European Union. Study Requested by the Directorate-General for Internal Policies.* Policy Department C: Citizens' rights and constitutional affairs, Civil liberties, Justice and Home Affairs. Brussels: European Parliament.

Coleman N. 2009. *European Readmission Policy: Third Country Interests and Refugee Rights.* Leiden: Brill.

Council of the European Union 2000. *High Level Working Group on Asylum and Migration.* Available from: www.consilium.europa.eu/uedocs/cms_data/docs/pressdata/en/misc/13993.en0%20ann.doc.html [17 April 2013].

Council of the European Union 2012. Synthesis on Member States' practical experiences based on delegations' responses to questionnaire discussed at the Working Party meeting on 1 February 2012. Brussels, 7 March 2012, 7260/12.

Council of the European Union 2013. *Joint Declaration Establishing a Partnership Between the Kingdom of Morocco and the European Union and Its Member States.* Brussels, 3 June 2013, 6139/13.

Dedja, S. 2012. The Working of EU Conditionality in the Area of Migration Policy. The Case of Readmission of Irregular Migrants to Albania. *East European Politics and Societies.* Vol. 26, No. 1, pp. 115–134.

Di Bartolomeo, A. Fakhoury T. and Perrin, D. 2009. *CARIM Migration Profile. Morocco.* Available from: www.carim.org/public/migrationprofiles/MP_Morocco_EN.pdf [17 April 2013].

Doctors Without Borders 2013. *Violence, Vulnerability and Migration: Trapped at the Gates of Europe. A Report on the Situation of Sub-Saharan Migrants in an Irregular Situation in Morocco.* New York, Doctors Without Borders.

Duhem, V. (2012). Maroc-Espagne, une relation qui s'intensifie. Jeune Afrique. [Online] Available from: www.jeuneafrique.com/Article/ARTJAWEB20121003173414/ [April 17 2013].

EPC 2012. *EU Readmission Agreements: Towards a More Strategic EU Approach That Respects Human Rights?* Policy Dialogue. March 21, 2012. Brussels.

ESI 2012. *Moving the Goalposts? Comparative Analysis of the Visa Liberalisation Roadmap for Kosovo and Other Western Balkan countries.* Available from: www.esiweb.org/pdf/Moving%20goalposts%20-%20A%20critical%20look%20at%20 the%20Kosovo%20visa%20roadmap%20%286%20July%202012%29.pdf [10 April 2013].

European Commission 2008. *Cap Vert-Communauté Européenne. Document de Stratégie Pays et Programme Indicatif National pour la période 2008–2013.* Available from: http://ec.europa.eu/development/icenter/repository/scanned_cv_csp10_fr.pdf [15 April 15 2013].

European Commission 2011a. *A Dialogue for Migration, Mobility and Security With the Southern Mediterranean Countries.* Communication COM(2011) 292 final. Brussels, 24 May 2011.

European Commission 2011b. *Communication from the Commission to the European Parliament and the Council. Evaluation of EU Readmission Agreements.* COM(2011) 76 final. Brussels, 23 February 2011.

European Union (2013). Roadmap. Towards a visa-free regime with Turkey. Available online at http://ec.europa.eu/dgs/home-affairs/what-is-new/news/news/docs/20131216-roadmap_towards_the_visa-free_regime_with_turkey_en.pdf. [15 April 2014].

Europol 2012. *Iraqi People Smuggling Network Dismantled.* [Online] Available from: www.europol.europa.eu/content/successful-action-against-people-smuggling-illegal-immigration [14 June 2013].

Europol 2013. *Successful Action Against People Smuggling and Irregular Migrants.* Available from: www.europol.europa.eu/content/successful-action-against-people-smuggling-illegal-immigration [14 June 2013].

Evin, A., Kirişci, K., Linden, R.H., Straubhaar, T., Tocci, N., Tolay, J. and Walker, J.W. 2010. *Getting to Zero. Turkey, Its Neighbors and the West.* Washington, DC: Transatlantic Academy, p. 30.

FIDH 2015. *Roundups of Migrants in Morocco – Is This the End of a Promising Political?* Available from: www.fidh.org/La-Federation-internationale-des-ligues-des-droits-de-l-homme/maghreb-moyen-orient/maroc/17016-rafles-de-migrants-au-maroc [9 June 2015].

Frenzen, N. 2010. EU-Turkey Readmission Agreement Negotiations Continuing. *Migrantsatsea*, 1 June.

Frontex 2012. *Annual Risk Analysis.* Available from: www.frontex.europa.eu/assets/Publications/Risk_Analysis/Annual_Risk_Analysis_2012_final.pdf, [17 April 2013].

GADEM 2010. *The Human Rights of Sub-Saharan migrants in Morocco.* Dakar: Justice Without Borders Project.

Ganghoff, S. 2003. Promises and Pitfalls of Veto Player Analysis. *Swiss Political Science Review.* Vol. 9, No. 2, pp. 1–25.

Halpern, C. 2010. Governing Despite Its Instruments? Instrumentation in EU Environmental Policy. *West European Politics.* Vol. 33, No. 1, pp. 39–57.

Hooghe L. and Marks G. 2006. The Neo-Functionalists Were (Almost) Right: Politicization and European Integration. In Crouch, C. and Streeck, W. (eds) *The Diversity of Democracy. A Tribute to Philippe C. Schmitter.* Cheltenham: Edward Elgar, pp. 205–222.

House of Commons 2011. *EU Readmission Agreements. European Scrutiny Committee – Twenty-sixth Report Documents Considered by the Committee.* London: Hansard

İçduygu, A. 2015. *Syrian Refugees in Turkey. The Long Road Ahead.* Washington: Transatlantic Council on Migration.

Jeune Afrique 2012a. *Manuel Valls a Rabat: la France souhaite faciliter l'obtention de visas pour les Marocains.* Available from: www.jeuneafrique.com/Article/ARTJAWEB20120726171912/france-maroc-immigration-cooperationmanuel-valls-a-rabat-la-france-souhaite-faciliter-l-obtention-de-visas-pour-les-marocains.html [17 April 2013].

Jeune Afrique 2012b. *Maroc-Espagne, une relation qui s'intensifie.* Available from: www.jeuneafrique.com/Article/ARTJAWEB20121003173414/ [17 April 2013].

Kassim H. and Le Galès, P. 2010. Exploring Governance in a Multi-Level Polity: A Policy Instruments Approach. *West European Politics.* Vol. 33, No. 1, pp. 1–22.

Kirişci, K. 2007. Turkey: A Country of Transition from Emigration to Immigration. *Mediterranean Politics.* Vol. 12, No. 1, pp. 91–97.

Knaus, G. 2013. Turkey, EU to Develop More Trust Through Plans for Visa-Free Regime. *Todays Zaman.* Available from: www.todayszaman.com/news-305251 [10 April 2013].

Köhler, J. (2011). What Government Networks Do in the Field of Migration: An Analysis of Selected Regional Consultative Processes. In Kunz, R., Lavenex, S. and Panizzon M. (eds) *Multilayered Migration Governance: The Promise of Partnership.* London: Routledge.

Kunz, R. and Maisenbacher, J. 2013. Beyond Conditionality Versus Cooperation: Power and Resistance in the Case of EU Mobility Partnerships and Swiss Migration Partnerships. *Migration Studies.* Vol. 1, No. 2, pp. 196–220.

Lahlou, M. 2007. *Migrations Transméditerranéeennes et Strategies Euro-Africaines. L'année 2006 sans L'espace Euroméditerranéen.* Barcelona : IEMED, CIDOB, p. 75.

Lagrand, T. 2010. *Immigration Law and Policy. The EU Acquis and Impact on the Order.* Nijemegan: Wolf Legal Publishers.

Lascoumes P. and Le Galès, P. 2007. Introduction: Understanding Public Policy Through its Instruments – From the Nature of Instruments to the Sociology of Public Policy Instrumentation. *Governance.* Vol. 20, No. 1, pp. 1–21.

Lavenex S. and Stucky R. 2011. Partnering for Migration in EU External Relations. In Kunz, R., Lavenex, S. and Panizzon, M. (eds) *Multilayered Migration Governance: The Promise of Partnership.* London: Routledge, pp. 116–142.

Majone, G. 1996. *Regulating Europe.* London: Routledge.

Ottaway, D. 2012. *Morocco's Islamists: In Power Without Power' Moroccan News Board.* Available from: http://moroccoboard.com/component/content/article/492-news- release/5670-morocco-s-islamists-in-power-without-power [17 April 2013].

Paçacı Elitok, S and Straubhaar, T 2012. *Turkey, Migration and the EU: Potentials, Challenges and Opportunities.* Hamburg: Hamburg University Press.

Political Programme 2011. *Programme électoral. Elections législatives du 25 nove,bre 2011. Pour un Maroc Nouveau.* Available from: http://fr.slideshare.net/AnasFilali/synthese-pjd-legislative-2011 [17 April 2013].

Renslow, N. 2010. Deciding on EU External Migration Policy: The Member States and the Mobility Partnerships. *Journal of European Integration.* Vol. 34, No. 3, pp. 223–239.

Roig, A. and Huddleston, T. 2007. EC Readmission Agreements: A Re-evaluation of the Political Impasse. *European Journal of Migration* Law. Vol. 9, No. 3, pp. 363–387.

Sedelmeier, U. 2011. Europeanisation in New Member and Candidate States. *6 Living Reviews in European Governance.* Available from: www.livingreviews.org/lreg-2011-1 [6 September 2013].

Stiglmayer, A. 2012. Visa-free Travel for Turkey: In Everybody's Interest. *Turkish Policy Quarterly.* Vol. 11, No. 1, pp. 99–109.

Strik, T. 2013. *Migration and Asylum: Mounting Tensions in the Eastern Mediterranean Committee on Migration, Refugees and Displaced Persons.* Report to the parliamentary assembly of the council of Europe (No.13106).

Thomas, D. 2011. *Making of EU Foreign Policy. National Preferences, European Norms and Common Policies.* London: Palgrave.

Tolay, J. 2012. Turkey's 'critical Europeanization:' Evidence from Turkey's immigration policies. In Paçacı Elitok, T. and Straubhaar, T. (eds) *Turkey, Migration and the EU: Potentials, Challenges and Opportunities.* Hamburg: Hamburg University Press, pp.54-68.

Trauner, F. and Kruse, I. 2008. EC Visa Facilitation and Readmission Agreements: A New Standard EU Foreign Policy Tool? *European Journal of Migration and* Law. Vol. 10, No. 4, pp. 411–438.

Tsebelis, G. 1999. Veto Players and Law Production in Parliamentary Democracies: An Empirical Analysis. *American Political Science Review.* Vol. 93, No. 3, pp. 591–608.

UNHCR 2015. Numbers of refugee arrivals to Greece increase dramatically, 18 August, www.unhcr.org/55d3098d6.html.

Wolff, S. 2012. *The Mediterranean Dimension of EU's Internal Security.* London: Palgrave.

Wolff, S. 2015a. Migration and Refugee Governance in the Mediterranean: Europe and International Organisations at a Crossroads. Roma, IAI, October, 23.

Wolff, S. 2015b. Integrating in Justice and Home Affairs: A Case of New Intergovernmentalism Par Excellence? In Bickerton, C. Hodson, D. and Puetter U. (eds.). *The New Intergovernmentalism: States and Supranational Actors in the Post-Maastricht Era.* Oxford: Oxford University Press.

Wunderlich, D. 2012. The Limits of External Governance: Implementing EU External Migration Policy. *Journal of European Public Policy.* Vol. 19, No. 9, pp. 1414–1433.

Appendix A: Interview coding

A.   confidential interview, DG Home official, 22 April 2013, Brussels
B.   confidential interview, DG Enlargement official, 22 April 2013, Brussels
C.   confidential interview, EEAS official, 23 April 2013, Brussels
D.   confidential interview, NGO expert, 22 April 2013, Brussels
E.   confidential interview, Moroccan officials, 19 July 2013, Brussels
F.   confidential interview, Permanent Representation, 23 April, Brussels
G.   confidential interview, European Commission official, 23 April, Brussels
H.   interview with Selim Yenel, Turkish Ambassador to the EU, 22 April, Brussels.

6 Europe's global approach to migration management

Doing border in Mali and Mauritania[1]

Stephan Dünnwald

Introduction

European migration management has been extended from the control of states' own territories to countries of origin and transit, thus creating new spaces of border surveillance and the disciplining of migrants and migration in Africa. This also remodeled the relationship between Europe and the sending or transit states concerned, and impacted the social, political and economic life of these countries, as well as the ways people conceive and organize mobility. Bargaining over bordering practices increasingly conditions development assistance, and soft tools like visa facilitation and circular migration schemes are used both as incentives and to keep targeted groups in their place. Though with the *Global Approach on Migration* Europe brought forward a comprehensive framework, the practical outcomes in terms of agreements, inter-state cooperation and consequences on the societal level show remarkable differences.

Remote control – this image could be used as a metaphor circumscribing European efforts to include and utilize African countries of transit and origin into a generalized defense against unwanted migrants. Yet this metaphor, which can be linked to the "police à distance" (Bigo and Guild 2005), suggests that this is working in a way that the European Union or its member states effectively control from afar what is happening in so-called third countries. In this contribution, I argue that this is not necessarily the case.

Taking the cases of Mauritania and Mali as examples, differences will be discussed by first drawing on the development of what is called externalization of European migration policy, then sketching out the process of how this externalization evolved both in Mauritania and in Mali.

In the case of Mauritania Europe achieved its goal: immigration from Mauritania to the Canary Islands is, since the measures began, about zero. However, taking the side effects of this Spanish–Mauritanian deal against migrants into account, an unbound harassment and hunt on migrants is taking place in Mauritania, which had been triggered by the European intervention, but is far from being under control. Furthermore, control is not remote: it has been the deployment of Spanish security forces, which finally led to an effective fight against migration 'on the beach.'

In the case of Mali, we see that European incentives to control the migration flux passing through and leaving the country were not strong enough to hamper the both open and integrative population politics in Mali. Even if we take Malian compliance and willingness to negotiate into account, the migration from and through Mali is far from being controlled by anyone. The developments of the past 15 years suggest that the inclusion of Mali into EU intended migration schemes would be a very long process.

Doing border

We should understand 'doing border'– as a way not only to do borders like border guards, patrolling, controlling, watching shift after shift, but, as a concept similar to 'doing gender,' as enacting and performing, and thus constituting and simultaneously questioning 'border.'

One of the most appropriate approaches to study and conceptualize bordering dynamics seems to be ethnographic border regime analysis. Ethnography provides a view which takes into account what is happening 'on the ground,' and border regime analysis frames ethnography with a mainly Foucaultian analysis of power relations. Borders are no longer understood as stable demarcation lines, but as spaces and fields defined, structured and negotiated between a multitude of actors (see Tsianos et al. 2009).

Approaching *doing border* methodologically means to focus on the application and negotiation of border practices, in addition to or sometimes even instead of a mere formal study of thematic documents and political decisions. Anthropological research on borders thus includes e.g. cultures of migration and cultures of bordering, the effects of gender roles or power relations, the biographical meaning or the multiple trajectories of migration.

This 'broader' view does not mean to neglect the role of the 'state.' Migration and even more migration control include negotiations on the national and international level. Yet the question why stopping migration to Europe worked in Mauritania and did not work in Mali cannot be answered on this level alone. Thus an anthropology of bordering processes is part of an anthropology of politics, at best including perspectives on micro, meso and macro levels in the analysis.

A global approach

Migration from Africa, predominantly conceived as migration of the poor, occupies European media and politics, though total figures are quite low (de Haas 2008). The 1990s saw a general tendency to restrict migration towards Europe applying a stricter visa regime and establishing enhanced border controls, readmission agreements with sending countries and a sharp rise in deportations. This did not stop migration from the south, but rendered it unauthorized and illegal. South–north migration was no longer a legitimate goal seen from the angle of European states and societies, but something to be controlled, and stopped. The notion of a 'Fortress Europe' denominates this European aim, and was illustrated by increased measures like fences, barbed wire, patrols and electronic surveillance along the southern and eastern borders of the European Union. Nonetheless, migration went on, although routes were shifting according to European attempts to close the holes in the fences.

As a landmark for the 'new,' externalized European migration policy we can see the incidents of Ceuta and Melilla, two Spanish exclaves on African territory, which in October 2005 were marked by the assault of sub-Saharan migrants trying collectively to overcome the heavy border fences surrounding the exclaves. Many Africans were severely hurt by the sharp wires and police clubs, with more than a dozen killed by gunfire. Moroccan security forces rounded up migrants still outside the fences and deported them to the desert or back to their home countries. All this happened under broad media coverage, and incented wild debates all over Europe as well as in African countries. In the aftermath of Ceuta and Melilla, while Spain fortified the fences surrounding the exclaves, the EU Council proposed the "*Global Approach to Migration*," a paper that stressed that controlling borders is not enough, but Europe needs a broader strategy including the countries of transit and origin into a concept of migration control and development.

European member states agreed upon a set of combined measures, decided on an informal meeting of the European Council at Hampton Court in October 2005, and aiming particularly at migration from Africa. These measures then emerged as Europe's *Global Approach to Migration* (European Commission 2006).[2] This *Global Approach* or *GAM* focused in the first instance on a dialogue with countries of transit and origin. Integrating the governments of these countries into a generalized migration and border management was the central message. This process can be traced back at least to the EU Council meeting at Tampere, Finland, in 1999 (Petrucci and Kalambry 2012), but after Ceuta and Melilla conferences and consultations between African and European politicians, it succeeded at an accelerated pace. Summits in Rabat and Tripoli in 2006, Madrid and Lisbon (2007) and Paris (2008) are the landmarks of a new African–European process of converging migration policies, surrounded by a swarm of informal meetings of high police, border guards and government officials.

The most important meeting between African and European heads of state was the 2006 summit at Rabat, Morocco, which resulted in the 'Rabat Process' coordinated by ICPMD, the International Centre for Migration Policy Development.[3] This process embraces a high number of single activities, most of them funded or co-funded by the European Commission.

In the externalization of the European migration management, which in the language of the EU Commission's *Global Approach* translates as "approaches on migration to optimize the benefits of migration for all partners in a spirit of partnership" (European Commission 2005, 5), African countries should take part in Europe's fight against irregular migration. Though this task is part of a more comprehensive approach, tackling irregular migration clearly is at center stage. While the EU identified the concentration on migratory routes as promising in stopping irregular migration from Africa, another, more general focus is the wish to combat the roots for migration through development measures, and to strengthen positive effects of migration through channelling remittances and assisting returnees. The last pillar then is the promise of opening routes for legal migration.

This model has been enacted most powerfully in the coastal states, especially Maghreb countries and Senegal, and could build on a number of bilateral processes that preceded the GAM. Italy had concluded agreements with Tunisia, and Libya, France and Spain already negotiated with Senegal, Morocco and Mauritania. By offering incentives like more options for legal migration, a rise in development assistance, and material and training in border issues, the EU was trying to establish a field for negotiations with the Maghreb countries and a number of West African states and to integrate them into a migration management scheme widely defined by the European Union.

Mauritania: fighting migration on the beach

Mauritania, scarcely populated and situated between the stronger neighbours Morocco and Senegal, was attributed for a few years the role of a window towards Europe. After the closing down of the transit routes via the Strait of Gibraltar and the Spanish exclaves Ceuta and Melilla in 2005, the Canary Islands became the predominant destination of some tens of thousands of mostly West African migrants. The islands belong to the Spanish kingdom, and are situated some 100 to 400 kilometres off the West African coastline. Soon Morocco controlled its Atlantic coast, and departure places shifted further south. Between 2005 and 2007, the Mauritanian towns of Nouadhibou and Nouakchott became prime departure zones. Dozens of small wooden boats, called *pirogues* or (in Spanish) *cayucos*, left the shores every night. Simultaneously, Spain and the European Union tried to influence the Mauritanian government to stop transit migration via Mauritania, and to take back migrants apprehended on sea. In Nouadhibou, these migrants were arrested, transferred to a detention camp and soon afterwards deported to the Malian and Senegalese borders.

Today the detention center in Nouadhibou is closed, the iron doors are shattered and the low buildings of the former school that were used to detain migrants are empty. The center was closed in May 2012, and the Spanish Red Cross, which had offered some assistance to detainees, has left. Nouadhibou's window towards Europe is closed. Standing in front of the detention center, the view back shall not only recall the years when thousands of migrants left the Mauritanian shores, thousands drowned or died of thirst, when many reached the Canary Islands and many others were detected on sea and turned back. It is also driven by the question of how the Mauritanian migration regime developed under Spanish influence, and what the consequences of this extension of European bordering processes to the African continent are for the everyday life of migrants and citizens.

Mauritania as a country of immigration

Long before Mauritania was known to be a country of transit for migrants heading for Europe it had been an immigration country. Almost all petty traders in the streets selling clothes, watches, sunglasses or mobile phones are foreigners, mostly from Senegal or Mali. Their life is not necessarily precarious: differences in prices, markets and trade connections guarantee a certain profit. Apart from this very visible presence in the streets, migrants keep whole industrial branches going. Immigration began under French colonial rule. In the 1950s, Mauritania still was a mostly nomadic society relying on livestock breeding and trade. Fish was not part of the diet; people say that Mauritanians live with their back to the sea. The development of fishing industries, building infrastructure and ore mining under colonial rule created a demand for qualified labour force, which was matched by immigrant workers from Senegal, Mali and some other West African countries.

Mauritania's membership in the Economic Community of West African States (ECOWAS) brought free movement within all member states, and even when Mauritania left ECOWAS in 1999 this free circulation of migrants continued, and was formally secured by bilateral treaties with Mali and Senegal. This was due not only to the need for skilled workers in Mauritania, but also to secure the status of about 200,000 Mauritanian migrants mostly working as traders and small shop owners in West African states. Estimations on migrant population from sub-Saharan Africa in Mauritania vary significantly between 65,000 and 200,000, mostly from neighbouring West African countries (Di Bartolomeo et al. 2010: 1, Profil Migratoire 2007: 113). The enormous variation again underlines that migration consists of mostly informal movements. Most migrants entered Mauritania legally and regularly. As the Mauritanian researcher Bensâad dryly states: "this immigration is neither formalized nor managed nor controlled by the state. It therefore is neither a trespass nor illegal, but simply left to the informal" (Bensâad 2008: 179). This is true for most West African countries, where migration and the presence of immigrants are normal elements of everyday life. Though free movement within the space of ECOWAS in fact faces a number of obstacles, irregular migration, as well as the classification of states as transit-countries, is a concept introduced by European and international efforts to organize and control migration (Fall 2007).

The relation between migrants and Mauritanians was never without tensions. West African immigrants met in Mauritania a population which had not yet overcome the history of a slave holder society. The distinction between 'black' (*haratin*) and 'white' (*beidane*) Moorish inhabitants, and *Halpulaar* and *Soninké* from the south are formative for social relations. Apart from slavery, informal dependencies seem to be more relevant today: 'Moorish' people are reluctant to do crafts and wage labour, 'black' Mauritanians and immigrants are kept from moving upwards the social ladder. Furthermore, tensions led to the displacement of tens of thousands of Mauritanians from Southern Mauritania in 1999. Reconciliation and return of the refugees from Senegal is proceeding slowly. Given this history, Mauritanian society has difficulties developing a common identity, and immigrants from the south further complicate matters, though for a long time they were mostly welcomed as a necessary work force. This was roughly the situation when, starting in 2005, a growing number of immigrants arrived in Mauritania to move on towards the Canary Islands.

Spanish borders in Africa

After 2000, Spain had introduced a number of measures to manage immigration, including regularization schemes, border surveillance, and the integration of countries of origin and transit into the fight against migration. In particular with the *Plan África* (published in 2006) Spain developed a narrow entanglement between migration governance and development policies, and opened a number of embassies in West Africa. In close connection to the European Union, Spain aspired to play a leading role in the formulation of European migration policies (Casas et al. 2010: 82f, Pinyol 2007: 3f). Spain developed an approach which rewards good conduct with assistance and additional development aid, quite close to what was adopted on EU level as the *Global Approach to Migration*.

In the wake of growing arrival numbers of boats with irregular migrants on the Canaries, Spain first concentrated on intensified sea patrols. Assisted by planes, Spanish marine and Guardia Civil vessels tried to detect migrant boats and divert them back to African shores. Since 2006 the EU border agency Frontex was coordinating these efforts. As a major part of the intercepted migrants had to be brought to the Canary Islands, these activities did not reduce the number of arrivals significantly. On the contrary, 2006 saw, with more than 30,000 migrants arriving on the islands, the biggest influx of sub-Saharan immigrants. This can be seen as the greatest incentive for Spanish authorities to extend migration control to the African mainland. In 2006 Spain concluded two treaties with the Mauritanian government. In the summer and autumn of 2006 the following measures were introduced: 250 Spanish police and Guardia Civil officers were garrisoned in Nouadhibou and Nouakchott. These forces brought along surveillance equipment including a helicopter with a night observation device. A surveillance plane was handed over to Mauritanian forces. The empty school No. 6 at Nouadhibou was transformed by the Spanish army into a detention center with the Spanish designation 'Centro de Estancia Temporal de los Inmigrantes, CETI' (Cruz Roja Espanola 2008: 14), but soon was called 'Guantanamito,' little Guantanamo, by the immigrants. Joint patrols of Spanish and Mauritanian security forces started in the harbors and coastal areas. Mauritanian officers on Spanish vessels guaranteed that these vessels can operate in coastal waters and return migrants to Mauritanian territory; Spanish police furthermore controlled Mauritanian coast guards to improve patrols and preclude bribing. Mauritanian forces were trained by Frontex surveillance standards. The Mauritanian coast guard was equipped with vessels, zodiacs (fast rubber boats), quads (small four wheel vehicles) and surveillance technology.

These measures were aimed at detecting migrants before they had reached open sea. Control was extended to the fishery harbors, and video surveillance and military exclusion zones were established. However not only were these securitized zones patrolled, also the quarters of the old town of Nouadhibou, which were inhabited mostly by migrants, came under surveillance. Migrants were arrested, transferred to the detention center and, after some days, deported in mini vans to the Senegalese or Malian borders.

With these provisions, transit migration from Mauritania towards the Canary Islands was practically stopped in 2007. In the document of a 'National Migration Management Strategy' this was celebrated as a 'spectacular' success (RIM 2010: 42). The close cooperation with third countries had proved to be decisive in the reduction of transit migration. However, stopping transit migration was not the end of the Spanish activities in Mauritania. Jointly with the European Union, Spain promoted a broad array of measures to integrate Mauritania into the more comprehensive approach of migration management, including the planning of a migration management strategy and a related action plan.

The Spanish-European negotiations with Mauritania were supported by the fact that, prior to the Spanish interventions, a Mauritanian migration policy practically did not exist. This was stated succinctly in a 'Migration Profile Mauritania,' a study financed by the United Nations High Commissioner for Refugees (UNHCR), the International Organization for Migration (IOM) and the European Union. It asserted that Mauritanian law on migration was sporadic and inconsistent. Authorities act indifferently, migration was rather tolerated than managed, and even the most relevant laws and directives on migration were not published frequently, and authorities did not know them (Profil Migratoire 2007: 127). Furthermore, the approach of offering material or financial incentives, mostly taken from development funds, in turn for cooperation in migratory issues, proved to be effective to enforce the collaboration with the Mauritanian government.

A further aspect favouring the fight against transit migrants in Mauritania was that there were hardly any Mauritanian citizens among the migrants heading for the Canaries. In contrast to the situation, for example, in Senegal or Mali, boat migration towards Europe did not include Mauritanian migrants, but consisted almost entirely of transit migrants from other West African countries.

It is mainly development aid with which the combat against migrants was bought. In 2006, Spanish payments of development aid rose rapidly, only to fall again drastically in 2009, when the boat journeys from Mauritania to the Canaries were stopped (Dünnwald 2014: 207). The European Union had more stamina. Directly or through Spain, and with considerable sums of money, it financed measures aimed at strengthening governance and, above all, the control and surveillance of the population and the borders. State-of-the-art electronics and numerous checkpoints were employed to register and prevent the entry of migrants – not only at sea borders but as well, and especially, at the long and porous borders with Senegal and Mali. Forty-seven new border posts were established, and crossing borders at these checkpoints became compulsory. Foreign nationals were required to carry identification and to consent to having their biometric data collected. Considerable pressure was exerted to push through these measures. Mauritania suspended some of the restrictions only after massive protests and threats by neighbouring countries. The country is economically dependent on several tens of thousands of migrants that live in Mauritania. Mauritania's economy started suffering under the number of migrants leaving the country; Mauritanian employers are unable to find migrants in Nouadhibou to work in the new fish processing plants, Chinese–Mauritanian joint ventures.

Most migrants in Mauritania entered legitimately and legally. All West African nations benefit from migration to a certain degree, either through their own nationals living in neighbouring states, or through immigration from those countries. With this wide-ranging liberality, the nations of West Africa have allowed an uncomplicated social attitude towards migration to flourish. This also conforms to the agreements made within the scope of ECOWAS, whose integration was actively promoted by the EU for many years. The presence of migrants is completely normal in all West African countries. That presence is hardly subject to any rules, and for that reason alone cannot be irregular. The fact that mobility acts as an economic motor for all West African nations is a further reason why ECOWAS is in favour of a liberal approach to migration. It is not only Mauritania that clearly benefits from the presence of Senegalese fishermen, Malian mine workers and construction workers from Guinea Bissau. The interference by the European Union and its member states undermines this liberality through the introduction of controls, administrative regulations and criminal offences that advance an arbitrary 'irregularization' and criminalization of migration and migrants.

By now, open xenophobia among the security forces and the general population had become commonplace. It was fuelled to such an extent by the fight against transit migrants that the Mauritanian human rights organization Association Mauritanienne des Droits de l'Homme (AMDH) warned of an increasingly racist approach to immigrants. Nevertheless, some migrants have become successful business people – as long as they were prepared to take on Mauritanian business partners and make adequate tribute payments. Well-to-do and long-established traders have advanced to the position of local community leaders. They had the necessary connections to the authorities and were able to help their compatriots in times of need. This did not always happen altruistically. In particular, arrivals of large groups of transit migrants brought a lot of money into town. Everyone made good business out of the passing migrants, and it was especially the newly arrived who had to buy the solidarity of the leaders of their own community.

Supported by the European Union and Spain, but also by the UNHCR, the International Organisation for Migration and the International Labour Organisation (ILO), Mauritania has since 2006 created a migration policy largely based on surveillance. Migrants are especially affected by the compulsory collection of biometric data crucial in obtaining residence permits and work permits. Judy can confirm this. In her capacity as leader of the Nigerian community in Nouadhibou she wanted to lead by example and made an effort to get papers early on. She paid the required 30,000 Oughiya – about €80 – for each member of her family and went to the capital Nouakchott on three occasions. She stated that "In the past, when they came round for controls they used to knock on the door; nowadays just they kick the doors down." I met Judy in the courtyard of the Catholic mission, a contact point not just for Christians but also other migrants in Nouadhibou, and one of the safe places for migrants in town. Repeated raids have unnerved the migrants – now many of them only leave their houses when it is necessary. In the past, says Judy, they used to meet up regularly, but today cohesion is crumbling; many are in constant fear and avoid the streets. The Nigerian community, as many other migrant communities in Nouadhibou, has also shrunk significantly. The Catholic mission run by Père Jérôme not only gives social support to all migrants but also essential medical treatment that would otherwise often be unaffordable: help for survival in a town that had turned into a dead end for migrants. Père Jérôme was here before in 2006, when he cared for stranded and sick migrants; many of those who were washed up dead on the beaches were buried by him in the mission's cemetery. Père Jérôme also works for stronger cohesion and cooperation between the various migrant communities, but the stubborn and pressurising control exerted by some of the community leaders has stood in the way of the necessary solidarity.

Pierre, a refugee from Rwanda, accompanied me to meet Mohamed, a Senegalese friend of his. He rented a small room in the center, where he shared a shower and toilet with three other lodgers. He commuted between Dakar and Nouadhibou, selling and buying. Mobile phones, imported by Mauritanian traders from the Gulf States, brought profits in Dakar, and clothes from Senegal sell well in Nouadhibou. Both Pierre and Mohamed were arrested in April 2012, and transported in pick-up trucks to the police prison in the center. The prison was over-crowded; they were not even offered drinking water, let alone food. Those who had money could order food from outside, but most people relied on relatives for water and food. While the men were kept in prison, the women had been deported almost immediately. On the second day, Pierre, who could provide a refugee card issued by UNHCR, was released. Mohamed was deported to Rosso, the Senegalese border town, on the third day after his arrest. He was lucky, he said, because he could return soon after and recover his goods. Often migrants lose their belongings and goods on deportation (see also Poutignat 2012).

David has worked a lot at Nouadhibou, as a fishermen, trader and building worker. Increasingly, he said, jobs were reserved for Mauritanians, as part of a new government strategy. Not only fishermen and taxi drivers, but also masons got work only when a certain number of Mauritanians had been hired as well. Often, employers not only asked for a residence permit, but for a work permit as well, a document which was obtained only on demand and for cash. Immigrants have to handle many obstacles. David was chairman of the small Gambian community, a duty which brings many troubles, he said. "Many of us wanted to go to Europe. Many still want to go," David said and grinned. We were sitting on a carpet in a dark room, together with two Gambian fellows. David came to Nouadhibou in 1992, when he was 18 or 19 years old. Four times, he tried the passage to Europe, and four times he failed. "We do not fear the sea. We know it is dangerous," he said. For a long time, there was no one who surmounted the controls. Even if Europe was still highly attractive, the distance became bigger. David told about some information campaigns on illegal travel and life in Europe, which were run to prevent migrants from leaving. Local associations had been paid by Europe to do the campaign, but he did not believe what was told to him: "Seeing is believing." Now pressure on migrants is growing. Many of his compatriots had been arrested, some deported in April 2012. Today, the Gambian community counts for about 300 persons at Nouadhibou; some years ago there were at least 2,000. Many had moved on since Nouadhibou became a dead end. David had stayed. He told me that he runs a restaurant now, and later I learned that he even owns two of them. In Nouadhibou, for migrants it is better not to show your wealth off (all accounts from field notes December 2012).

Remote control? Europe's dubious success

European funds and Spain's skilful negotiations have created a situation in Mauritania which left migrants largely to their own devices. Mauritania has seen the introduction of a controlling regime in which migrants are under general suspicion. Since the detection of would-be transit migrants is hardly possible before they actually embark, control, suspect and repression target all immigrants, and even black Mauritanians. This was accomplished relatively easily as a political aim because there are comparatively few Mauritanians among the migrants willing to leave. The criminalized image of a transit migrant is that of a black foreigner. The control of migrants fuels racism in a society whose identity between Maghreb and sub-Saharan Africa has always been fraught with tension. Not only has the transit been halted. Many West African migrants have left Nouadhibou, leaving behind a serious demand on the labour market. Others feel intimidated; while the surveillance programmes initiated by Europe are spreading even beyond Mauritania's southern borders, a dubious authoritarian regime profits from European funds to control its own population.

So far, the European intervention has triggered processes that are far from being under control. This was, as has been argued, not only the effect of European negotiations with Mauritania, investing much development assistance in surveillance equipment, staff and training. Eventually successful was this intervention only by the garrison of several hundred Spanish security forces in the towns of Nouakchott and Nouadhibou. Thus not remote control, but close surveillance of territory and the work of Mauritanian forces was the key to success. This, by the way, still seems to be true: in 2012, a unit of about 50 Spanish Guardia Civil was based at Nouadhibou, frequently patrolling harbor and beaches.

The Mauritanian migration analyst Ali Bensâad warns that Mauritania meanwhile runs the risk of creating an artificial irregularity by aligning itself with European demands and migration policy, all the while disregarding traditional migrational relations between Mauritania and its neighbours; he notes that "Apart from social and regional tensions Mauritania risks to disturb its proper socio-economic system based largely on immigration" (Bensaâd 2008: 2).

Mali, playground for migration management

Mali, in contrast to Mauritania, belongs to the 'hinterland' of migration towards the European Union. Mali has a long tradition as a north–south trading link between the Maghreb and the states along the Gulf of Guinea as well as a country with manifold cultures of regional migration. Migration is an important aspect in Malian cultures. Migration is an *aventure*, and persons returning from other countries are praised not only for bringing back wealth, but also for having seen the world. Though labour migration is increasingly a necessity for many families, migration in Mali is a social and cultural phenomenon that cannot be reduced to earning money abroad. The most important migration link to Europe was established by the ethnic group of Soninké from the western region of Kayes already in the 19th century, first going to French harbours like Marseille or Le Havre, and since the 1950s gathering in the industrial belt of Paris (Manchuelle 1997). During the 1990s, migration towards Europe enhanced, comprising different ethnic groups also from other regions in Mali, and many migrants went to Spain instead of France, working in the booming construction and agribusiness. Migration and migrant remittances are important factors in Mali, but most migrants tend to return after some years abroad. This explains why for example the Malian community in France comprises only some 80,000 persons, those without legal residence included. Nonetheless, France, like other European immigration countries, has stopped its liberal migration policies since the 1970s, gradually restricting the issuing of visas and extending deportation measures.

Mali is at the same moment a country of origin and an important transit country for migrants from Cameroon, the Congos and a number of other African states, with the capital Bamako being an important hub for changing passports, gathering information and arranging further travel.

Increasingly Bamako developed into a place not only for the departure, passing through and return of migrants, but also for activities targeting these flows. Bamako, among other West African cities, became a focal point for international organizations dealing with migration issues. Different organizations were trying to establish knowledge resources about migration, to distribute information, and further integrate African societies into the fight against the formerly unknown figure of the *illegal migrant*.

Again, the events of Ceuta and Melilla represent a turning point. In 1996, the *sans papiers* occupying the church of St. Bernard in Paris were arrested and deported. Once in Bamako some of them gathered with other deportees, mostly from African countries like Angola, and formed the *Association Malienne des Expulsés* (AME). The aims were at first instance to help members to reintegrate, and to support other deportees through social assistance and political activities. Without proper financing, the organization stayed small and did not develop. Over the years, Bamako saw a constant flux of returnees, but it was not until October 2005 that these processes accelerated and structures began to evolve. When, following mass attempts of sub-Saharan migrants to overcome the fences of the Spanish exclaves Ceuta and Melilla, within one week in October more than 400 deportees from Morocco arrived at Bamako airport, a broad public debate started in Bamako about what was actually happening.

Already some years earlier, Mali had developed governmental structures to address the Malian diaspora and assist migrants abroad. However, the Ministry for Malians Abroad and African Integration (MMEIA), and its technical branch, the Délégation Générale pour les Maliens de l'Extérieur (DGME), never had much influence and effect on migration or migrant communities. The same is true for the HCME, the High Council of Malians Abroad, which was founded to be the umbrella organization for Malian migrants' associations, but suffers from inadequate financial support and quarrels over leadership (see Whitehouse 2012: 155ff, for a more detailed account).

The rather poor performance and detached standing of Malian state institutions became clear with the massive arrival of deportees from Morocco, when the DGME and other branches, as well as the *protection civile*, were passive, and did not provide assistance to the returnees. It was civil society, especially Aminata Draman Traoré, a central figure in the Malian anti-globalization movement, who offered shelter and food to the returnees, organized the collection of their testimonies and coordinated, together with the AME and other groups, public hearings and protest. This protest, linked closely to the history of the Malian-French relationship and the broader context of Africa in times of globalization, further developed when, in the spring of 2006, the decentralized World Social Forum at Bamako gathered activists and organizations from Africa, Europe and elsewhere. Migration and deportation was the central theme of the forum, displayed in migrants' testimonies and theatre sketches, and Malian associations linking up with human rights associations, transnationalizing the protest against deportation, and against European efforts to block migrants. Out of these events and activities a number of returnee associations arose, and the AME managed to get stable funding by the French church based CIMADE, the German non-government organizations (NGOs) Medico International and Pro Asyl, and thus rapidly extended and transnationalized its work and increased its international visibility.

Dealing with, dealing over migration

As a regional hub of migration to and from Europe and as a strategic place within the European *Global Approach*, Bamako also became a key site for European migration management.

Against the background of a complex mobility of persons, mainly circulating within West Africa, but also heading north towards the Maghreb and Europe, we see a broad alliance of political and security forces trying to prevent exactly this *flux migratoire* going north. This alliance, seen from a Bamako perspective, comprises the European Union, together with some single member states standing out as prime new or old destination countries, like France, Italy and Spain. The International Organization for Migration and United Nations bodies like UNHCR, United Nations Development Programme (UNDP) or ILO take part in this alliance as well. The common denominator among such actors is an understanding of the abovementioned migratory movements as problematic, as irregular or illegal, because it is lacking authorization from distant destination countries. As indicated above, most migration in West Africa is not illegal or irregular; what is illegal about this migration is that it (in some cases) takes the Maghreb or Europe as destination, which only then, beginning with the attempt of irregular border crossing, makes this migration unauthorized (Carling 2007). Nonetheless, to my surprise even the UN refugee organization UNHCR addressed (at a human rights conference I attended in Bamako in 2006) migration as a problem and as irregular. Talking about migration in terms of illegality or irregularity became frequent as well for Malian government officials when meeting with Europeans, but Mali's political elite could well distinguish when it was appropriate to follow the European discourse and when to praise the migrant – no matter whether regular or irregular – as an almost constitutive part of the Malian society. So all relevant institutions – embassies, IOM or ILO – were eager to arrange conferences, to install special attachés close to the Malian government and organize workshops and regular meetings on migration management issues. Looking back on these activities, they can best be characterized by a sentence of the then head of the IOM-office at Bamako,

Kapirovsky, who complained that everyone has its workshops, and all people are meeting frequently, but talking things over and over again does not change anything.[4]

At least since the EU agreed on a *Global Approach to Migration*, not only the fight against irregular migration, but also the fostering of (circular) migration for development was among the pillars sustaining this cooperation.

Sometimes (especially in critical accounts on EU migration policy) the focus was put on the measures driven by the EU or member states' efforts to control and reduce irregular migration in West Africa (Casas et al. 2010, CIMADE 2010). However, implementing migration management outside the European Union meant collaborating with third countries, such as Mali, and convincing the respective government and authorities that such collaboration is useful, necessary and potentially beneficial for the third country. Thus bordering practices target control measures, readmission agreements or co-development projects, and even comprise efforts to establish a new view on migration on the level of the government, civil society and the population as a whole. This disciplining of migration (Geiger 2013) shifts the focus from the mere political measure to the bargaining over it, and the responses of Malian authorities and civil society.

Mali's position regarding European activities to extend migration management to West Africa can be described as twofold. On the one hand, Europe, France and – at least during the period of 2006 to 2012 – Spain were major donors of development assistance and economic partners. Since 2006, cooperation in migration issues emerged as a condition to sustain this flow of money, and also to attract additional funding.[5] On the other hand, Mali is one of the main integrating forces within ECOWAS, which includes and promotes free circulation of citizens. Similar to Mauritania, Mali depends to a certain degree on highly skilled external labour force, mostly from Senegal and Côte d'Ivoire. As almost a quarter of the Malian population is residing in mostly West African countries, Mali's interest to invest in migration control is limited, as neighbours could impose similar measures on Malian citizens. Furthermore, migrants in France exert not only economic power (through remittances and business investments), but constitute a force that can execute substantial social and political pressure as well. When it came to the signature of a readmission agreement, demonstrations both in Paris and in Bamako largely influencing public opinion could be seen as one of the major reasons why the then Malian president Amadou Toumani Touré finally refrained from concluding the treaty (Soukouna 2011). It is thus a double interest and double discourse on migration and migration management that marks the position of the Malian government.

France and Spain as main actors in migration management

Both France and Spain, as EU member states most affected by immigration from West Africa, tried to influence Malian politics of migration management.

France, as the former colonial power, is seen as a central actor in Mali and the 'ennemi' targeted by migrant and human rights organizations. While up to the mid-1970s, Malian nationals did not even need a visa to enter France, immigration regulations have become increasingly restrictive since. The fight against irregular migrants, the *sans papiers*, and deportations already accelerated during the 1990s, and after Ceuta and Melilla France intensified its efforts to conclude readmission agreements with major West African states, including Mali.[6] But, while almost all surrounding states gave in and accepted readmission, Mali resisted. A first attempt by Nicolas Sarkozy failed in 2003 (Panapress 2003), followed by longish negotiations. The Malian government wanted France to agree upon a fixed number of 5,000 regularizations per year in return for signing the readmission agreement. With this procedure, the estimated 20,000 Malians as irregular migrants would gain residency in France over a four year period. France initially offered only 1,500 regularizations, and insisted on a case-to-case examination, which seemed to be just the procedure France was already following (Soukouna 2011: 56). Though France increased its offer slightly, after a number of attempts, in late 2009 the consultations eventually failed, and Brice Hortefeux, then minister of integration, who had visited Bamako personally, had to leave without the Malian signature on his agreement.

For Mali, this failure had a number of immediate negative impacts. First, France stopped the *codéveloppement* assistance, and official development assistance decreased from 2008 to 2009 from 60 to 53 million Euros. Furthermore, France declared almost the complete Malian territory as a zone threatened by terrorism, with immediate negative effects on trade, tourism and travel of French citizens to Mali in general (Soukouna 2011, 59ff). At the street level and in the media people celebrated the refusal to sign the readmission treaty as a victory, as a sign of support for Malians abroad and as a symbol of resistance towards the French and European interests in migration control.

Spain arrived in Bamako only in 2006 with the opening of an embassy in the framework of the *Plan África* initiative to combat irregular migration and intensify development assistance addressing the root causes of migration (Gobierno de Espana 2006). Spain concluded a number of almost identical agreements with West African governments, the so-called Second Generation Agreements, which are in line with the European policy of the *Global Approach*, and have a focus on readmission. While readmission is clearly defined in an annexed document, legal migration is formulated in vague terms and always depending on the labour market situation in Spain (Serón et al. 2011: 35). With Mali, Spain signed a similar agreement, including the acceptance of up to 800 legal temporary migrants annually (Doumbia 2012). Characteristically, the scheme for circular migration never was put to work properly. Only once, in 2009, a number of 26 Malian migrant workers could go to Spain for a four months period. Afterwards, Spain cancelled the program.

Here, the *Plan África*, in line with the European *Global Approach*, should be read as comprehensive and addressing Spanish efforts for development rather than mere restrictive migration matters. In the past years, a broader public as well as a number of publications critically targeted the *Plan África* for its reluctance in human right matters and its failure regarding development activities (Martinez Bermejo and Rivero Rodriguez 2008; Romero 2008). In comparison to France, Spanish diplomats acted much more carefully. Though imposing migration constraints on Mali, they offered a number of incentives, even if the realisation failed. Adepoju also suggested that Spain's negotiations in Africa regarding migration control were more efficient because Spain did not have the burden of colonial ties with West African states (Adepoju et al. 2009: 66). This is probably an important aspect, if we compare the complex and ambiguous relationship between France and Mali, especially regarding migratory issues (Quiminal 2012). Furthermore, Malian migrants arrived in greater numbers in Spain only during the late 1990s, and thus there are no long lasting transnational relations or established Malian migrant communities. Also, deportations from Spain were never criticised publicly in Mali. Protests against expulsions from Ceuta and Melilla targeted the European Union rather than Spain in particular, and charter deportations from Spain were hardly noticed at all. Furthermore, while the French government cut co-development funds substantially in 2009, Spain more than doubled its development assistance for Mali in the context of readmission negotiations (Cimade 2009, 9; Serón et al., 2011, 44).

The CIGEM

The European Union, present with a delegation office at Bamako, was not an important actor in the field of managing migration in Mali until 2008. At the Rabat summit in 2006, the then commissioner for development issues Louis Michel agreed, together with representatives from Mali, Spain, France and ECOWAS, upon establishing a *Center for Information and Management of Migration*, the CIGEM, at Bamako. The center was equipped with 10 million Euros and opened in 2008. Due to the relatively small budget, the CIGEM tried to bundle different, partly already existing projects like the TOKTEN program of the UN Development Program. The center's primary task, that is, organizing and mediating labour migration from Mali to Europe, was rejected immediately by a number of EU member states, as they wished to keep full sovereignty over national labour markets. Divested of its main goal, the CIGEM concentrated on the remaining tasks: the information of potential migrants about the risks of irregular migration, assisting return migrants and assisting the Malian government with the draft of a comprehensive migration policy strategy. None of these activities was entirely successful. For example, in 2011 the migration strategy still was a draft of about a dozen pages, and never concluded. The integration of return migrants into the Malian labour market failed, as the cooperation with Malian institutions proved to be difficult and a formal labour market hardly exists. Information campaigns directed towards persons willing to go to Europe were not successful, either. As Isaie Dougnon points out, Malian migrants are well aware of the risks of migration, as risk is an essential part of the concept of migration, in the culture of most Malian ethnic groups (Dougnon 2012).

Reviewing the European efforts to establish a border management in Bamako over the past years, it is astonishing that the outcome is so meagre. The description of European activities could be extended, for instance by mentioning the equipment of 17 Malian border posts financed by Spain, or a number of Spanish co-development activities, but the result would not change the overall picture. As far as I can rely on my informants, it is very questionable that the Spanish money for border enforcement ever arrived at border posts, and crossing Malian borders is still much the same as before. Only few Spanish development projects were realised at all, and those I could visit suffered from severe problems and were not finalised after years.

The situation in Bamako is different from the cases of Libya or Mauritania. There is no clear European will to invest into the integration of Mali into an externalized European border management system. No member state stands out taking the lead; instead, French and Spanish officials seem to handle the subject without much empathy. The EU itself did not seem to invest a lot into the process: once the CIGEM was established, it was not given much attention. Furthermore, agreements about migration were not accompanied by greater investment schemes or a significant rise in development assistance (as in the case of Spain and Mauritania). As migration control works out in the countries on the Atlantic as well as on the Mediterranean coast, already strongly diminishing the influx of migrants, the pressure to come to terms with the Malian government was low. One of the main reasons why migration management did not work out in Mali seems to be the ambivalent and very diplomatic stance Malians took while negotiating over migration issues. Malian diplomats were eager to show up on conferences or to agree to position papers. However, when it came to the realisation on the ground, nothing quite seemed to work.

Looking at the Malian side, we can detect a number of more miscellaneous reasons why Europe did not succeed in its intent. Among these reasons is the ambiguous relationship with France, characterized by a counter-hegemonic position towards France of great parts of the Malian society. Furthermore, Mali's quality as a country of origin and transit, closely embedded in the West African sub-region, is adherent to the imposition of control measures. While Arab countries could target 'black' immigrants as intruders, in Mali this is neither possible nor on anyone's agenda. Setting up control measures, the repulsion of potential transit migrants at Malian borders, or their detention would clearly oppose the Malian overall policy in the region, and perhaps would foster similar reactions in neighbouring countries. Furthermore, the strong Malian diaspora in France has a significant influence on migration politics, and is not reluctant to interfere.

Establishing an effective European system of migration management in Mali thus will remain a challenging task, and, apart from practical political interests, is confronted by Malian self-conceptions of being a welcoming society. Finally, taking into account the low capacity and complex interests of Malian authorities, it will need more than counsellors to impact a change on Malian governmental institutions.

In December 2015, the CIGEM closed without notice, its website shut down. Bamako will remain a challenging place for European aspirations towards migration management.

Conclusion: effects of bordering Africa

In the conceptual framework of EU externalization of the control of movement we tend to think of borders predominantly in that way: Europe establishes borders to prevent migrants from proceeding towards Europe. From a European perspective, this process and its consequences are studied under the aspects mainly of technology, law and human rights. This hides the fact that this establishment of bordering processes is more comprehensive, affects and is affected by the societies concerned. Borders in Africa often are not the sort of fixed territorial lines we know from elsewhere. The borders, often still relics from the division of colonial spheres of interest, more often than not cut ethnic groups and traditional landscapes. Cross border mobility is everyday practice, and often vital to the social and economic welfare of those living on or crossing the border. Therefore, many borders are spaces where mobility is only loosely controlled, and control posts on roads created a plethora of ways to circumvent them (see Mechlinski (2010) for a vivid account).

Mauritania now, in the framework of the National Action Plan on Migration, established new border posts, equipped with electronic data processing facilities on its Senegalese and Malian borders. If this works out, it will fundamentally change the character of the African border. It affects not only the life and relations of the ethnic groups living on both sides of the borders, e.g. the Soninké, but also the supply of migrant workers used to seasonal migration, or seeking work in the mines and harbors of Mauritania. Control measures on migrants and deportations already had a negative impact on the Mauritanian economy, creating a shortage of labour force in fish processing industries in Nouakchott and Nouadhibou.

Bordering is going on. Apart from the processes which directly link up to the EU border programs, as they are for instance collected in national action plans in third countries or listed as single projects in the Rabat Action Plan website run by ICMPD, in West Africa we find a number of projects and processes which are more or less closely intertwined with them. This applies first of all to the projects to fight and survey activities linked to Islamic terrorism. Europe, single member states like especially France, but also the United States have more or less concealed activities running in the whole

Sahara/Sahel zone. This applies also to projects to reinforce African borders, as in the framework of the African Union Border Program (AUBP). Here, the German development agency GIZ is conducting a project to measure and fix border lines, establish border posts and enhance control and border processing capacities. Both lines of activities, counter-terrorism and border enforcement, work in different political spheres, and are not included into the EU program of fighting irregular migration, but both are definitely part of a broader process to establish more generalized surveillance schemes and population control in West Africa.

Notes

1 A first version of this chapter was published as Wolff, S. (2014). 'The Politics of Negotiating EU Readmission Agreements: Insights from Morocco and Turkey'. *European Journal of Migration and Law*, 16:1, 69–95 and we thank Brill Publishers for granting us the permission to publish this updated version in this book.
2 The GAM had been reconfirmed and extended to the *Global Approach to Migration and Mobility* in 2011, see European Commission (2011).
3 See http://processusderabat.net/web/index.php.
4 Alexander Kapirovsky, interview December 2010.
5 Aminata Traoré for instance stated, when asked why Mali agreed to the CIGEM, that, "unfortunately, for the Malian government 'a bad project is more valuable than no project at all'" (Pour l'Etat malien malheureusement 'un mauvais projet vaut mieux que pas de projet du tout') (Herrou 2008).
6 Readmission, that is the readmission of own nationals and third country nationals, is one of the central features of migration management, as it is precondition for the effective removal of unwanted immigrants.

Bibliography

Adepoju, A., F. Van Noorloos, and A. Zoomers. 2009. Europe's Migration Agreements with Migrant-Sending Countries in the Global South: A Critical Review. *International Migration*. Vol. 48, No. 3, pp. 42–75.

Bensaâd, A. 2008. L'"irrégularité" de l'immigration en Mauritanie: une appréhension nouvelle, sonséquence d'enjeux migratoires externes. Fiesole, Robert Schuman Centre for Advanced Studies.

Bigo, D. and E. Guild. 2005. Policing in the Name of Freedom. In Bigo, D. and Guild, E. *Controlling Frontiers. Free Movement Into and Within Europe*. Aldershot: Ashgate.

Carling, J. 2007. Unauthorized Migration from Africa to Spain. *International Migration*. Vol. 45, No. 3, pp. 3–37.

Casas, M., S. Cobarrubias, and J. Pickles. 2010. Stretching Borders Beyond Sovereign Territories? Mapping EU and Spain's Border Externalisation Policies. *Geopolitica(s)* Vol. 2, pp. 71–90.

CIMADE. 2009. *French Agreements concerning the concerted management of migration flows and co-development*. Paris, La Cimade.

CIMADE. 2010. *Prisoners of the Desert. Investigation into the Consequences of European Migration Policies at the Mali–Mauritania Border*. Paris, La Cimade.

Cruz Roja Espanola. 2008. *Migraciones africanas hacia Europa. Estudio cuantitativo y comparativo. Anos 2006-2008. Centro no. 6 de Nouadibou, Mauritania.* Madrid, Cruz Roja Espanola.

de Haas, H. 2008. The Myth of Invasion: The Inconvenient Realities of African Migration to Europe. *Third World Quarterly.* Vol. 29, No. 7, pp. 1305–1322.

Di Bartolomeo, A., T. Fahkoury, and D. Perrin. 2010. *CARIM Migration Profile Mauritania.* Fiesole, Robert Schuman Centre for Advanced Studies.

Dougnon, I. 2012. Migration as Coping with Risk: African Migrants' Conception of Being Far from Home and States' Policy of Barriers. In Kane A. and Leedy T. H. (eds.) *African Migrations: Patterns and Perspectives.* Bloomington: Indiana University Press, pp. 35–58.

Doumbia, C. 2012. Mali-Espagne: un accord de "dupes" sur l'immigration. *Le Challenger.* Bamako.

Dünnwald, S. 2014. No Transit. Mauretanien und die Externalisierung europäischer Migrationskontrolle. In Pro Asyl, Brot für die Welt (ed.) *Im Schatten der Zitadelle. Der Einfluss des europäischen Migrationsregimes auf "Drittstaaten."* Karlsruhe: von Loeper, pp. 171–212.

European Commission. 2005. *Priority actions for responding to the challenges of migration: First follow-up to Hampton Court.* Brussels, European Commission.

European Commission. 2006. *The Global Approach to Migration One Year on: Towards a Comprehensive European Migration Policy.* Available at: http://eur-lex.europa.eu/LexUriServ/LexUriServ.do?uri=COM:2006:0735:FIN:EN:PDF.

European Commission. 2011. *The Global Approach to Migration and Mobility. European Commission.* Brussels, European Commission.

Fall, P. D. 2007. La dynamique migratoire ouest africaine entre ruptures et continuités. In *African Migrations Workshop IMI*, pp. 24. Accra, Ghana. Available at: www.oecd.org/fr/dev/pauvrete/44806149.pdf.

Geiger, M. 2013. The Transformation of Migration Politics: From Migration Control to Disciplining Mobility. In Geiger M. and Pécoud A. (eds.). *Disciplining the Transnational Mobility of People.* EBasingstoke: Palgrave Macmillan, pp. 15–40.

Gobierno de Espana. Ministry of Foreign Affairs and Cooperation. 2006. *Plan África 2006–2008.* Madrid, MAEC.

Herrou, A. 2008. "Aminata Traoré: Le CIGEM est "le machin" de Louis Michel," in Afrik.com.

Manchuelle, F. 1997. *Willing Migrants: Soninke Labor Diasporas 1848–1960.* Athens, Ohio: Ohio University Press.

Martinez Bermejo, E. and J. Rivero Rodriguez. 2008. *El Plan REVA y La Ayuda Espanola. Migraciones y Cooperación.* Madrid, Grupo de Estudios Africanos UAM/Covenio MAEC.

Mechlinski, T. 2010. Towards an Approach to Borders and Mobility in Africa. *Journal of Borderland Studies.* Vol. 25, No. 2, pp. 94–106.

Panapress. 2003. *Mali Would Not Help France Deport Its Own Citizens. Panapress.* Dakar: Panapress.

Petrucci, F. and A. Kalambry. 2012. *Evaluation of Visibility of EU External Action. Final Report, Volume 7, Thematic Report on Migration, Part 1, Mali.* Particip GmbH et al. Available at: http://ec.europa.eu/europeaid/how/evaluation/evaluation_reports/reports/2012/1307_vol_7_en.pdf.

Pinyol, G. 2007. The External Dimension of the European immigration Policy: A Spanish Perspective. In AA.VV. *The Euro-Mediterranean Partnership (EMP): Perspective from the Mediterranean EU Countries.* Rethimnon, Crete.

Poutignat, P. 2012. Migration at the Level of Individuals. In Streiff-Fénart, J. and Segatti, A. (eds) *The Challenge of the Threshold.* Maryland: Lexington.

Profil Migratoire 2007. *Profil Migratoire de la Mauritanie.* UNHCR, Europäische Union, IOM.

Quiminal, C. 2012. Ambiguous Europe: Repertoires of Subjectivation among Prospective Migrants in Bamako, Mali. In Streiff-Fénart, J. and Segatti, A. (eds) *The Challenge of the Threshold.* Maryland: Lexington.

RIM, Republique Islamique de Mauritanie. 2010. *Document de Stratégie nationale pour une meilleure gestion de la migratoire.* (Unpublished document.)

Romero, E. 2008. El Plan África, la política migratoria espanola de 'nueva generación' y la guerra contra los pobres. *Virus Fronteira Sur, 14 August 2008.* Available at www.rebelion.org/docs/74895.pdf.

Serón, G., A. Jolivel, M. Serrano Martín, and J. L. Gázquez. 2011. *Coherencias de políticas Espanolas hacia África: Migraciones.* GEA-Grupo de Estudios Africanos. Madrid, Línea de Investigación en Migraciones.

Soukouna, S. 2011. *L'Échec d'une Coopération Franco Malienne sur les Migrations: Les logiques du refus malien de signer.* Université Paris I Panthéon Sorbonne.

Tsianos, V., S. Hess, and S. Karakayali. 2009. *Transnational Migration. Theory and Method of an Ethnographic Analysis of Border Regimes.* Sussex Centre for Migration Research. Working Paper No 55.

Whitehouse, B. 2012. *Migrants and Strangers in an African City. Exile, Dignity, Belonging.* Bloomington and Indianapolis: Indiana University Press.

7 At a distance

The European Union's management of irregular migration in Eastern Europe

Lyubov Zhyznomirska

Introduction

In the course of European integration, the European Union (EU) has become an active policy-maker and trend-setter in inter-state migration relations. Since the Tampere Justice and Home Affairs (JHA) Council (held in 1999) proclaimed the EU to be the area of freedom, security and justice (AFSJ), the salience of migration and internal security issues in the EU's relations with its neighbouring countries has increased. Various mechanisms of remote control that contributed to the overall externalization of migration regulation to non-EU countries have provided a rationale and blueprint for the overall construction of the AFSJ grounded in the dynamic of the Self/Other interaction (Balzacq 2008; Zhyznomirska, 2011). The linking of security reforms to the ideas of good governance, democracy and the rule of law in the EU's external relations has served to expand the realm of the applicability of the EU's normative power. Through the construction of the AFSJ, the EU differentiates itself from its periphery and conveys its supreme standing in relation to other countries in the region.

Whereas the policy measures of remote migration control have been extensively analysed from the EU's perspective, the impact of these measures on migration management policies and practices in the countries of transit and origin of migrants has been less investigated. This chapter problematizes the role of non-EU countries in establishing and maintaining the EU's regime of migration controls by examining the responses of the EU's eastern neighbouring countries (in particular, that of Russia and Ukraine) to the EU's pressure concerning the prevention and control of unregulated migration towards its territory. Although the EU acts as the facilitator and promoter of norms in the area of migration, ultimately, the decision to pursue – or not – certain policy directions is made and justified by the government of a given country, either in accordance with its domestic interests (in a *quid-pro-quo* manner) and/or under external pressure. Also, the relationship between the EU and a given non-EU country is dynamic, rather than simple, one-way process in which the all-powerful EU imposes its interests upon neighbouring countries. The impact hence differs depending on the possibility of applying various remote control dimensions (succinctly described by Zaiotti, this volume) to a given country or region and upon the domestic politics of a country under the EU's external pressure. The examination of what policy measures are being applied and to what extent allows us to deepen our understanding of why the inter-state system of migration management in Europe has been developing the way it has been.

The empirical material is drawn from the cases of the EU–Russia and EU–Ukraine relations on migration management since the 1990s until the end of 2013. Such historical perspective enables the author to highlight the dependency of bilateral and international migration co-operation on the larger, macro-political dynamic of relations.[1] This chapter evaluates the cumulative effect of "Europe" on the cases under study in the area of international migration management where the activities and ways of doing things are perceived by local elites and expert communities as associated with "European" experiences, traditions and practices.

The chapter proceeds as follows. First, I analyse the effects of the EU's policy of the externalization of migration management on the international migration governance, followed by the examination of factors that determine the EU/third country migration co-operation. Then I analyse the consequences of the EU's migration diplomacy for the case studies under consideration. Finally, prior to concluding this chapter, I evaluate the institutionalization of the readmission mechanism in Europe.

Evaluating the EU's externalized management of migration

Although the EU's migration regime impacts the countries of origin and other countries of destination by virtue of the EU being a magnet for migrants (the changes made in the EU's migration regime affect the migration flows, depending on the permissiveness or restriction of the EU's migration policy) (Lavenex and Uçarer 2004), the EU has also had a deliberate strategy of managing migration flows indirectly, by influencing third countries' capacities to manage migration to, from and through their territories.[2] Since the 1990s, EU policy makers have attributed great importance to migration issues in external relations, making migration one of the dimensions of the EU's foreign policy towards third countries. Migration management clauses have been inserted into the EU's co-operation frameworks with the Euro-Mediterranean region, with the countries covered by the European Neighbourhood Policy (ENP), with Latin American countries, as well as with 77 countries grouped under the ACP title (i.e., African, Caribbean and Pacific). The Union conducts regular exchanges on immigration management with the USA, Canada and the Russian Federation. It also has Europe–Asia dialogue on migration and includes migration matters in its dialogue with India and China.

The EU's *Global Approach to Migration* (GAM), introduced by the European Council in December 2005 to address the "seriousness of illegal migration" to the EU, was a first comprehensive expression of the EU's external migration relations. Later modified into the *Global Approach to Migration and Mobility* (GAMM), it advanced an idea of dialogues with non-EU countries from which immigrants to Europe originate. It put together the issues of migration, development, foreign policy and other policies in the Union's external relations (Commission 2006). According to the official rhetoric, the EU promotes global migration management and seeks to foster regional co-operation among states facing similar challenges with emigration and immigration, irregular migration, and human trafficking and smuggling. The GAMM provides enough flexibility to incorporate tools and practices deemed beneficial to all parties involved (i.e., states of origin, destination, transit and possibly migrants themselves). Content-wise, it serves as a convenient rhetorical platform that consumes governmental technologies (e.g., biometrics and surveillance for border protection) and social spaces (e.g., detention centres, airport security zones, local border traffic regimes) to maintain migration and mobility controls on a regional, if not global, level. It normalizes the extension of the EU's migration governance practices beyond its territory and gives legitimacy to its attempts to regionalize the regulation of migration flows. It also allows the EU to institutionalize transnational migration partnerships directed at the regulation of labour, humanitarian and irregular migration flows, increasing its influence on setting the agenda of international migration governance.

Having linked migration to development (largely, in response to the pressure from the countries of origin of migrants), the EU has increased its leverage in negotiations with third countries on migration co-operation extending its "soft power" influence beyond into the territories of the neighbouring countries and/or countries of origin of migrants. Thus far, this regime has largely focused on control and prevention of migration flows. As Lavenex and Kunz (2008) convincingly argue, the EU has approached development as a strategy to reduce and prevent migration into its territory, rather than as a strategy to create greater development *per se* in the countries of origin.

Clearly, the EU's migration politics in non-EU countries is advanced by such international organizations like the International Organization for Migration (IOM), the *International Centre for Migration Policy Development* (ICMPD) and the United Nations High Commissioner for Refugees (UNHCR) as they implement the EU-funded projects in these countries (see also Lavenex 2007). In particular, the IOM has been actively promoting the agenda of "global migration management", and it has been assisting governments in non-EU countries with the introduction of European and international standards in migration management (Georgi 2010; Geiger 2010; Hess 2010). ICMPD, in turn, has been acting as a Secretariat to and a member of the Core Group within the *Prague Process* that acts to promote close migration partnerships between the states of the European Union/Schengen area, Western Balkans, Eastern Partnership and Central Asia, as well as Russia and Turkey, in line with the EU's *Global Approach to Migration and Mobility*.[3] Since the European Commission funds those projects and determines the guidelines for implementation, these organizations may be viewed as agents of Europeanization of the countries where they implement EU projects and follow the EU's overarching framework for migration co-operation with third countries (see Geiger 2010 on IOM efforts in Albania).

The EU also facilitates and creates what has been called "multi-level migration governance" (Betts 2008, 2011) or "multilayered migration governance" (Kunz et al. 2011) by expanding the venues where states can engage in creating ad hoc or permanent migration partnerships on bilateral, regional or inter-regional bases. While internationally we have been observing an emergence of fragmented and incoherent governance of migration in specific categories (e.g., refugees, workers, irregular migrants, tourists) and issue areas (e.g., remittances), in Europe this regime is more structured thanks to the activities and interests of separate states (such as the Switzerland-initiated Geneva Process or the Sweden-initiated Söderköping Process) and organizations like the EU and the Council of Europe. This all has effects on the international politics of migration.

The EU's strategy of externalizing migration controls has caused the expansion of the migration control regime to the countries that partake in the common logic of management of migration to Europe. In the global migration system there exists a hierarchy of power among nation states with regard to the capacities to control population movements, to maintain the integrity of their borders and to shape the structure of international migratory flows (Held et al. 1999). As non-EU countries adopt the practices and norms of the Union, we see the spatial extension of the EU's political authority beyond the perimeter of its member states towards neighbouring countries. There, the EU's power is expressed in 1) policy transfer, 2) expertise and knowledge transfer, and 3) operational co-operation with the relevant government authorities that elect, or are induced, to follow the European practices of migration and border control. The Union's governance combines both inclusionary and exclusionary techniques in relation to its partner countries and their populations.

While co-operation (or conflict) between states on the regulation of migration flows is not something historically new, what is new in the case of the European Union is the "attempt to relate and co-ordinate action at the EU level and – in certain policy areas – formulate common policies" (Geddes 2008, 20). What is also unique to the European Union is the attempt to assist and guide non-EU countries in how their migration regulation should be instituted. Moreover, the EU's innovation lies in the continuous learning and adjustment to how non-members of the Union may be of use to the EU MSs in stemming the flows of unwanted population into the EU territory; all the weight of the European Union as an international actor and a normative power is used to extract the compliance of third countries. EU governments have formally chosen to manage migration into the EU territory with the assistance of third countries with whom they co-operate.

The EU has sought to increase the capacities of governments in transit and source countries to better control irregular movement: 1) of their populations across borders (such as through better identification and protection of the documentation and exit-entry controls), and 2) of the foreigners transiting through or entering their territory (for example, by means of technically equipped border protection services). Instrumental linking of the countries into 'migratory routes' generated a possibility for an organized action aimed at solving the 'common' problem of irregular migration. Representation of irregular migration as a threat to all societies serve to mobilize governments 'along the migrant routes' into action, as none of the countries wished to look as if it were covering up or supporting 'illegal' migration. Due to their geographical proximity to the (prosperous) EU, co-operating with the EU – particularly for the transit countries – became obvious and expected because of the wide-spread understanding of the problem being imported.

Notably, the EU agenda expanded from irregular immigration to include the issues of migrant integration, freedom of movement within third country territories and external relations on migration. Such conditionalities became possible within the context of the visa liberalization negotiations with the non-EU countries, considering the latter's interest in advancing towards a visa- free regime with the EU.

Another cultivated 'common sense' belief was the representation of illegal migration as a problem that required *concerted* actions by governments, both internally within the EU and outside of its membership confines. Since it received the competence to conduct external relations in the immigration area, the Commission has been advocating for transnational solutions to Europe's migration problem. The EU, as a result, has acted to facilitate various bilateral and multilateral co-operation platforms in order to solve the irregular migration problem through co-operation. The EU's 'government at a distance' takes place through the inculcation of its ideas and norms in non-EU countries, popularized by the Commission's support of such mechanisms as the sharing of expertise and best practices, shaping of the governments' capacities, the spread of the EU's training manuals (as in the case of the Schengen border code trainings) or the practical on-the-job training (for example, of the border and migrant officials in non-EU countries). Instruments such as study missions and expert exchanges are other governmental technologies enabling the EU to produce and accumulate 'knowledges' about a given country's migration situation and later to claim that it offers a 'tailor-made' approach with a partner country's migration challenges taken into account when it proposes its solutions to the 'common' problem of irregular migration. Other examples are the activities of intergovernmental platforms like the Söderköping Process or the Budapest Process, and the activities of the European Union Border Assistance Mission (EUBAM) whose main objectives are to popularize the information or to assist the participating countries in improving their internal policies with regard to migration and borders.

In this respect, it is useful to analyse the tools and techniques the EU has been using to "mainstream" (the Commission's term) immigration into its external relations while at the same time paying attention to the geopolitical ordering of space and the creation of "buffer zones" around the EU.

The political rationalities and tools employed by the EU in relation to its accession countries and non-members differ, and reflect the construction of an inside/outside dichotomy: the accession countries are potential insiders, whereas the neighbouring countries (viewed as transit countries) are viewed as both 'in' and 'out', with fuzzy borders and borderlands constructed around the Union to meet its geopolitical goals. In the enlargement context, the transfer of governance models of migration controls happens in the context of accession negotiations (e.g., financial, technical and expert knowledge assistance in creating the capacities of accession countries in migration and border management). In the case of the Western Balkan countries, membership prospect (provided to these countries by the European Council in 2003) offers a similar logic. In addition to financial and technical assistance, a roadmap to a visa-free regime with the Union has motivated these governments to undertake the required policy changes in migration controls in line with the EU's demands (Trauner 2009a, 2009b).

In the case of neighbouring countries with no membership perspective, political maneuvering by EU institutions and MSs has been important to exercising the EU's authority and influence. In the relations with the eastern European neighbours of the Union (i.e., Ukraine, Moldova, Belarus, Georgia, Armenia, Azerbaijan), the EU's experience with the Western Balkans provides a map for possible future action. Similarly – as was the case with the Western Balkans – visa dialogue and a roadmap to a visa-free regime function as leverage for the EU to demand compliance with its norms and rules, and hence allow it to intervene into third states' governance capacities in migration and border management. Facilitating mobility for the neighbouring countries' nationals (that was restricted in the first place due to the introduction of strict visa rules by the EU for these potentially "risky" nationals) emerges as a tool to secure third countries' co-operation on irregular migration. This co-operation includes undertaking policy changes to align security and mobility rules in line with European ones, with roadmaps for a visa-free regime used as a potent foreign policy tool with "willing" partners, who are themselves interested in a liberalized travel regime to the EU for their citizens. What happens in the external dimension, then, is representative of the processes used by the EU to balance its security interests with third countries' sensitivities and interests.

As the EU's response to the 2015 migrant tragedies in the Mediterranean Sea shows (see European Council 2015), the multilayered migration management regime in Europe remains focused on control and policing, with government resources dedicated in the first place to irregular migration and other migration-related phenomena associated with (in)security (such as counter-trafficking). The control- and security-oriented migration governance regime that the EU has established internally is being expanded beyond its borders through co-operation with non-EU countries (see also Düvell 2011).

Factors determining EU/third country migration co-operation

It has been suggested that "a relatively high level of compliance can be expected" in the area of Justice and Home Affairs co-operation (to which migration issues belong) with European Neighbourhood Policy (ENP) countries because

> by taking aim at organized crime and terrorism, many objectives endeavor to strengthen state control, which is accepted as legitimate by the ENP participants. JHA co-operation with the EU is an obvious interest of ENP countries whose governments are interested to increase the capacities of their law enforcement agencies pointing out to the leverage the EU has over these countries.
>
> (Occhipinti 2007, 114–15)

However, the EU is also perceived to be in an asymmetrical, disadvantaged position in relation to third countries concerning its internal security or issues related to human rights (Occhipinti 2007, 115). When their security or economic well-being are at stake, the argument goes, third countries may not be willing to accept the EU's rules and may resist the externalization of controls and may heavily insist upon compensation for something that can potentially jeopardize their domestic situation (for example, readmission) (Lavenex and Wichmann 2009). Clearly, there are mutual dependencies, and power positions shift depending on the issue under discussion. States' elites are socialized in accordance with the dominant discourse that portrays challenges as transnational, requiring concerted actions on the part of various states to deal with them. Some states, however, have less say in determining international regulations or accepted 'best practices' in addressing those security challenges. Rather, they are under external pressure to set priorities domestically in accordance with the international agenda in order to be perceived as 'good' and 'responsible' actors in the system.

The EU's migration relations with its immediate neighbours to the east focus, first of all, on control measures designed to curb irregular migration flows, both of their own citizens and third country nationals transiting towards the EU territory. Despite the various dimensions of the GAMM, neither labour nor development aspects have yet been introduced into the EU's migration co-operation with Ukraine and Russia. The emphasis has been squarely on *mobility* of their nationals (mainly envisioned as an increase in travel opportunities for their nationals for tourism and/or cultural exchanges), but not on *labour* migration opportunities per se. Ukrainian and Russian nationals are rewarded with greater access to the Schengen territory once the security of travel of foreigners and citizens is guaranteed by their governments. This situation is paradoxical considering that the political rationality of the EU's comprehensive approach to migration management formally includes an increase in legal channels for labour migration to counter irregular migration. Labour migration has been linked to irregular migration and has been presented as one of the tools in the EU's migration control strategies (Walters 2010).

How had Ukraine and Russia responded to EU pressures with regard to international migration management in the 1990s–2000s? In this section, I provide the explanations that are to be found in the international/foreign policy dimensions of these countries' politics.

Ukraine and Russia are two cases to which different political frameworks and expectations on the part of the EU have been applied – the EaP in the case of Ukraine, and 'strategic partnership' and the four common spaces in the case of Russia. Both political platforms have built-in expectations concerning legal approximation to EU norms. However, relations on migration and border management are characterized by more operational and interagency co-operation and have less emphasis on legal approximation in both cases. Despite differences in Ukraine's and Russia's positions towards the EU, their responses to the EU's migration diplomacy have been similar. Factors such as the state's interest in greater control capacities over immigration and borders, the dominant position of the security-oriented domestic elites, the international environment of states 'fighting' 'illegal' migration, and developed countries' interest in increasing the capacities of 'weaker', new states to control their borders and territories, are some of the observed factors that have caused the similarity of outcomes in these two differently positioned cases.

The initial willingness of the eastern neighbouring countries to co-operate with the EU on migration management was spurred by the reality of increasing migration flows through their territories in the 1990s due to their geographic proximity to the EU. Whether and how co-operation evolved was also influenced by the state's foreign policy orientation; the organizational and political culture of the state (in particular, its openness to foreign influences); the acuteness of the perception of sovereignty issues in relations with the EU and the predisposition of domestic politics to the EU's influence.

The EU's leverage over a given country is one important variable that determines that country's response to EU pressure. This leverage can be expressed in economic (e.g. trade relations, financial and technical assistance), political or diplomatic terms. Additional leverage exists if a country has EU membership aspirations. It has also been argued that variables such as intra-EU coherence on a specific issue and on its foreign policy stance, the structure of incentives between the EU and its neighbours, and mutual perception of legitimacy are important in determining the degree of policy convergence intended by a third country (Barbé et al. 2009b). Moreover, policy convergence can have a different basis than the EU's norms; namely, it can be a convergence towards international and bilaterally developed norms (Barbé et al. 2009a).

The ENP has an expectation of policy convergence on the part of the ENP countries. The principle of "sharing everything with the Union but institutions", proclaimed by then European Commission President Romano Prodi, implied close co-operation with neighbouring countries in exchange for their legislative and regulatory approximation (Prodi 2002). More recently in its relations with the EaP countries the EU has been subscribing to the principle of "more for more" as the way to create additional incentives for EaP countries to express willingness to undertake economic and political reforms both in line with their interest for greater co-operation and EU demands. Such principles notwithstanding, the foreign policy orientation of the governing elites affects the EU's ability to exert influence over a given country (see Schmidtke and Chira-Pascanut 2011, for the analysis of Moldova's case).

In the case of Russia until 2012, we observed regulatory approximation in the four common spaces that resembled the EU's influence over ENP countries. Despite a lack of direct EU leverage, there was partial adaptation of Russia to the EU's normative framework (e.g., see Potemkina 2010). This was particularly evident in cases where European and international norms coincided, or when what the EU required from Russia resembled, or was institutionalized through, other international platforms in which Russia participated (Hernández i Sagrera 2010). On irregular immigration co-operation, the ability of the EU to promote its norms at the international level increases its chances of being successful in pressuring some third countries, which otherwise might not be willing to co-operate with the EU.

On the other hand, as foreign policies fluctuate and develop in relation to each other, the scope of what the EU has to offer and how it corresponds to what a given country expects from the EU plays an important role in determining the climate of co-operation and the willingness of a particular government to comply with EU pressure. This directly affects the sensitivity to the question of sovereignty in the relations between the EU and a particular third country on issues such as governing borders and control of population movement.

A given state's response to the EU's migration diplomacy is also dependent on the domestic situation regarding migration, and the priority that the government gives to migration-related questions. Ukraine, despite the pressures from the EU, historically was not complying with all the migration-related requirements of the EU Action Plan on justice and home affairs but sought closer co-operation on 'illegal immigration' and border management – the priorities aligned with the government's domestic agenda. In the case of Russia, the level and intensity of co-operation with the EU increased once the security communities in Russia acquired a greater degree of control over the political agenda in the country. Arguably, the response to the external EU pressure increased once similarly-minded individuals – i.e. law-enforcement agencies and national-security-oriented individuals – were able to steer domestic attention to questions of policing and law enforcement in relation to foreigners/immigrants (Zhyznomirska 2012).

Another important factor is the role of non-governmental sector and expert communities, as well as of individual citizens, in speaking the language of the EU. Experts play an important role in enabling the transfer of EU norms and practices. As Miller and Rose (2008, 34–35) point out, "language ... plays a key role in establishing these loosely aligned networks, and in enabling rule to be brought about in an indirect manner". The existence of "shared vocabularies, theories and explanations", while letting each participant remain autonomous, enables governing "at the distance". If the actors speaking the EU's language are present and influential, more pressure is placed on the government to introduce reforms that the EU requires. This is a dimension that differs between the two cases examined. While in Ukraine expert communities sought to hold the government accountable for fulfilling the requirements of the bilateral agreements (such as the Visa Liberalization Action Plan), no similar pressure was visible in Russia prior to 2012.

The consequences of the EU's migration diplomacy for Ukraine and Russia

This section highlights some of the consequences of the EU's migration diplomacy for third countries' migration, foreign and security policies. There are some institutional, administrative, discursive and normative influences of the EU on its eastern neighbouring countries (for a short overview, see Table 7.1). From a comparative perspective, until 2012 Ukraine and Russia showed a varying degree of European influence in the development of their migration policies. Nevertheless, both were under the normative and institutional influence of 'European' practices of migration management through EU-funded training programs for civil servants for migration and border control authorities, counter-trafficking programmes, foreign expert knowledge transfer and readmission policy development, to name a few areas. Russia even copied some elements of the EU's migration policies into its own migration regulation framework, in particular, by giving a greater role to migration questions in its external relations (see Korneev and Leonov, this volume; Potemkina 2010).

We can talk about direct/indirect and intentional/unintentional impact of the EU on third countries, as well as about voluntary and involuntary adaptation on part of non-EU countries (Lavenex and Uçarer 2004). The indirect impact of the EU is associated with the externalities of the European integration process and its impact on the countries outside of the EU's territory, and it has been expressed in the neighbouring countries' voluntary and necessitated reactions to the policy changes that have taken place in the EU. The most obvious example of such impact were the legislative and practice-oriented changes in Ukraine and Russia caused by the expansion of the Schengen order towards the accession countries of Central and Eastern Europe. The EU's immigration control policies caused its eastern neighbours to follow its lead, insofar as they, too, did not want to bear the pressure of immigration unwanted in the EU. These policy changes occurred prior, or in parallel, to Ukraine's and Russia's negotiation of

Table 7.1 Europe's impact on Ukraine and Russia's border control policies

	Ukraine	Russia
EU-influenced institutional changes in migration area	No and yes, in terms of institutional architecture. No and yes, with regard to capacity-building of existing institutions. Strong internal disagreements and no political will to create workable institution(s) in charge of migration; no external motivation provided by the EU to create one comprehensive institution in charge of migration. More recently, such motivation was provided by the EU in the context of the visa negotiations with Ukraine. The creation of an institution in charge of migration was listed as one of the conditions Ukraine must meet to move closer to the visa free regime for its citizens.	No, in terms of institutional architecture. Yes, in terms of assisting with capacity-building of existing institutions.
Assistance with drafting legislation	Yes – both direct and indirect (through international organizations the IOM and UNHCR, with the activities funded by the EU).	Yes – indirect (through international organizations like the IOM and UNHCR, with the activities funded by the EU).
Administrative capacities of migration agencies	Yes – the law enforcement and border protection agencies. Indirect assistance with refugee issues – through funding of projects of domestic and international non-governmental organizations working with refugees and asylum seekers in Ukraine, and to human rights organizations.	Yes – migration agency, the law enforcement and border protection agencies. Indirect assistance with refugee issues – through funding of projects of domestic and international non-governmental organizations working with refugees and asylum seekers in Russia, and to human rights organizations.

	Ukraine	Russia
Technical assistance and equipment (for borders guards and police)	Yes	Yes – technical assistance for border management.
Knowledge transfer through personnel training and expertise sharing	Yes	Yes
Assistance with the construction of physical migration-related infrastructure	Yes	Yes, but limited.
Presence in the national discourses	Yes	Yes

readmission agreements with the EU, and they were meant, among other things, to minimize the expenses associated with detention and deportation of irregular migrants. These countries also had turned to restrictive migration control policies in order to decrease their attractiveness as transit zones for EU-bound human smugglers and irregular migrants in the late 1990s to early 2000s.

In turn, the direct impact has taken place through migration diplomacy and internal security conditionality attached to other spheres of co-operation, and its intention usually extends to policy transfer and the off-loading of migration responsibility for the migration flows to other countries. The impact – that is, approximation of norms and practices in the third country towards the European ones – is greater when the country's and the EU's interests coincide or are two-directional (i.e. can be satisfied simultaneously through closer co-operation), or when the EU has a political (i.e. membership prospect) or policy leverage (i.e. visa).

The cases show that the Europeanization of non-EU countries consists of processes, rather than outcomes. The notion of "Europeanization" includes both "domestic impact of Europe" and creative usage of Europe for domestic consumption (Jacquot and Woll 2003). Europeanization takes place in a creative manner and not as a unidirectional process, and both politics and policies are Europeanized. Both the 'EU' and 'Europe' are used in national discourses in whatever ways to justify the policy choices of the domestic actors promoting change. On the one hand, politicians might reject the pressure from the EU; on the other hand, they might use Europe creatively to legitimate change they seek. Whereas policy actors in EU member states can use the perceived threat of action by the European Court of Justice against their country to shape the domestic debate on a given policy issue (Radaelli and Exadaktylos 2010), policy actors from non-EU countries may use the argument about the credibility and importance of their relations with the EU, or the necessity to appear as a credible partner in bilateral relations in order to steer the policy debate and legislative processes in a desired direction (e.g. the debate about adopting readmission and visa facilitation agreements in the Ukrainian parliament, or the border-related reforms in Russia). Europeanization takes place incrementally (Ladrech 1994), and it is not necessarily a progressive process with no return.

Establishing a direct causal relationship is difficult, considering the complexity of Europeanization processes, as well as the impact of both global processes and domestic forces also seeking reforms. Irregular migration, border protection and biometric surveillance technologies are the latest obsession of states in their attempts to show that they are security providers to their populations, and to create the illusion of being in control of various flows, including human flows. However, the EU is not the only agent of Europeanization. There is also the Council of Europe, other European intergovernmental fora, *international governmental organizations (*IGOs), and individual states that collectively or unilaterally act to advance their policy agenda or solve a given problem at hand. Arguably, we may also suggest that separate EU member states co-operating with non-EU countries may greatly shape the process of policy learning and policy transfer in a given non-EU country (for example, the Ukrainian authorities learning from the Polish example of the accession process), thus impacting the Europeanization processes there (see Featherstone 2003). Even in the cases of bilateral influences, the EU arguably provides the architectures and the procedures for reforms, because it was the EU that affected the practices and processes in justice and home affairs in EU member states in the accession process in the first place.

The institutionalization of readmission practices in Europe

By now, we could talk about the spread of the readmission principle in Europe. Facing the Union's pressure, Ukraine and Russia incorporated the principle of readmission into their relations with third countries and sought to sign readmission agreements with the major source countries of unauthorized migration. With its international influence and 'big power' status, Russia has been especially active in adapting the principle of readmission in its foreign policy.

Besides normative adaptation, there is a material side to the functioning of the readmission space in Europe. Thanks to EU assistance, third countries develop their readmission infrastructure (e.g. the erection of reception centres, designation and training of the personnel) and adopt new procedures in order to be able to deal with their now international readmission obligations. Although the EU plays the role of a motivator and facilitator for the establishment of the readmission realm in Europe, third countries also play their role in how this space functions as they introduce policies and practices to implement the readmission. Therefore, all states in Europe – both the sending and the receiving ones – play their role in establishing the readmission space. The transit countries invest (whether from their own budget funds or with funding from the EU) in the development of physical structures and training of the personnel of the centres where third country nationals are kept before they are sent further. Even in their sovereign realm of internal policy implementation, they are affected by European readmission practices, because they follow – or are informed by – the European 'best practices' manuals compiled especially for them (for example, by the IOM or ICMPD). These manuals contain detailed information concerning the proper functioning of the centres and 'correct' detention of foreigners, based on practices of detention and reception centres in EU member states. Ultimately, such readmission co-operation maintains the international system in which states – both sending and receiving – seek to increase their sovereign power over mobile subjects and claim control over their territories and populations.

In inter-state relations, the understanding of readmission is quite technical. Readmission agreements specify the procedure by which nationals and third country citizens are transferred. For the country signing a readmission agreement with the EU, the priority is to determine what to do with a returned foreigner. How the readmission cases will be handled is thus determined by the domestic experts participating in the negotiations, in accordance with the sensitivities and imaginaries they might have with regard to the application of sovereign power to these foreigners. The discussion may centre on questions of institutional and financial capacity to accept such individuals. However, disagreement usually centres upon the acceptable pieces of identification of one's transit through, or origin from, a given country (e.g. bus or airplane tickets, a piece of clothing made in a country). For a requesting country, the number of positive responses to its readmission applications and the consequent transfers of individuals to a requested country measure the success of readmission. Effectiveness is also measured by decreased interceptions of irregular migrants and/or traffickers in a particular country, or a list of countries, within one 'route'.

Readmission and visa facilitation agreements institutionalize the European practices of mobility regulation. At the core is the EU's approach to countering the 'illegal' movement of foreigners into its territory. This leads to strengthened state capacities, bound into the EU's migration regime, to regulate and track the movement of their own citizens through new systems of identity documents and databases, entry and exit system, etc. These different practices shape the EU-centred zone of security that is regulated by EU demands regarding the security of travel documents (e.g. biometric passports that are read by the machines installed at the EU external borders), acceptable border technologies (e.g. infrared detectors of human movement) and tools (e.g. interviewing at border crossing points) to detect 'unwanted,' potentially 'illegitimate' travellers. Countries co-operating with the EU must ensure compliance, lest their citizens not be able to travel into the EU. Security professionals in third countries are thus motivated to increase their capacities and modernize their own migration control technologies in order to monitor the movement of foreigners and separate the flows into 'good' and 'bad' ones.

Evidently, visa facilitation agreements are instruments to counter the negative perceptions of the exclusiveness of the "EU club" that is perceived or experienced by third country nationals facing EU consulates. Nevertheless, they do not erase the outsider status of these nationals; rather, this status is modified in a way that removes just one layer of the insecurity and "threat" perception attached to these nationals. Simultaneously in the EU, we observe the implementation of technological innovations at the borders to compensate for the removal of the visa/consulate pre-screening as an important security mechanism. Borders are becoming biometric, with "dataveillance" (Bigo 2011) used as a compensatory measure – a process that simply transfers the power to exclude from EU consulates in foreign countries to border agencies. The burden of screening is partially shifted to the countries of origin that have to introduce these biometric passports and the upkeep of databases that contain personal data. The policing function increases in the case of countries of origin, and hence the state's responsibility for its own citizens and the potential security threats coming from them to other states increases. The states of origin bear the pressure for creating systems of population management that would allow screening against any potential danger coming from their individual citizens. The trade-off for better travel options is a suspicious treatment of everyone, and restricting travel for those deemed risky or dangerous. In the domestic debates, some of the reforms in the internal security sector are justified by the EU's requirements as part of visa negotiations (in Russia's case) and/or as part of the country's EU integration strategy (Ukraine's case). The conditionality that the EU attaches to lifting visa requirements plays an important role in stimulating third countries to undertake reforms in line with requirements of the European Union. In practice, this leads to what William Walters (2011) dubs "the zones of security" that are composed of the states that maintain a certain standardized level of security and co-operate with each other. These "zones" share similar border technologies and the knowledge of how these technologies should be used to monitor borders, and the bodies and goods crossing them. They also include the standardization of personal identification documents, the absence of which may prevent citizens of a given country from entering the space where a certain degree of document protection is practised and expected.

Conclusions

Andrew Geddes (2008, 3) rightly argues that international migration is an integral process to

> the development, consolidation and transformation of European states; the underlying motives for European economic and political integration; the international system within which European integration is embedded; the ways in which organizational and institutional processes at national, European and international levels 'make sense' of the complex human material that comprises international migration flows.

Importantly, it is in relation to the borders that the categories associated with international migration (for example, as "legal" or "illegal", a "threat" or "opportunity", "bogus" or "genuine") are being constructed (Geddes 2008). What happens to borders, and how borders get shifted to where the difference is constructed, are matters affecting the political landscape of inter-state relations in Europe.

The changing meaning of borders on the European continent points to the emergence of a multilevel system of governance of population – one directed at citizens and non-citizens alike, with actors at various levels exercising their influence over how mobility is being governed. This occurs when various levels of government participate in governing human mobility for the sake of bona fide travellers – both citizens and foreigners alike. The mobility of individuals is being managed through increased control over both immigration and emigration. This takes place, at least partially, because of the pressure placed by the EU upon sending and transit countries to control immigration into the EU territory. Importantly, non-EU countries that choose to co-operate also have their interests and play active roles in managing migration towards Western Europe. The model is being presented as a form of a 'global' migration management, shaped by concerted government actions on both sides of the borders and is employed towards or against non-nationals. How states deal with international migration illuminates the changing territorial and sovereignty boundaries of contemporary nation states.

In response to the EU's pressures to increase their control of irregular migration flows, third countries politicize the absence of freedom of movement for their nationals while downplaying the restrictive measures introduced and implemented to control the movement of foreigners towards the EU territory, and to regulate the mobility of their own citizens. Loss of control over people's movements at state borders is interpreted as a threat to national sovereignty and security. Such discourse constructs the need to re-insert the state's control over its territory and borders by investing into new surveillance technologies and the discretionary power of border guards to prevent a foreigner from entering the territory. As a result, the sovereign logic is being re-inserted at the borders. In such an environment, migration control and assumption of responsibility for migration control by non-EU countries happens both voluntarily, and due to the EU's pressure. The governmental project of migration management around the EU core contains a logic of strengthening sovereign power over populations in the countries to which migration controls are being extraterritorialized. In addition, in its push for return and re-integration policies in the countries of origin, the EU stimulates third countries to be 'good' sovereigns (especially towards those citizens who were in irregular status in EU countries) and to invent policies that will keep their populations in.

Notes

1 Since 2014 EU–Russia relations have deteriorated due to Russia's annexation of Crimea and its aggression in the east of Ukraine causing factual freezing of relations on migration management with Russia. This case shows how important a degree of "friendliness" in relations is to co-operation on migration.
2 Considering the reluctance of EU member states to release their national prerogatives over labour migration and residency policies (Collett 2009), the level of co-operation on irregular migration in the EU has been particularly noticeable since the late 1990s–early 2000s. There have been significant changes in the area of irregular migration and border controls – the area where the EU is perceived to

have an "added value" to MSs' policies, and we observed the shifting of migration control mechanisms "up and out" (Lavenex 2006) – that is, to the EU level and into the realm of the EU's relations with non-EU countries.

3 ICMPD. *The Prague Process* ICMPD online, www.icmpd.org/Prague-Process.1557.0.html (accessed 17 June 17 2015).

Bibliography

Balzacq, Terry. 2008. *The External Dimension of EU Justice and Home Affairs: Tools, Processes, Outcomes.* CEPS Working Document No. 303. Brussels: Centre for European Policy Studies. Accessed November 7, 2008, www.ceps.eu/book/external-dimension-eu-justice-and-home-affairs-tools-processes-outcomes.

Barbé, E., Costa, O., Herranz, A. Surralles, and Natorski, M. 2009a. Which rules shape EU external governance? Patterns of rule selection in foreign and security policies. *Journal of European Public Policy.* Vol. 16, No. 6, pp. 834–52.

Barbé, E., Costa, O., Herranz, A., Johansson-Nogués, E. and Sabiote, M.a. 2009b. Drawing neighbours closer ... to what? Explaining emerging patterns of policy convergence between the EU and its neighbours. *Cooperation and Conflict.* Vol. 44, No. 4, pp. 378–99.

Betts, A. 2008. Global Migration Governance. *Working Paper 43.* Oxford: Global Economic Governance Programme. Available at: www.globaleconomicgovernance. org/wp-content/uploads/BettsIntroductionGEGWorkingPaperFinal.pdf [10 May 2012].

Betts, A. 2011. Introduction: global migration governance. In Betts, A. (ed.) *Global Migration Governance.* Oxford: Oxford University Press, pp. 1–33.

Bigo, D. 2011. Freedom and speed in enlarged borderzones. In Squire, V. (ed.) *The Contested Politics of Mobility: Borderzones and Irregularity.* London and New York: Routledge, pp. 31–50.

Collett, E. 2009. "Beyond Stockholm: Overcoming the Inconsistencies of Immigration Policy." European Policy Centre Working Paper No. 32, December 2009.

Commission of the European Communities. 2006. Communication "The Global Approach to Migration one year on: Towards a comprehensive European migration policy." COM (2006) 735 final, 30 November. Düvell, F. 2011. Irregular migration. In Betts, A. (ed.) *Global Migration Governance.* Oxford: Oxford University Press, pp. 79–108

EU Presidency. 2009. Prague Ministerial Conference 'Building Migration Partnerships'. *Joint Declaration.* Available from: http://ec.europa.eu/home-affairs/doc_centre/immigration/docs/bmp_ jointdeclaration_en.pdf [10 September 2012].

European Council. 2015. *European Council meeting (25–26 June 2015): Conclusions.* EUCO 22/15, Brussels. 26 June 2015.

Featherstone, K. 2003. Introduction: in the name of 'Europe'. In Featherstone, K. and Radaelli, C.M. (eds) *The Politics of Europeanization.* Oxford: Oxford University Press, pp. 3–26.

Geddes, A. 2008. *Immigration and European Integration: Beyond Fortress Europe?* 2nd edn. Manchester: Manchester University Press.

Geiger, M. 2010. Mobility, development, protection, EU-integration! The IOM's national migration strategy for Albania. In Geiger, M. and Pécoud, A. (eds) *The Politics of International Migration Management.* Basingstoke, New York: Palgrave Macmillan, pp. 141–59.

Geiger, M. and Pécoud, A. 2010. The politics of international migration management. In Geiger, M. and Pécoud, A. (eds) *The Politics of International Migration Management.* Basingstoke, New York: Palgrave Macmillan, pp. 1–20.

Georgi, F. 2010. For the benefit of some: the International Organization for Migration and its global migration management. In Geiger, M. and Pécoud, A (eds) *The Politics of International Migration Management.* Basingstoke, New York: Palgrave Macmillan, pp. 45–72.

Held, D., McGrew, A., Goldblatt, D. and Perraton, J. 1999. *Global Transformations, Politics, Economics and Culture.* Stanford, California: Stanford University Press.

Hernández i Sagrera, R. 2010. "The EU-Russia readmission-visa facilitation nexus: an exportable model for Eastern Europe?" *European Security* 19 (4): 569–584.

Hess, S. 2010. 'We are facilitating states!' An ethnographic analysis of the ICMPD. In Geiger, M. and Pécoud, A. (eds) *The Politics of International Migration Management.* Basingstoke, New York: Palgrave Macmillan, pp. 96–118.

Jacquot, S, and Woll, C. 2003. "Usage of European Integration – Europeanization from a Sociological Perspective." *European Integration online Papers* (EIoP) vol 7, No. 12. Available at http://eiop.or.at/eiop/pdf/2003-012.pdf (last accessed October 26, 2015). Kunz, R., Lavenex, S. and Panizzon, M. 2011. Introduction: governance through partnerships in international migration. In Kunz, R., Lavenex, S. and Panizzon, M. (eds) *Multilayered Migration Governance: The Promise of Partnership.* London and New York: Routledge.

Ladrech, R. 1994. Europeanisation of domestic politics and institutions: the case of France. *Journal of Common Market Studies.* Vol. 32, No. 1, pp. 69–88.

Lavenex, S. 2006. "Shifting Up and Out: The Foreign Policy of European Immigration Control." *West European Politics* 29 (2): 329–350.

Lavenex, S. 2007. The external face of Europeanization: third countries and international organizations. In Faist, T. and Ette, A. (eds) *The Europeanization of National Policies and Politics of Immigration: Between Autonomy and the European Union,* New York: Palgrave MacMillan, pp. 246–64.

Lavenex, S. and Kunz, R. 2008. The migration–development nexus in EU external relations. *European Integration.* Vol. 30, No. 3, pp. 439–57.

Lavenex, S. and Uçarer, E.M. 2004. The external dimension of Europeanization: the case of immigration policies. *Cooperation and Conflict: Journal of the Nordic International Studies Association.* Vol. 39, No. 4, pp. 417–43.

Lavenex, S. and Wichmann, N. 2009. The external governance of EU internal security. *Journal of European Integration.* Vol. 31, No. 1, pp. 83–102.

Miller, P. and Rose, N. 2008. *Governing the Present: Administering Economic, Social and Personal Life.* Cambridge: Polity Press.

Occhipinti, J. D. 2007. Justice and home affairs: immigration and policing. In Weber, K., Smith, M.E. and Baun, M. (eds) *Governing Europe's Neighbourhood: Partners or Periphery?* Manchester and New York: Manchester University Press, pp. 114–33.

Potemkina, O. 2010. "EU-Russia cooperation on the common space of freedom, security and justice – a challenge or an opportunity?" *European Security* 19 (4): 551–568.

Prodi, R. 2002. "A Wider Europe - A Proximity Policy as the key to stability." Speech delivered at "Peace, Security and Stability International Dialogue and the Role of the EU" Sixth ECSA-World Conference, Brussels, 5-6 December 2002. Available at   http://europa.eu/rapid/press-release_SPEECH-02-619_en.htm   (accessed

June 30, 2015).Radaelli, C. M. and Exadaktylos, T. 2010. New directions in Europeanization research. In Egan, M., Nugent, N. and Paterson, W.E. *Research Agendas in EU Studies: Stalking the Elephant.* New York, London: Palgrave Macmillan, pp. 189–215.

Schmidtke, O. and Chira-Pascanut, C. 2011. "Contested Neighbourhood: Toward the 'Europeanization' of Moldova?" *Comparative European Politics* 9 (4–5): 467–485.

Trauner, F. 2009a. "Deconstructing the EU's Routes of Influence in Justice and Home Affairs in the Western Balkans." *Journal of European Integration* 31(1): 65–82.

Trauner, F. 2009b. "From membership conditionality to policy conditionality: EU external governance in South Eastern Europe." *Journal of European Public Policy* 16(5): 774–790. Walters, W. 2010. Imagined migration world: the European Union's anti-illegal immigration discourse. In Geiger, M. and Pécoud, A. (eds) *The Politics of International Migration Management.* Basingstoke, New York: Palgrave Macmillan, pp. 73–95

Walters, W. 2011. "Rezoning the global: Technological zones, technological work and the (un-)making of biometric borders." In *The Contested Politics of Mobility: Borderzones and Irregularity*, ed. Vicki Squire, 51–73. London and New York: Routledge.

Walters, W. 2012. *Governmentality: Critical Encounters.* London and New York: Routledge.

Zhyznomirska, L. 2011. "The European Union's 'Home Affairs' Model and its European Neighbours: Beyond the 'Area of Freedom, Security and Justice'?" *Comparative European Politics* 9 (4/5): 506–23.

Zhyznomirska, L. 2012. "The European Union's Migration Co-Operation with its Eastern Neighbours: The Art of EU Governance Beyond its Borders." Unpublished PhD manuscript, University of Alberta.

8 Eurasia and externalities of migration control

Spillover dynamics of EU–Russia cooperation on migration[1]

Oleg Korneev and Andrey Leonov

Introduction

Various analyses of externalization of European Union (EU) migration controls to the east usually stop at the borders of Eastern Partnership countries and Russia. Even the number of scholarly works comprehensively analysing EU–Russia cooperation on migration is surprisingly limited (for notable exceptions see Hernandez i Sagrera 2010; Potemkina 2010; Trauner et al. 2013). This level of scholarly interest does not correspond to the level of EU–Russia migration cooperation and to the importance attributed by both parties to the intertwined issues of migration and external polices, as well as to the changing roles of these two actors in their respective 'near abroads' (Charillon 2004).

Moreover, research that addresses EU–Russia cooperation within the overarching project of the Common Space of Freedom, Security and Justice (FSJ) often looks at this 'space' as if it were regarded by both parties as a naturally coherent field, where cooperation had to be developed evenly in all of the sub-fields. Consequently, this misperception often leads to the conclusion that EU–Russia cooperation on issues of justice and home affairs is rather unsuccessful because, on one hand, there is still no visa-free regime (Potemkina 2010) and, on the other hand, developments in the sphere of human rights and judicial reforms in Russia are far from being satisfactory (Ehin 2009). This chapter argues that in order to better appreciate the results of EU–Russia cooperation in the field of internal security, one should not analyse it in connection with EU attempts to promote human rights and democracy in Russia. Such an idealist view has prevented many researchers from a more pragmatic assessment that would take into account some real achievements of this cooperation that fit well with the commonly defined interests and goals in the sphere of internal security and migration in particular.

Finally, rare existing studies of EU–Russia cooperation on migration almost exclusively focus on negotiation processes and eventually leave out the implementation phase as well as the broader regional impact of this cooperation. In other words, what happens further afield is traditionally escaping the attention of those dealing with issues of conditionality and policy transfer as applied to EU relations with its 'neighbours'. One of the most important issues that is largely neglected is the effect that EU–Russia cooperation on migration has had on Russia's relations with its neighbours in the Commonwealth of Independent States (CIS), namely with countries in Central Asia. Scholars have already paid attention to the EU efforts aimed at the regulation of migration issues in Central Asia (Gavrilis 2009; Laruelle and Peyrouse 2010). Nevertheless, they mostly take into account only the EU's direct involvement in the region. Issues that relate the EU, Russia and Central Asia in terms of migration have only sporadically been addressed in the existing literature; no elaborate research has been conducted to investigate them profoundly.

This chapter focuses on EU–Russia cooperation on migration and extends this analysis geographically and conceptually – to include countries of origin and transit for migrants coming to Russia and to trace the impact of EU migration policies far beyond its immediate neighbourhood. It shows that externalization of EU migration policy significantly affects countries that do not have the same degree of direct involvement with the EU as its European Neighbourhood Policy (ENP) partners and involves problematic externalities[2] for the EU normative engagement in Eurasia. First, the chapter provides a brief picture of the developments in EU–Russia cooperation on migration control. Second, the chapter analyses the place of the readmission agreement in this cooperation. Third, it traces the milestones of Russia–Central Asia relations on readmission and demonstrates the role of the EU in shaping Russia's priorities, as well as regional migration management dynamics. The chapter argues that the EU–Russia cooperation on migration control has triggered significant spillover dynamics in the wider Eurasian region, as well as in the Middle East, South and South East Asia. Attention to what happens beyond the framework of the EU–Russia readmission agreement can help us get better insights about the degree of EU success in the externalization of its migration control approaches. Finally, the chapter points to several important normative issues and contradictions – externalities of EU policies – stemming from the spread of readmission agreements in the region where human rights of migrants and persons seeking international protection are not guaranteed.

EU–Russia cooperation on migration

Since its inception, intra-EU evolution of the 'area of justice, freedom and security' has been reinforced by the development of projects aiming to guarantee its viability in the volatile international environment. Such projects have found their way through some of Mediterranean policies (Pastore 2002), through enlargement negotiations (Mitsilegas 2002), European Neighbourhood Policy and Eastern Partnership (Lavenex 2004), as well as through 'common spaces' between the EU and Russia. In the context of the Eastern enlargement, it has been argued that the EU needs a 'buffer zone' to keep soft security challenges, including uncontrolled migration, as far as possible from its somewhat harmonious internal space (Potemkina 2002). The EU used to have such a buffer zone in the east – quite naturally provided by the countries of Central and Eastern Europe. However, the geopolitical reality has changed and this zone has not only ceased to exist but has become a constitutive part of the EU itself, which is now bordering with a rather problematic region in terms of soft security risks. As a reaction to this, the EU while delimiting its own 'area of justice, freedom and security' has been also trying to create another buffer zone to consistently safeguard its status of a security community in relation to the outside world (Korneev 2007). Migration-related security concerns have become top priorities of EU Justice and Home Affairs (JHA) in cooperation with the neighbouring countries in the east. Manifestations of this dynamic have been quite obvious, for example, through the EU actions in the framework of the European Union Border Assistance Mission to Moldova and Ukraine (Kurowska and Tallis 2009). Another example of such practices is the case of the EU orchestrating management of the Ukrainian-Russian border through, among others, 'privatization' of this sphere (Gatev 2008). In the same logic of its remote control strategy, the European Union has been trying to shape Russian migration policy. A major policy change – Russian immigration policy developing an external dimension – coincided with the steep intensification of the EU–Russia cooperation in justice and home affairs in 2000s.

The EU–Russia Road Map for the Common Space of Freedom, Security and Justice was adopted in May 2005 together with the road maps for three other 'common spaces' (EU–Russia Summit 2005). Some scholars have argued that EU–Russia 'common spaces' can be considered as frames for potential regimes encompassing various spheres of the EU–Russia relations (Alexandrova-Arbatova 2006). The internal structure of the four road maps indeed reflects some sort of balance between the EU's and Russia's interests and, prevailingly, the EU's values. With regard to the latter, a member of the Working Group on Eastern Europe, South Caucasus and Central Asia of the Council of Ministers (COEST) has emphasized that

> values are more important for the EU than for Russia, but interests are important for the both parties ... the whole ideology of the four common spaces is based on the assumption that they include issues that are more important for the EU than for Russia and vice versa.
> (Interview at the Finnish Permanent Representation to the EU, 21 May 2007)

However, some of the priority areas for cooperation identified in the road maps are apparently more interest-driven than values-oriented. This definitely applies to the common space of freedom, security and justice. The Road Map openly claims that 'cooperation between the EU and Russia in the area of Freedom, Security and Justice is already advanced and has become a key component in developing a strategic partnership between the parties' (EU–Russia Summit 2005: 21). In other words, the importance of JHA cooperation for the EU–Russia relations has been finally recognized at the highest level and the 'commitment of the parties to further strengthen their strategic partnership on the basis of common values' has been reconfirmed. Thus, the parties paid tribute both to the omnipresent 'strategic partnership' vitally important for Russia and to the 'common values' defended by the EU. The Road Map also states that cooperation 'must reflect the necessary balance between security, on the one hand, and justice and freedom, on the other' (21–22). However, shortly after the emergence of the Road Map, some EU officials expressed opinion that even though neither party would officially admit prioritizing one of the aspects in this cooperation, it existed in practice, and the sector of 'justice' was clearly underdeveloped (interview with Wouter van de Rijt, 16 May 2007). The preamble of the Road Map puts the emphasis on 'adherence to common values' and declares 'equality between partners', but the rest of the document – defining the goals and actions to be taken – is more pragmatic.

The Road Map defines a whole set of concrete (even though non-legally binding) guidelines for unilateral and bilateral actions aimed at the creation of this common space. The first part of the Road Map is devoted to 'freedom', which implies that the partners aim to 'facilitate human contacts and travel' while effectively combating 'illegal migration'. The conclusion of a readmission agreement appears as an issue of major importance for cooperation on migration. Overall, despite numerous objectives and actions envisaged in the Road Map, it is clear that the main concern of the EU, and thus defined as a key challenge by the Russian authorities as well, is irregular migration.[3] The need to work together against 'illegal immigration' has been on the EU–Russia cooperation agenda since their relations were institutionalized by the PCA. This idea has been later reiterated in other strategic documents guiding cooperation, such as Common Strategy of the European Union on Russia (European Council 1999) and Russian Mid-term Strategy for Relations with the EU (The Government of Russia 1999). The Road Map has simply restated this need, this primary common interest, and integrated it in a larger framework. That is why it would have been more logical to put 'the fight against illegal migration' under the following heading of the Road Map – "security". The latter has, however, put an emphasis on the fight against terrorism, as well as against trafficking in human beings. This makes the whole Road Map even more security-driven and clearly confirms that migration-related concerns constitute the top priorities of the EU JHA cooperation with Russia.

The section on 'justice' has its place at the end of the list of objectives and this creates an impression that it entered the Road Map only because the EU needed to emphasize its values-based identity and its values-oriented foreign policy, thus striking the balance between security-related issues and human rights concerns. Nevertheless, the need for such 'injections' is questionable. This situation is well described by Emerson (2005, 2) who has claimed that

> the de-democratizing Russia of President Putin manifestly could not embark on negotiations on a common space of democracy. Yet the EU could not ignore the subject. The result is token inclusion of a few lines in this common space for FSJ.

Apparently, the EU was not interested in pushing the human rights agenda too much, since it would jeopardize common efforts in a much more important field of internal security where common threats of irregular migration, terrorism and drug trafficking were identified by both parties.

Two years after the Road Map was agreed on, both the EU and Russia have characterized the Common Space on FSJ as the best functioning one (Interview at the Russian Permanent Representation to the EU. Brussels, May 2007; Interview at the DG RELEX, the European Commission, June 2007). Nevertheless, there still existed a number of factors that hindered cooperation. Thus, both parties have underlined that some of the issues that form the common agenda of cooperation in the JHA are significantly politicised and this prevents a lot of positive developments (among the examples are the issue with asylum standards in Russia and the problem of Chechen asylum seekers in the EU, etc.). They have also expressed the opinion that when issues move from the political level to the technical one, then problems are solved faster and in a mutually beneficial manner (Interview at the DG Justice, Liberty and Security, the European Commission, May 2007; Interview at the Russian Permanent Representation to the EU, May 2007). Indeed, not all of the areas of cooperation included in the FSJ Road Map enjoy the same success as the sphere of migration management.

The fact that countries on the way of migrants to the EU are becoming not only transit but destination countries as well explains why the EU is willing to invest in asylum facilities and immigration infrastructure in Russia and in the Eastern Neighbourhood – the evidence for such investments is provided by specific projects financed fully or partially by the EU and implemented through mediation of various governmental and non-governmental international as well as domestic structures (Hernandez i Sagrera and Korneev 2012). However, other more sensitive issues such as improvements in the field of justice in Russia or Russian policy and practice in the field of asylum, the problems of border management, and the rights of the Russian speakers in Latvia and Estonia are still on the margins of the actual cooperation either because of Russia's or EU Member States' reluctant positions respectively. A nice summary of this complex relationship has been given by a representative of the European Commission Delegation in Russia who admitted: 'In general I see both the expansion of activities and the expansion of rhetoric. Public discussion tends to focus on the statement "the relation is in trouble". The rhetoric makes it more difficult to move forward' (Sean Carroll, Head of

Press and Information Section of the European Commission Representation to the Russian Federation, 29 April 2007). At the same time, as it has been bluntly emphasized by a Council representative, 'in security issues it is easier to develop operational measures together' (Interview with Wouter van de Rijt, 16 May 2007).

Evaluating the progress in the development of the EU–Russia Common Space of FSJ, Ehin rightly notes that 'interaction between the European Union and Russia has been characterized by selective cooperation in areas where interests coincide' (Ehin 2009, 68). In this assessment, however, she views the whole of the common space of FSJ as one of such areas, without any differentiation inside this field. This *prima facie* evaluation leads to a distorted picture of cooperation, as Ehin pays attention only to the issues of human rights and judicial reforms, which, in her view, are largely ignored by Russia. Contesting these claims, we argue that it is important to distinguish between declared and actual policy goals of the partners. In the Road Map virtually all goals that do not directly relate to security issues and focus on promotion of human rights and democracy fall under the category of 'declared goals' – not only for Russia, but also for the EU. Therefore, it is meaningless to evaluate the success of cooperation in these areas with the same degree of rigidity as applied to clearly defined goals in soft security cooperation. The later belong to actual goals of the European Union and are, thus, the only goals shared by both partners, whereas some of the declared goals are important only for the EU and others – only for Russia.

This said, one should admit that the project of this common space bears some influence of the value-driven approach to building a partnership with Russia. In a way, it has been designed in order to foster at least some degree of socialization of the Russian internal security system with EU norms and best practices. Nevertheless, coherent socialization of Russia into the EU system of values in the field of justice and home affairs was not the major intention of those who drew the Road Map – neither from the Russian, nor from the EU side. The analysis of the Road Map clearly shows that this 'space' is driven by security concerns of the both parties. One of the most prominent elements in this field has been defined by the partners as 'the fight against illegal immigration'. The instrument that has attracted most practitioners' and scholars' attention in this regard is the EU–Russia agreement on readmission.

The EU–Russia Readmission Agreement: implementation and implications

Despite many implicitly negative formulations and an instructive tone inherent in some of EU documents related to Russia, as well as frequently one-sided media coverage of EU–Russia relations, the cooperation in the field of JHA has produced some results positively evaluated by both sides (Revenko 2010; European External Action Service 2011). Undoubtedly, the most visible of them was the conclusion and the ratification of the two EU–Russia agreements on visa-facilitation and readmission, which were signed in 2006. Since then, the successful implementation of the Agreement between the Russian Federation and the European Community on Readmission (2006) has been defined by the European Commission as one of its top priorities, as well as a crucial pre-condition for talks on visa-free regime (Van Elsuwege et al. 2013). The eventual transfer of readmission mechanism to the very core of the Russian migration policy is probably the most prominent case of successful EU policy transfer to Russia in this field. Dialogue on readmission has come to dominate the agenda of EU–Russia cooperation on migration since the move towards the development of the Common Space of Freedom, Security and Justice in 2003. The priorities defined in the Road Map set the facilitation of travel while 'fighting illegal immigration' as the first policy objective. Whereas the EU in a very normative manner has always emphasized the urgent need to "facilitate human contacts and travel between the EU and Russia" (EU–Russia Summit 2005), the security concerns have finally prevailed and the Union has *de facto* blocked Russian visa-free initiatives, instead pushing forward a lower profile offer to conclude a visa-facilitation agreement coupled with a readmission agreement (Interview with Wouter van de Rijt, 16 May 2007).

Difficult negotiations culminated with the signature and ratification of the agreements that the Russian Presidential Aid optimistically defined as 'a milestone on the way to a visa-free regime' (Yastrzhembsky 2007). However, although subsequently the visa-free track of EU–Russia cooperation has repeatedly faced a deadlock, this does not mean that the entire cooperation on migration is a failure. Potemkina notes that 'the lack of optimism concerning the prospects for visa exemption can be made up for with enthusiasm in cooperation against illegal migration' (2010: 555). One could also argue that successful cooperation on the fight against irregular migration confirms its place among the actual priorities of the bilateral cooperation shared by the both partners, whereas the vaguely defined goal of 'the visa-free regime in the long-term prospective' is still mostly a preoccupation of the Russian negotiators. The strive of the Russian authorities for the recognition of Russia's special place and role among EU partners has defined much of what the Russian government has done so far to obtain a visa-free regime with the EU. Importantly, many recent expert assessments point to the fact that reaching a visa-free agreement is not only in Russia's but also in EU's interests, in particular if one adopts a political economy perspective (Mananashvilli 2013; Van Elsuwege et al. 2013).

However at the time of double track negotiations with Russia, EU officials, while confirming that 'both agreements are what the EU was looking for', expressed worries about potential problems with the implementation of the readmission agreement. One of them has characterized it as 'the first readmission agreement with a major partner country' (Interview at the DG RELEX, the European Commission, 6 June 2007). Another EU representative bluntly stated that 'both agreements are what the EU was looking for.

The EU badly needed this, especially the re-admission agreement. It would be very important that Russia implements them properly' (Interview at the DG Justice, Liberty and Security, the European Commission, 19 April 2007). The problem of implementation has always been the major concern for the EU. Implementation of the readmission agreement by Russia could indeed be a problem. During the negotiations and especially after the signature, the EU–Russia Readmission Agreement was often qualified by Russian officials and many independent – also EU-based – experts as an obvious burden for Russia. One of the most prominent public commentators of EU–Russia relations Timophey Bordachev has even argued that

> Russia has exchanged an elephant for a small dog, giving in to the European Union and getting in exchange only a symbolic dividend – visa-facilitation procedures that might eventually jeopardize Russian interests simply because the parties will continue to move on the track of further visa-facilitation leaving out a possibility of a visa-free regime.
>
> (Bordachev 2006)

Quite to the contrary, the implementation of the Agreement shows that these views were too pessimistic and that the actual 'burden' for Russia is not so heavy.

In line with the Readmission Agreement, Russia and the EU are obliged to readmit their nationals who are staying irregularly in the territory of the other party, as well as those third country nationals who transited through their territory. A special clause of the Agreement has given Russia a three-year transitory period, when the country had to readmit only its own nationals as well as the citizens of those countries with whom Russia itself had already concluded readmission agreements. During this transitory period Russia was supposed to take measures to secure its territory from migrants planning to use it for transit to the EU, as well as to prepare for readmission of irregular migrants from the EU. Such a preparation implied the establishment of centres for readmitted migrants, improved and strengthened infrastructure for migrants' accommodation and border infrastructure, concluding readmission agreements with countries on the perimeter of Russia (Interview at the Russian Permanent Representation to the EU, May 2007), as well as concluding with individual EU Member States implementation protocols – de facto detailed technical and procedural schemes – indispensable for the whole-scale implementation of the EU–Russia readmission agreement.

Importantly, as Potemkina notes, 'Russia had to put much effort into modernising its legal basis, because the term "readmission" appeared in Russia's legislation only in 2006 with the adoption of the respective federal law and the additional legal acts' (Potemkina 2010, 556). The first steps in this direction, however, were made already in 2003 after the conclusion of the readmission agreement between Russia and Lithuania (Agreement between the Government of Russia and the Government of Lithuania on Readmission

2003), which was described as 'an important step along the road of implementing the joint statement of the Russia–EU summit of November 11, 2002, on transit between the Kaliningrad Region and the rest of the territory of the Russian Federation' (Russian MFA 2003), but more generally was regarded by both the EU and Russia as a pilot project allowing them to test capacities of bilateral cooperation on readmission (Chizhov 2005). In other words, both the agreement between Russia and Lithuania and the subsequent agreement between Russia and the EU have constituted important phases in the process of readmission policy transfer from the EU to Russia.

Major innovations in legislation introducing the readmission mechanism have also triggered changes in relevant, although non-directly linked, spheres of Russian legislation. Russia has started issuing biometric passports using the standards of the International Civil Aviation Organization (ICAO). The respective Presidential Decree No. 1709 'On passport of citizen of the Russian Federation, confirming identity of citizens of the Russian Federation abroad, containing additional biometric personal data of passports' holders on electronic chip' (Presidential Decree No.1709 2012) entered into force in January 2013. The Russian Parliament has also introduced several important legislative changes to the normative acts regulating migration control measures in Russia. The Federal Law No.320-FZ 'On amendments to the Federal Law on the legal status of foreign citizens in the Russian Federation', the Federal Law No.321-FZ 'On amendments to the Article 26 of the Federal Law on the rules of exit from and entry to the Russian Federation' entered into force on 11 January 2013. These legislative changes were in line with the prescriptions of the Joint Monitoring Committee and evaluations produced by the European Commission (later – by the European External Action Service).

In this context, the European Commission made it clear that a smooth implementation of the EU–Russia readmission agreement means securing all necessary changes in legislation and infrastructure relating to readmission procedures, as well as in practices of all the relevant Russian governmental bodies (Interview at the European External Action Service (EEAS), 15 February 2011). There was a need for both technical assistance and substantial policy transfer through specific projects implemented in daily cooperation with the major Russian counterpart – the Federal Migration Service (FMS). For the purposes of better implementation of the EU–Russia readmission agreement, the EU initiated a special financial project 'Assistance to the Government of the Russian Federation in Establishing a Legal and Administrative Framework for the Development and Implementation of Readmission Agreements (2006/120-282)' for the period from February 2007 to January 2009 to be implemented together with the International Organization for Migration (Korneev 2014). The Russian Federal Border Service has paid specific attention to the improvement of infrastructure and border management schemes on the southern part of the Russian border in recent years.

Russia started implementing the Readmission Agreement in October 2007 – even in the absence of implementation protocols with EU Member States. It was the second country after Albania to do so. To date, Russia has concluded implementing protocols with all EU Member States that are part of the readmission agreement. By July 2008 EU Member States had determined only about 100 cases eligible for readmission (Lahti 2008). There was quite a steady increase afterwards, and by February 2009 the Russian Federal Migration Service (FMS) had received some 1500 applications for readmission from EU Member States. Since October 2007 Russia has received 4715 readmission requests from 20 EU Member States. More than 3500 requests have been examined and 2214 out of them have been accepted as eligible for the readmission procedure. By November 2010, 793 persons had been readmitted (Arestova 2010), including some looked for by the police and even some related to terrorist activity, and cases of false documents have been discovered as well (Yakovlev 2009).

The FMS has established good working relations with its EU-based counterparts and cooperation in migrants' identification and removal is progressing smoothly (Hernandez i Sagrera and Potemkina 2013, 15–16), even though approximately 50 per cent of all readmission applications from EU Member States to Russia are rejected by the FMS as ill-founded. Importantly, as indicated by a representative of the European Commission 'Russia's adherence to the EU-level readmission agreement has been instrumental in stimulating EU Member States to use this legal mechanism instead of other bilateral means' (Coleman 2012).

Prior to the entry into force of the EU–Russia readmission agreement, EU Member States and Russia, similar to Eastern Partnership countries (Brunarska et al. 2013), used various legal mechanisms in their cooperation for expulsion purposes, not limiting themselves to readmission procedures. Since the EU–Russia readmission agreement entered into force, Russia has been consistent in requesting EU Member States to cooperate under the agreement instead of using other schemes for 'return'. The lack of consistency in EU Member States' use of EU-level readmission agreements has been emphasized by the European Commission as a serious problem for the EU's readmission policy in its evaluation report in 2011 (European Commission 2011). Therefore, the insistence of the Russian Federal Migration Service on the use of the EU-level readmission agreement has been important for the development of the EU's readmission policy.

Quite naturally, EU–Russia cooperation on readmission is also linked to cooperation on border management and to the development of detention facilities for irregular migrants in Russia. By the end of 2013, several centres for temporary accommodation of readmitted third country nationals were established with expert assistance from the International Organization for Migration (IOM) and financial assistance from the European Commission (Korneev 2014). In its first official evaluation of the Common Steps, the European Commission has emphasized: some of the centres for migrants awaiting readmission have been built with the support of the EU project AENEAS following the conclusion of the EU–Russia readmission agreement. The centres for irregular migrants awaiting readmission that have been visited by the EU experts (Moscow region, Rostov-on-Don and Pskov) have been found to be in line with the general EU standards (European Commission 2013: 14).

Notwithstanding this evaluation given by the European Commission, such centres, unfortunately, replicate 'the best practices' of similar facilities with often inhumane conditions on the EU territory (Valluy 2005; Chappart et al. 2012). The freedom of people that get into such centres is limited to the extent that they become de facto imprisoned there,[4] despite the fact that they are not criminals and that their offence – if there is any at all – is of an administrative nature. Human rights organizations are unambiguous in their evaluation of such practices (Human Rights Watch 2005, 2010).[5] The situation may change after recent developments in the Russia federal legislation, in particular with the adoption of the federal law aimed to improve detention conditions of foreign citizens placed in temporary detention facilities (Federal Law No.14-FZ 2015). However, active copy-pasting of the EU experience casts a shadow not only on Russia's pledges for human rights protection, but also on the EU declarations about its human rights promotion mission. The EU readmission agreement with Russia that could create important rights-oriented context for migration cooperation does not provide any sufficient safeguards relating to rights of potentially readmitted migrants (Trauner et al. 2013). Russia has been eagerly embracing control-oriented norms and mechanisms from the EU, shifting the burden of unpopular measures onto its partners. This 'Europeanisation' (Lavenex 2004) of Russia unfortunately implies the transfer of EU norms and practices of ambiguous character. The same applies to the developments further afield.

A web of readmission agreements in Eurasia

Evaluating the implementation of the readmission agreement over the three transitory years, both Russian and EU officials underline the fact that there was identified a very limited number of Russian citizens falling under the readmission procedure.[6] However, after these three transitory years (by the summer 2010) Russia has assumed responsibility for all irregular immigrants entering the EU from Russian territory. Russia did not have readmission agreements with migrant origin or transit countries before the signature of the readmission agreement with the EU. Since then, politically, concluding such agreements has become part of the implementation tool-kit as interpreted by the European Commission. The Commission has explicitly stated on many occasions that implementation of the EU–Russia agreement will also be judged based on Russia's success in concluding readmission agreements with important countries of origin and transit, which is much more than just an effective implementation of the EU–Russia agreement per se. Technically, they were also necessary in order to diminish expenses that Russia would bear in the readmission procedures. This double motivation partly explains the eagerness of the Russian government to start negotiations that focused on major migrants' origin and transit countries such as Vietnam, Turkey, Pakistan and Afghanistan, but primarily targeted the Central Asian region.

It has been claimed that Russia's role as a transit country for migrants from Central Asia, for whom the EU is the final destination, is bound to grow (interview at the EEAS, 15 February 2011). The EU is, therefore, directly concerned by Russia's ability to manage its migration flows. The perceived salience of this challenge for both the EU and Russia is emphasized by the following statement made by the Head of EU Delegation to Russia:

[We have] fears that a visa-free regime with Russia would lead to a large inflow of irregular migrants to the European Union, as well as to an increase in criminal activities. This is not really about [the fear of] Russian citizens. However, the transparency of your [Russian] southern borders creates a risk of increase in human and drug trafficking.

(Valenzuela 2010, 2)

Even though this argument is quite disputable, because for the moment the migrants from the five Central Asian states do not constitute major migration pressure for the EU, the particular importance of these countries is explained by several interrelated factors. First, their migration potential coupled with unstable economic conditions, internal conflicts and environmental problems in the region cannot be neglected (Foresight 2011). Second, these countries are major transit routes not only for migrants from the larger Asian region, but also for human and drug trafficking arriving from neighbouring Afghanistan and Pakistan. The perception of threat by the EU – apparently also shared by Russia – has been decisive for an intensive externalization of the Russian immigration policy.

The Russian government seemed to gradually realise the need for closer cooperation with these countries, taking into account this multiplicity of regional 'push' factors and regional border problems. Some experts have argued that for Russia the security of the southern borders of Central Asia is seen as a question of domestic security, not out of 'imperialism', but of pragmatism, because some 7000 kilometres of the Russian border with Kazakhstan are nearly impossible to securitise (Laruelle 2009). Indeed, the Russian authorities claim to have been paying specific attention to the situation on this part of the Russian border for several years (Strekha 2010). However, according to independent assessments, the lengthy and porous border with Kazakhstan is still in a precarious state (Olekh 2008). Such a border, of course, cannot serve as an effective barrier either for irregular migration, or for human or drug trafficking. The modernisation of the border infrastructure needs time. This situation requires that the irregular flows are better controlled downstream, which confirms Central Asia's role as a buffer zone for Russia itself (Laruelle 2009).

However, integrating Central Asian states in the network of readmission agreements has proven to be an extremely difficult task for Russia which has been emphasized by the Head of Department on Readmission of the Russian Federal Migration Service:

> Unfortunately, Russian proposals to activate the readmission dialogue do not always get a positive reaction from the CIS countries. Russia has repeatedly asked them to speed up negotiations on readmission agreements, but ... Kyrgyzstan and Tajikistan keep low profile ... So called 'package deals' may be a good solution for this problem. In this case Russia can condition the signature of any international treaty important for these states by their signature of readmission agreements with Russia. In order to involve these countries in the readmission dialogue, it is necessary to use the potential of such international organizations as CIS, CSTO, EAEC, IOM, OSCE, as well as the EU capacities.
>
> (Yakovlev 2009)

It is indeed the conclusion of the readmission agreement with the EU that has been used by the Russian government as the leverage in similar negotiations with Central Asian republics. Moreover, Russian negotiators have made use of successful negotiation tools previously employed by the EU towards Russia, namely 'package deals' involving positive conditionality.

In some cases, Russia was very active; in others almost no démarches were taken. Quite unexpectedly, the first success came with the signature of the agreement with Uzbekistan.[7] By the summer of 2007 Russia had managed to conclude the readmission agreement with this most populated Central Asian country, when a serious package deal was used by the Russian side as leverage. In addition to the readmission agreement, an agreement on labour activity and the protection of rights of citizens of the two countries was also signed (Uzbekistan Daily 2007). The Head of the Russian Federal Migration Service, Konstantin Romadanovsky acknowledged that the signature of the readmission agreement with Uzbekistan would play a positive role in introducing a visa-free regime in the EU–Russia space (ibid). Russia was then expecting a quick chain reaction, waiting for other Central Asian states to express a willingness to sign similar agreements, but had to wait for five more years for a new agreement.

Already in July 2007 Russian authorities declared that the agreement with Kazakhstan was being finalized. Kazakh authorities, however, have long been very vague on this issue, delaying the finalization of the agreement which was eventually signed only in June 2012. It still has not entered into force. The major reason for this non-ratification and, consequently, non-implementation of the agreement seems to be its ambitious legal scope – the final text includes provisions on readmission not only of citizens of the parties, but also of third country nationals (Agreement between the Government of Russia and the Government of Kazakhstan on Readmission 2012). In general, the main difficulty with concluding readmission agreements with Central Asian countries is related to the 'third country nationals' clause. Not only competent Russian bodies, but also the relevant authorities of Central Asian states, understand that many of those who would be eventually identified as irregular migrants in Russia would come not from but through Central Asia, arriving mostly from such countries as Bangladesh, Pakistan, India, Sri-Lanka and some South-East Asian countries (Olekh 2008). Other Central Asian countries share this view, but the chances of dealing with readmission of third country nationals are much higher for Kazakhstan; due to its geographic position, developed transport infrastructure and booming economic situation it is a key country for transit migration in the region.

For quite a long time, Kyrgyzstan seemed to be waiting for its neighbours who have the biggest migratory pressures on Russia to sign the agreements first. But eventually, due to its increased political and economic dependency on Russia after the new political turmoil in 2010, Kyrgyzstan gave in to Russian demands. The agreement between the two countries was signed in October 2012 in a package deal providing for preferential treatment of Kyrgyzstani labour migrants in Russia. This readmission agreement, in contrast to the one between Russia and Kazakhstan, entered into forced relatively quickly with the ratification by both countries finalised in the summer of 2013. The reason for this rather unproblematic implementation can be explained by its restricted legal scope – the agreement covers only citizens of the parties leaving beyond its remit third country nationals (Agreement between the Government of Russia and the Government of Kyrgyzstan on Readmission 2012). This, however, does not mean that the agreement holds just symbolic value – in contrast to Kazakhstan where the issue of concern for Russia was indeed its transit role, Kyrgyzstan was much more a country of massive emigration to Russia. This agreement, therefore, provides the Russian authorities with a sanction-like tool that can be used as a threat to Kyrgyzstani authorities fearing social turmoil that is likely to take place in the case of massive migrant returns.

To date, Tajikistan remains the only country in the region that does not have a readmission agreement with Russia.[8] In December 2009, the main official Tajik media – National Information Agency of Tajikistan 'Khovar'– claimed that Tajikistan was ready to conclude a readmission agreement with Russia (Khovar 2009). A year later, in October 2010, the Tajik president met with the Head of the Russian Federal Migration Service, which is itself an extraordinary event (Federal Migration Service 2010). Then, in December 2010, Tajik and Russian Ministers of Interior devoted special attention to this issue during their meeting in Dushanbe (Deutsche Welle 2010). However, no agreement has been signed yet and the parties continue negotiations and informal discussions on various levels. Apart from the mentioned fear of becoming responsible for readmission of third country nationals, the Central Asian governments' attitude towards the readmission is conditioned by other internal and external factors. First, they do not want to see their countries on the 'black list' representing 'a migration threat'. Second, heavily depending on migrants' remittances (World Bank 2011) and being aware that a readmission agreement in force may negatively affect many thousands of migrant-workers overstaying their legally defined periods of sojourn in Russia, Central Asian governments do not want to put them at risk of being sent back home where they would most probably be unemployed and pose a challenge for social stability. Finally, the Central Asian states fear that the signature of a readmission agreement with Russia might weaken their positions in similar negotiations with other parties, first and foremost with the EU.

Paradoxically, the EU has never tried to directly include Central Asian states in its network of readmission agreements, since it is not ready to propose any visa-facilitation in exchange (Interview at the EEAS, 15 February 2011), as in the cases of Russia, Moldova and Ukraine. Nevertheless, there is now a potential for these states to conclude readmission agreements with Russia and between themselves. This has been triggered by the dynamics of EU–Russia cooperation on readmission. Therefore, one might argue that 'transformative power of Europe' (Boerzel and Risse 2009) does not always need to directly involve those whose policies it wants to transform. The role of efficient intermediaries (such as Russia) is important in this case. This explains why the EU–Russia readmission agreement has been characterized by an EU official as the only one that has been properly used in EU readmission policy (Coleman 2012). This has to do with the need to extend the network of readmission agreements further from EU direct neighbours/partners to other countries in the world. A thorough analysis shows that EU–Russia cooperation on readmission has brought about a diffusion of the readmission mechanism that has been not only successfully implemented in the EU–Russia relations, but has eventually been incorporated into migration and external policies of Russia itself. Russia has become an active transferring actor as regards readmission mechanisms in relations with its partners and major migrants' origin/transit countries in Central Asia, the Middle East, as well as in South and South East Asia, thus fostering EU promotion of readmission in various regions of the world.

Enforced 'emulation' (Boerzel and Risse 2009) of some of EU migration management practices by Russia might be explained by pragmatic calculations of such key policy actors as the Russian Federal Migration Service and Russian Ministry of Foreign Affairs and might reflect Russia's intention to employ efficient policy solutions for problems similar to those of the EU. Just as the EU has been shifting responsibility for migration management to the east, Russia has adapted a similar strategy towards its southern neighbours. In so doing, Russia has been trying to use the EU's interests in migration management as leverage in its own negotiations with third countries. Subsequently, the need to conclude readmission agreements with third countries has even entered the new Migration Policy Concept of the Russian Federation adopted by a presidential decree in 2012.[9] This policy line represents one of those cases where the EU security interests coincide with the security interests of Russia or, more accurately, with the securitarian agenda of the Russian government. Last, but not least, cooperation on migration with the EU provides the Russian government with leverage and symbolic power vis-à-vis its neighbours in the CIS countries. It is particularly the case for Russia's relations with the countries of migrants' origin and transit in Central Asia. Russian authorities learn not only to emulate EU measures in the field of migration control, but also to exploit justifications that have already been used by the EU in the process of construction of its own migration policy. The EU, thus, provides a sort of legitimisation for Russian policy, most probably being aware of this situation. A major feature of the EU policy towards Russia is the intention to adapt institutional and content-specific characteristics of Russia's migration-related policies to EU needs. The EU experience or, more precisely, multiple references to this experience, play an important role in the construction of Russian migration policies – in particular in their advanced securitisation. The relevant agreements (those already in place or those being negotiated) seem to be part and parcel of the EU 'externalization' strategy implemented through the policies of its partners, in this particular case, through Russia's policies towards its 'near abroad'. We clearly observe a dynamic of 'diffusion' (Boerzel and Risse 2012) that is not contained within the framework of the EU–Russia cooperation, but has acquired a broader regional dimension.

By developing cooperation with Russia and, consequently, socialising countries from the Eurasian migration system (Ivakhnyuk 2008) with the norms and practices of readmission, the EU has created scope conditions for more active promotion of its migration management standards in the region. Some of the EU Member States, such as Hungary, the Czech Republic, Latvia and Lithuania, have moved towards closer migration cooperation with Central Asian countries. Kazakhstan, aiming to get visa-facilitation preferences with the EU, is certainly the leader in this process (Newskaz 2015). This strategic policy aim of the Kazakhstani government, in a similar vein to Russia's migration cooperation with the EU, explains its efforts to conclude readmission agreements with Kyrgyzstan and Tajikistan (Kazinform 2015). Moreover, apart from the development of bilateral cooperation on readmission, many countries of the region are involved in multilateral cooperation that has started within the framework of the CIS. In 2012, the Parliamentary Assembly of the CIS drafted a model readmission agreement in order to promote coherent readmission practices among CIS Member States. All negotiation and implementation problems notwithstanding, this dynamics testify to a certain success in the diffusion of readmission practices in the vast post-Soviet region.

Introducing the mechanism of readmission agreements in its 'wider neighbourhood' can be a way for the EU to pressure for reforms in justice and home affairs in countries of the region and this view can hold if we remember that in order to implement readmission agreements properly countries indeed need to upgrade their related standards and technical capacities. However, it would be misleading to believe that such positive adaptation would automatically follow the conclusion of readmission agreements. Positive changes in this area, if they ever take place, need time. In the immediate terms, quite to the contrary, exporting the mechanism of readmission to Central Asia goes against promotion of some core rights of migrants and refugees, as well as of victims of human trafficking. For example, when the OSCE is trying to persuade Tajikistan to establish national referral mechanism for protection of victims of human trafficking, Russia's simultaneous efforts to conclude a readmission agreement contribute to a rather negative image of a migrant, as well as of potential victims of human trafficking that are looked at through the prism of illegality.

Conclusion

Cooperation on migration-related issues is important for both the EU and Russia. The importance of this cooperation for the EU is higher than success in the sub-field of human rights within the same common space, while Russia also benefits, which is why migration management continues to be one of the most dynamic and successful fields of cooperation, regardless the pending visa-free negotiations. The implementation of the readmission agreement has shown that fears of both the EU and Russia with regard to potentially high numbers of irregular migrants were largely exaggerated. Instead, this agreement has proved to be beneficial for Russia insofar as its own migration strategy in the region is concerned. By engaging in this type of migration management efforts with the EU, the Russian government has created for itself a legitimation platform for further similar restrictive migration management cooperation with those countries that 'source' migrants to Russia. More generally, the steep intensification of the EU–Russia cooperation on migration has contributed to the rapid development of the external dimension of Russia's migration policy.

The EU, through this cooperation with Russia, has acquired additional channels to promote one of its most internationally visible instruments of migration management – readmission agreements – in Central Asia where its direct involvement had not been successful (Emerson et al. 2010). This cooperation has produced a significant domino effect in the Eurasian migration system. Russian mediation has played an important role in EU policy transfer leading to the introduction of cooperation on readmission in the inter-state relations in the region far from the EU borders. Unfortunately, the externalities of EU policies creating a web of readmission agreements, enforcing readmission practices and developing relevant detention infrastructure without prior establishing proper legislative and practical safeguards can potentially endanger the position of migrants in Eurasia – the region where human rights of migrants and persons seeking international protection are already at risk.

Notes

1 This chapter is a significantly revised, updated and extended version of the following article: Korneev, O. (2012) Deeper and Wider than a Common Space: EU–Russia Cooperation on Migration Management, *European Foreign Affairs Review*. Vol.17. No.4. Pp. 605-624.

2 By 'externalities' we mean (often unintended) side-effects of particular policies as primarily understood in economics (see Lavenex and Uçarer (2002) for a discussion of the use of this term in studies of EU migration policies).

3 Consider the aggregated data on Russian nationals irregularly staying in the EU and apprehended by the Member States provided by Eurostat (European Commission 2011). There were 10 375 Russian nationals apprehended in different EU Member States in 2009. FRONTEX, which uses a different data collection methodology, provides different numbers: 9 526 for 2009, 9 471 for 2010 and 10 314 for 2011 (FRONTEX 2012). Apart from these divergent and non-comprehensive data, there is no reliable statistics on irregular migrants in the EU and on irregular migrants from Russia, in particular. However, one of the concerns of the EU is a high number of asylum seekers from Russia, since often refused asylum seekers fall under the category of irregular immigrants and thus need to be returned. In 2011, Russia ranked second as the 'country of origins' of asylum seekers in the EU (18 200 applicants) after Afghanistan (EMN Bulletin 2012).

4 One of such centres in Hungary, next to the Austrian-Hungarian border, was visited by Oleg Korneev in spring 2006, during a study trip organized by Central European University (Budapest).

5 See also multiple comments on detention conditions for migrants in Russia by various Russian NGOs.

6 Reliable data on readmission applications from EU Member States to Russia, as well as on numbers of positive replies and refusals by the Russian authorities (for the years 2007–2009) can be found in the Annex 2 to the 'Evaluation of EU Readmission Agreements' (European Commission 2011).

7 This success is however relative, since the agreement with Uzbekistan does not cover third country nationals.

8 More accurately, the same is also valid for Turkmenistan, but due to extremely restrictive exit control policies enforced by its government, the country is not perceived as a major source of migrants and, thus, is not on the list of priorities for Russia.

9 The need to conclude readmission agreements with countries of origin figures already in the previous Migration Policy Concept adopted in 2003 (related to the signature of the readmission agreement with Lithuania). However, only the new Concept defines such agreements as key instrument of Russia's international cooperation on migration issues.

Bibliography

Agreement between the Government of Russia and the Government of Kazakhstan on Readmission (2012) [in Russian]. http://mid.ru/bdomp/spd_md.nsf/0/ CA0341557D774D7E43257E4C00246FFD

Agreement between the Government of Russia and the Government of Kyrgyzstan on Readmission (2012) [in Russian]. www.mid.ru/BDOMP/spd_md.nsf/0/73DC6C84 91313D4A43257E4C002471B7

Agreement between the Government of Russia and the Government of Lithuania on Readmission (2003) [in Russian]. www.mid.ru/BDOMP/spd_md.nsf/0/ E154A619FCB345F943257E4C00246E19

Agreement between the Russian Federation and the European Community on Readmission (2006) [in Russian]. www.russianmission.eu/userfiles/file/agreement_on_readmission_2006_russian.pdf

Alexandrova-Arbatova, N. 2006. Russia-EU Relations: Still at the Crossroads. *The EU–Russia Review.* Vol. 2, pp. 17–21.

Arestova, L. 2010. Intervention at the Training School 'EU Immigration and Asylum Policies, Border Security: State of Play and Prospects of Russia–EU Cooperation on Migration'. MGIMO, Moscow, 25–29 October.

Belguendouz, A. 2005. L'Europe des Camps: La Mise à l'Écart des Étrangers. *Cultures & Conflits: Sociologie Politique de l'International.* Vol. 57, Paris: L'Harmattan.

Boerzel, T. and Risse, T. 2009. The Transformative Power of Europe: The European Union and the Diffusion of Ideas. *KFG The Transformative Power of Europe.* Working paper No. 1. May 2009.

Boerzel, T. and Risse, T. 2012. From Europeanisation to Diffusion: Introduction. *West European Politics.* Vol. 35, No. 1, pp. 1–19.

Bordachev, T. 2006. *Russia has exchanged an elephant for a small dog* [in Russian]. www.svoboda.org/content/transcript/158928.html

Brunarska, Z., Mananashvili, S. and A. Weinar 2013. *Return, Readmission and Reintegration in the Eastern Partnership Countries: An Overview.* CARIM-East RR 2013/17, Robert Schuman Centre for Advanced Studies, San Domenico di Fiesole (FI): European University Institute.

Cassarino, J.-P. 2010. *Unbalanced Reciprocities: Cooperation on Readmission in the Euro-Mediterranean Area.* Washington DC: Middle East Institute.

Chappart, P., Charles, C. and Schulmann, L. 2012. Politique européenne de readmission: cooperer pour mieux renvoyer. In Clochard, O. (ed.) *Atlas des Migrants en Europe: Geographie critique des politiques migratoires.* Paris: Armand Colin.

Charillon, F. 2004. Sovereignty and Intervention: EU's Interventionism in its 'Near Abroad', in Cox, M. (ed.) *E.H. Carr. A Critical Appraisal.* London and New York: Palgrave.

Deutsche Welle 2010. Tadjikistan and Russia Cannot Reach an Agreement on Readmission [in Russian], *Deutsche Welle*, 8 December. www.dw-world.de/dw/article/0,,6309830,00.html.

Ehin, P. 2009. Assessment of the Common Space of Freedom, Security and Justice, in Nikolov, K. Y. (ed.) *Assessing the Common Spaces between the European Union and Russia.* Sofia: Bulgarian European Community Studies Association, pp. 68–88.

Emerson, M. 2005. *EU–Russia Four Common Spaces and the Proliferation of the Fuzzy,* CEPS Policy Brief, No. 71, May, p. 2. Brussels: CEPS.

Emerson, M., Boonstra, J., Hasanova, N., Laruelle, M. and Peyrouse, S. 2010. *Into EurAsia: Monitoring the EU's Central Asia Strategy.* Brussels: CEPS.

European Migration Network (2012) *EMN Bulletin: A Report from the European Migration Network for the period January to May 2012.* June 2012.

EU–Russia Summit 2005. Road Map for the Common Space of Freedom, Security and Justice. *Annex*, Vol. 2, pp. 21–34.

European Commission 2011. *Evaluation of EU Readmission Agreements.* COM (2011) 76 final. Brussels: European Commission.

European Council 1999. *Common Strategy of the European Union on Russia.* http://trade.ec.europa.eu/doclib/docs/2003/november/tradoc_114137.pdf

European External Action Service 2011. *EU–Russia Common Spaces Progress Report 2010.* Brussels, March. http://eeas.europa.eu/russia/index_en.htm.

Federal Migration Service 2010. *President of Tajikistan and the Director of the Russian Federal Migration Service Konstantin Romodanovsky had a working meeting in Dushanbe on the 27th of October 2010* [in Russian]. www.fms.gov.ru/press/news/news_detail.php?ID=39132

Foresight 2001. *Migration and Global Environmental Change: Final Project Report.* London: The Government Office for Science.

FRONTEX 2012. *Annual Risk Analysis 2012.* Warsaw: FRONTEX.

Gatev, I. 2008. Border Security in the Eastern Neighbourhood: Where Bio-politics and Geopolitics Meet. *European Foreign Affairs Review*, Vol. 13, No. 1, pp. 97–116.

Gavrilis, G. 2009. Beyond the Border Management Programme for Central Asia (BOMCA). *EU-Central Asia Monitoring (EUCAM)*. Policy brief No. 11. November.

Chizhov, V. 2005. We Do Not Want to Transform Russia Into a 'Transit Camp' [in Russian]. www.vremya.ru/print/130680.html.

Coleman, N. 2012. Questions and answers session, 'The Management of the External Borders of the EU and its Impact on the Human Rights of Migrants: The Italian Experience. A Consultation between the UN Special Rapporteur on the Human Rights of Migrants, Mr François Crépeau Civil Society, and Academia'. European University Institute, Florence, 3 October, 2012.

European Commission 2013. First Progress Report on the implementation by Russia of the Common Steps towards visa free short-term travel of Russian and EU citizens under the EU-Russia Visa Dialogue. COM 923 final. Brussels: European Commission.

Government of Russia 1999. *Midterm Strategy for EU–Russia Relations: 2000-2010* [in Russian]. www.mgimo.ru/fileserver/2004/kafedry/evro_int/reader4meo_3-6.htm.

Hernandez i Sagrera, R. 2010. The EU–Russia Readmission – Visa Facilitation Nexus: An Exportable Migration Model for Eastern Europe? *European Security*. Vol. 19, No. 4, pp. 569–584.

Hernandez i Sagrera, R. and Korneev, O. 2012. *Bringing EU Migration Cooperation to the Eastern Neighbourhood: Convergence Beyond the Acquis Communautaire?* European University Working papers, RSCAS 2012/22. Florence: European University Institute.

Hernandez i Sagrera, R. and Potemkina, O. 2013. *Russia and the Common Space on Freedom, Security and Justice: Study.* Brussels: European Parliament.

Human Rights Watch. 2005. *Ukraine: On the Margins. Rights Violations against Migrants and Asylum Seekers at the New Eastern Border of the European Union.* Vol. 17, No. 8(D). New York: Human Rights Watch.

Human Rights Watch. 2010. *Buffeted in the Borderland. The Treatment of Asylum Seekers and Migrants in Ukraine. December 2010.* New York: Human Rights Watch.

Ivakhnyuk, I. 2008. *Eurasian Migration System: Theoretical and Political Approaches* [in Russian]. Moscow: MAX Press.

Jileva, E. 2002. Larger than the European Union: The Emerging EU Migration Regime and Enlargement, in Lavenex, S. and Uçarer, E. M. (eds.) *Migration and the Externalities of European Integration.* London: Lexington Books.

Kazinform 2015. *Kazakhstan Plans to Sign Agreements for Return of Illegal Migrants with Russia, Belarus, Kyrgyzstan and Tajikistan* [in Russian]. www.inform.kz/rus/article/2746858.

Khovar 2009. *Tadjikistan Is Ready to Conclude a Readmission Agreement* [in Russian]. www.khovar.tj/index.php?option=com_content&task=view&id=17047.

Korneev, O. 2007. *The EU Migration Regime and Its Externalization in the Policy Toward Russia.* InBev-Baillet Latour Working Paper No.31. Leuven: Catholic University of Leuven.

Korneev, O. 2013. EU Migration Governance in Central Asia: Everybody's Business – Nobody's Business? *European Journal of Migration and Law.* Vol. 15. No. 3, pp. 301–318.

Korneev, O. 2014. Exchanging Knowledge, Enhancing Capacities, Developing Mechanisms: The Role of IOM in the Implementation of the EU–Russia Readmission Agreement. *Journal of Ethnic and Migration Studies.* Vol. 40, No. 6, pp. 888–904.

Kurowska, X. and Tallis, B. 2009. EU Border Assistance Mission: Beyond Border Monitoring? *European Foreign Affairs Review.* Vol. 14, No. 1, pp. 47–64.

Lahti, T. 2008. Head of Political Section, European Commission Permanent Representation in Moscow. *Presentation at the EU Study Weekend,* Tomsk, July.

Laruelle M. 2009. Russia in Central Asia: Old History, New Challenges? *EU-Central Asia monitoring (EUCAM).* Working paper No.3.

Laruelle, M. and Peyrouse, S. 2010. *L'Asie centrale à l'aune de la mondialisation. Une approche géoéconomique.* Paris: Armand Colin.

Lavenex, S. 2004. EU External Governance in 'Wider Europe', *Journal of European Public Policy.* Vol. 11, No. 4, pp.680–700.

Lavenex, S. and Uçarer, E. M. (eds) 2002. *Migration and the Externalities of European Integration.* London: Lexington Books.

Mananashvili, S. 2013. *Access to Europe in a Globalised World: Assessing the EU's Common Visa Policy in the Light of the Stockholm Guidelines.* Migration Policy Centre Working Paper 2013/74. San Domenico di Fiesole (FI): European University Institute.

Mitsilegas, V. 2002. The Implementation of the EU Acquis on Illegal Immigration by the Candidate Countries of Central and Eastern Europe: Challenges and Contradictions. *Journal of Ethnic and Migration Studies.* Vol. 28, No. 4, pp. 665–682.

Newskaz 2015. *Kazakhstan has ratified the agreement on readmission with Hungary* [in Russian]. http://newskaz.ru/politics/20150218/7606848.html.

Olekh, G. 2008. Siberian Part of the Russian Border with Kazakhstan: State of Affairs and Perspectives [in Russian]. *CAMMIC Working Papers No.1,* Center for Far Eastern Studies, University of Toyama, April.

Pastore, F. 2002. Aeneas's Route: Euro-Mediterranean Relations and International Migration, in Lavenex, S. and Uçarer, E. M. (eds) *Migration and the Externalities of European Integration.* London: Lexington Books.

Potemkina, O. 2002. *Russia's Engagement with Justice and Home Affairs: A Question of Mutual Trust,* CEPS Policy Brief No.16, March 2002.

Potemkina, O. 2006. EU–Russia Cooperation in Justice and Home Affairs, *The EU–Russia Review.* Vol.2, pp. 39–45.

Potemkina, O. 2010. EU–Russia Cooperation on the Common Space of freedom, Security and Justice – A Challenge or an Opportunity? *European Security.* Vol. 19, No. 4, pp. 551–568.

President of Russia 2012. *Concept of the State Migration Policy of the Russian Federation through to 2025.* www.kremlin.ru/news/15635.

Revenko, N. 2010. Deputy Head, Russian Mission in the EU. Intervention at the Training School 'EU Immigration and Asylum Policies, Border Security: State of Play and Prospects of Russia–EU Cooperation on Migration', MGIMO, Moscow, 25–29 October.

Russian Ministry of Foreign Affairs 2003. *Press-Release 'Russian-Lithuanian Readmission Agreement Signed'.* www.mid.ru/bl.nsf/062c2f5f5fa065d4c3256def00 51fa1e/ce42b317e7e2e11043256d25002b9311?OpenDocument.

Strekha, M. 2010. Head of Strategic Planning Department, Russian Federal Border Service. Intervention at the Training School 'EU Immigration and Asylum Policies, Border Security: State of Play and Prospects of Russia–EU Cooperation on Migration', MGIMO, Moscow, 25–29 October.

Trauner, F., Kruse, I. and Zeilinger, B. 2013. Values Versus Security in the External Dimension of EU Migration Policy: A Case Study on the Readmission Agreement with Russia. In Noutcheva, G., Pomorska, K. and Bosse, G. (eds) *The EU and Its Neighbours: Values vs. Security in European Foreign Policy.* Manchester: Manchester University Press, pp. 201–217.

Uzbekistan Daily 2007. *Russia, Uzbekistan sign four agreements including deal on migration,* 4 July. www.uzdaily.com/articles-id-698.htm.

Valenzuela, F. 2010. Interview. *Komsomolskaya Pravda* [in Russian], 6 May, p. 2.

Valluy, J. (ed.) 2005. L'Europe des Camps: La Mise à l'Écart des Étrangers, *Cultures & Conflits,* No. 57.

Van Elsuwege, P., Fomina, J., Korneev, O., Sembaeva, A. and Voynikov, V. 2013. *EU–Russia Visa Facilitation and Liberalization: State of Play and Prospects for the Future.* Berlin: EU–Russia Civil Society Forum.

World Bank 2011. *Migration and Remittances Factbook,* 2nd edition. Washington D.C.: The World Bank.

Yakovlev, V. 2009. *The Road to Visa-Free Agreements* [in Russian]. www.fms.gov.ru/press/publications/news_detail.php?ID=27272.

Yastrzhembsky, S. 2007. *Russia and the EU Will Have Visa Free Regime in 15 Years* [in Russian]. www.rosbalt.ru/2007/10/12/421974.html.

Personal interviews

Directorate General Justice, Liberty and Security, the European Commission, 19 April 2007.

Directorate General RELEX (Russia unit), the European Commission, Brussels, 6 June 2007.

European External Action Service, Brussels, 15 February 2011.

Finnish Permanent Representation to the EU, Brussels, 21 May 2007.

Russian Permanent Representation to the EU, Brussels, 20 May 2007.

Sean Carroll, Head of Press and Information Section of the European Commission Representation to the Russian Federation, the 'EU study weekend', Pushkin (Russia), 29 April 2007.

Wouter van de Rijt, Principal Administrator, DG JHA, Council of the European Union, Brussels, 16 May 2007.

Part III

Externalizing migration management in North America

9 The US Visa Waiver Program and the management of mobility across the Atlantic

Rey Koslowski[1]

Introduction

Managers of a multinational corporation call an emergency meeting in London and a New York businessman hops on a flight to Heathrow; a German couple books a last-minute vacation and flies to Florida; a Boston woman breaks her hip and her Irish sister comes from Dublin to care for her. Such spontaneous transatlantic travel has become a common occurrence made possible by international cooperation enabling visa-free travel – cooperation that is taken for granted until the relative ease of travel is threatened, as it was after the attacks of September 11, 2001.

The Visa Waiver Program (VWP), which permits visa-free travel to the United States (US) for nationals of states such as the United Kingdom (UK), France, Germany and Ireland, emerged from obscurity after the 9/11 attacks, when it became clear that Zacarias Moussaoui, the so-called "20th hijacker," had entered the US using just his French passport. Then British national Richard Reid boarded a transatlantic flight in December 2001 with only his passport and tried to detonate a bomb in his shoes. Citing such examples, Robert Leiken described the risks of "a passport-carrying, visa-exempt mujahideen coming from the United States' western European allies" (Leiken, 2005).

After 9/11, the US Congress considered abolishing the Visa Waiver Program but then only stiffened its requirements while adding new members was put on hold. Meanwhile, excluded states like Bulgaria, the Czech Republic, Hungary, Poland, Romania, Slovakia and the Baltic states were joining the US-led 'coalition of the willing' to fight the 'war on terrorism' in Iraq and Afghanistan and they expected US reciprocity with visa-free travel. Radek Sikorski, now Poland's Foreign Minister, once poignantly noted that British and French citizens, whose ranks included Al-Qaeda terrorists detained in Guantanamo, were allowed to travel to the US without a visa, whereas not a single Polish national had been identified as a terrorist but visas were still required of Poles (Sikorski, 2004). Polish soldiers returning from a tour of duty in Iraq could not take their families to Disneyworld without considerable extra costs and hassles, even if they managed to get a visa.

The US Visa Waiver Program excluded all but one of the ten new member states that joined the European Union (EU) in May 2004. Given that the EU has a common visa policy, asymmetries in US treatment of 'old' and 'new' EU member states raised a thorny transatlantic problem. According to the EU's common visa policy at the time, any EU member state from which the US requires a visa could reciprocally require visas of US nationals and invoke a solidarity clause that would, in turn, result in US nationals needing visas to travel to all EU member states. Were this to occur, the US would require visas of all EU citizens and visa-free transatlantic travel would end, resulting in State Department consular service costs in the hundreds of millions of dollars, tourism losses in the billions and incalculable ill-will among the traveling public.

This US–EU conflict occurred in a changing environment of diplomatic practices that Anne Marie Slaughter describes in her book, *The New World Order*, which draws attention to the networks of government officials who "increasingly exchange information and coordinate activity to combat global crime and address common problems on a global scale" (Slaughter 2005, 1). Rather than thinking of the world in terms of unitary states interacting through the head of state or foreign ministry, Slaughter argued that the state has become "disaggregated" with the accelerating growth of "transgovernmental relations" (Keohane and Nye 1974) between parts of states, including interactions of finance ministries, environmental ministries and interior ministries.

Slaughter's depiction of growing transgovernmental relations and governmental networks is supported by the fact that the US Department of Homeland (DHS), its counterpart interior ministries in the EU and the European Commission are not only implementing international agreements negotiated by foreign ministries, they are increasingly developing their own diplomatic capabilities and negotiating agreements governing visa policies and border controls. For example, the DHS established an international section of its Office of Policy and posted a DHS attaché to the US Missions to the European Union and NATO in Brussels. Similarly, the European Commission's Directorate General for Justice, Freedom and Security established an 'External Relations and Enlargement' unit and an 'International Aspects of Migration and Visa Policy' unit as well as posting an official with responsibility for justice and home affairs in the European Commission's delegation in Washington. Due to painstaking diplomacy between the DHS and its European counterparts, a revision of the EU's common visa policy and reform of the US Visa Waiver Program that enabled entry of seven EU member states, transatlantic visa-free travel has survived. The diplomatic impasse, however, still festers given that the US still requires visas of nationals from several EU member states. The political compromise of US visa reforms may also have unforeseen consequences for international travel that raise new issues for policymaking and diplomacy. In some cases, agreements between interior ministries and law enforcement agencies are reversing longstanding international norms of state-to-state relations going back to the League of Nations. Finally, there are few realistic options for the Obama Administration to address this issue other than fully implementing legislation already passed and pressing Congress for sufficient resources to do so.

Passports, visas and visa-free travel

The modern passport and visa developed toward the end of the 19th century and at a 1920 Paris Conference sponsored by the League of Nations, when states standardized passport and visa formats, adopting the now familiar multi-page book passport (Lloyd 2003; Salter 2003). At the time, states considered making issuing states responsible for vetting passport applicants for their criminal records and their admissibility to other states but decided to only require that the passport signifies that an individual is the national of the issuing state. Destination states remained responsible for investigating the credentials presented by travelers and making entry decisions. To help control their borders, states increasingly relied on the visa, an authorization given by a state to the nationals of another state to travel and present themselves to authorities for inspection at ports of entry. Visa applications typically involve submission of identity documents, return tickets, bank statements, immunization records and an interview with consular officials abroad who then issue visas by stamping them in the prospective traveler's passport.

As the volume of international travel increased with the introduction of large passenger jets, many states eventually eliminated visa requirements for short-term visits on a bilateral reciprocal basis. For example, the US Visa Waiver Program permits travel to the US for purposes of business or pleasure for up to 90 days without a visa by nationals of states that similarly permit visa-free travel by US nationals. Begun as a pilot program with the UK and Japan in 1988, the Visa Waiver Program became permanent in 2000, when 17.6 million travelers entered under the program accounting for over half of overseas visitors (GAO 2002, 21). The program was made permanent largely because it saved billions of dollars in costs that would have been incurred processing visas and it facilitated significant growth in international tourism to the US during the 1990s. The program grew to 29 members in 1999, but dropped to 27, when Argentina's and Uruguay's memberships were terminated. As of March 2015, the Visa Waiver Program includes 38 countries.

The visa was initially developed as a tool of immigration law enforcement but it also became a tool of diplomacy. States use the issuance or denial of visas to individuals, certain groups or all nationals of particular states in efforts to influence other states' policies (Stringer, 2004). As will be made clear below, foreign policy considerations have been crucial in changing US visa policies and, in the process, collided with immigration law enforcement.

Security concerns vs. economic benefits

In response to the 9/11 attacks, Congress passed the USA PATRIOT Act requiring states in the Visa Waiver Program issue machine-readable passports by 2003. After the 'shoe-bomber' incident, members of Congress called for the elimination of the US Visa Waiver Program altogether. The Government Accountability Office then estimated that eliminating the program would initially cost the State Department up to $1.28 billion for consular facilities and staffing and generate ongoing annual costs of up to $810 million (roughly 11 percent of the State Department's entire $7.4 billion 2002 fiscal year budget). A Commerce Department study estimated that, over a five-year period, eliminating the program could mean a loss of three million visitors, $28 billion in tourism receipts and 475,000 jobs (GAO 2002, 22–23).

After fully considering these costs, Congress retained the program but passed legislation in 2002 that required members of the Visa Waiver Program to issue passports with biometrics on radio frequency identification (RFID) chips. Then in 2006, UK officials uncovered a plot of over 20 British nationals of Pakistani origin, who planned to board US-bound flights and blow them up with liquid explosives. Congress held hearings where the Director of National Intelligence testified that Al-Qaeda was recruiting Europeans because they could travel to the US with just a passport (McConnell 2007). Once again, members of Congress introduced legislation to eliminate the Visa Waiver Program.

US–EU diplomacy and political pressures for change

Other members of Congress and President Bush viewed the Visa Waiver Program differently and began advocating its expansion in response to domestic pressures, changing foreign policy agendas and EU enlargement. The Visa Waiver Program included all 15 members of the EU before the May 2004 enlargement (except Greece) but only one of the ten new member states (Slovenia). US citizens enjoy visa-free travel to all EU member states under its common visa policy, but after enlargement nationals of ten EU member states did not enjoy visa-free travel to the US. As enlargement approached, then Director General for Justice, Freedom and Security of the European Commission, Jonathan Faull (2004), argued that the US should allow visa-free travel to citizens of all EU member states. Nevertheless, the US resisted such arguments and persisted in bilateral arrangements that bypassed the EU's common visa policy.

Visa-free travel became a top priority of Central and Eastern European foreign policy towards the US. For example, during his January 2004 visit to the US, Polish President Aleksander Kwasniewski put President Bush on the spot during a photo-op and asked him to drop the US visa requirement (Kamen, 2004). Members of Congress with large Central and East European ethic constituencies also took up the cause. Noting that 9.2 million

Americans of Polish ancestry live in the United States, then Congressman Rahm Emanuel introduced legislation in April 2004 to include Poland in the Visa Waiver Program. The cause spread through Congress leading the 2006 Comprehensive Immigration Reform Act to include provisions that would establish a probationary admission to the Visa Waiver Program for EU member states "providing material support to the United States or the multilateral forces in Afghanistan or Iraq" (US Senate, 2006).

US visa policy reform efforts and US–EU diplomacy occurred under the shadow of a 'nuclear option' that would eliminate transatlantic visa-free travel and could be triggered by a single EU member state invoking a solidarity clause in the common EU visa policy that would, in turn, result in US nationals needing visas to travel to all EU member states. In early 2004, EU officials warned Washington of this scenario as new member states were about to join. As tensions peaked at the end of 2004, the Bush Administration developed a 'Road Map' initiative to clarify requirements for joining the Visa Waiver Program and 13 countries seeking admission joined the process, including new EU member states and future EU member states, Bulgaria and Romania. In June 2005, the EU took the nuclear option off the table by amending the solidarity mechanism (European Council 2005) and instituting a requirement whereby the European Commission issues regular progress reports on visa reciprocity by third countries, like the US. Lack of progress can be grounds for imposing visa restrictions. For example, if the Visa Waiver Program did not expand to at least some new member states by the end of 2008, the EU promised to impose temporary visa requirements on US nationals holding diplomatic and official passports.

Visa Waiver Program reform

Congress eventually struck a political compromise between the two extremes of eliminating the Visa Waiver Program and adding new states to the existing program by opting to reform the program with Section 711 of the *Implementing Recommendations of the 9/11 Commission Act of 2007*. The biggest obstacle to expanding program membership has been its 3 percent visa refusal rate requirement. The visa refusal rate is the percentage of visa applications from a country's nationals that are rejected by consular officers and it largely depends on officers' judgment of whether applicants are likely to comply with the terms of their visas. The *Implementing Recommendations of the 9/11 Commission Act of 2007* authorized the Secretary of Homeland Security to waive the 3 percent visa refusal requirement and accept countries with refusal rates of between 3–10 percent, thereby opening the door to several EU member states (see Table 9.1).

Congress initially required the 3 percent refusal rate in order to minimize the arrival of travelers who enter the US legally but overstayed their visas, as had been the case with an estimated 30–40 percent of the 11.7 million illegal migrants in the country (Passel et al. 2013). It would have made sense to

Table 9.1 Visa refusal rates of states joining "road map" initiative

Country	Visa refusal % per fiscal year		
	FY 2007	FY 2008	FY2009
Romania	37.7	25.0	26.3
Poland	25.2	13.8	13.5
Bulgaria	14.3	13.3	17.8
Lithuania	12.9	9.0	17.6
Slovakia	12.0	5.3	8.3
Latvia	11.8	8.3	19.5
Hungary	10.3	7.8	21.1
Czech Republic	6.7	5.2	6.9
South Korea	4.4	3.8	5.5
Estonia	4.0	3.9	6.2
Malta	2.7	2.5	3.8
Cyprus	1.8	1.7	1.4
Greece	1.6	1.5	2.0

Source: FY2007–2009 tables posted on US State Department (n.d.)

require a maximum visa overstay rate but exit data was deemed too inaccurate to calculate reliable overstay rates. Exit data are only collected from airline manifests and the traveler's I-94 arrival/departure card, with half of the card collected upon entry at passport controls, with the other half subsequently collected upon departure. All too often, I-94 forms have been lost or not properly entered into databases. Through the 'beyond the border' pilot program, Canadian border officials at certain border crossings have shared their entry data to serve as exit data from the US but no similar arrangement exists along the border with Mexico where there is no exit process and exit data is not collected. The DHS really does not know for sure who leaves the US and will not until it fully implements the exit capabilities of the automated biometric entry-exit system, US-VISIT.

This helps explain why Congress conditioned its authorization for DHS to admit countries with 3–10 percent visa refusal rates on several DHS actions and applicant state cooperation with the US on counterterrorism initiatives. The legislation specifically required: implementation of an Electronic System for Travel Authorization (ESTA), development of DHS capacity to verify the departure of those travelers who entered the US and information sharing agreements between the US and Visa Waiver Program members on "known and suspected terrorists," "preventing and combating serious crime" and "lost and stolen passports" (US Congress 2007).

Congress required the DHS to put in place an Electronic System for Travel Authorization (ESTA) similar to that used by Australia. ESTA requires travelers to submit biographical data found in their passports through a website at least 72 hours in advance of departure. Beginning in 2009 all Visa

Waiver Program travelers must use ESTA and those denied will be directed to apply for a visa at a US consulate. Initially, there was no automated system that informed airline staff issuing boarding passes for US-bound flights whether or not a traveler received authorization through ESTA to board but the verification system went live in 2010 and, in that year, airlines managed to comply the requirement to verify ESTA approval for almost 98 percent of passengers traveling under the VWP before boarding (GAO 2011).

Congressional authorization of Visa Waiver Program expansion also required that states with 3 to 10 percent visa refusal rates could only join the Visa Waiver Program after the DHS could certify the departure of 97 percent of international air travelers. In order for DHS to maintain this authority, the 9/11 Act further requires that departure of those who enter under the Visa Waiver Program is verified by biometric exit controls at airports by June 30, 2009 – a deadline that was not met. Once a biometric air exit process is in place, the 9/11 Act requires the DHS to set a maximum visa overstay rate for membership in the Visa Waiver Program.

The DHS considered three options for collecting biometric exit data: at airlines' departure check-in, at the Transportation Security Agency (TSA) security checkpoint or at the departure gate. The DHS initially decided in April 2008 to require airlines to collect biometrics but after intensive airline lobbying, Congress required that DHS retest its options. DHS then effectively abandoned plans to implement a biometric air exit process and focused on improving collection of biographic exit data from airline manifests. Members of Congress repeatedly called on DHS to complete biometric exit controls and, in 2013, several Senators made their support of Comprehensive Immigration Reform legislation conditional on the inclusion of provisions to make biometric exit data collection mandatory. They succeeded in getting a provision in the bill that passed in the Senate, which mandates biometric air exit capabilities at the ten US airports with the highest international travel volume (US Senate 2013, Section 3303).

When ESTA is combined with the requirement that Visa Waiver Program states share criminal data and terrorist information regarding their citizens, US authorities can grant visa-free travel on the basis of individual screening instead of a traveler's nationality. Information sharing agreements have been negotiated by the DHS and corresponding ministries of new Visa Waiver Program countries and, subsequently, with the balance of those countries previously in the program.

The texts of most agreements were not made public but some, such as the Preventing and Combating Serious Crime (PCSC) agreement between the US and Estonia, have. This agreement states that Estonia and the US may

> even without being requested to do so, supply the other Party's relevant national contact point ... with personal data .... (which) shall include, if available, surname, first names, ...date and place of birth, current and former nationalities, passport number, numbers from other identity

documents, and fingerprint data, as well as a description of any conviction or of the circumstances giving rise to the belief ... that the data
subject(s) ... will commit or has committed a serious criminal offense
... (or) ... will commit or has committed terrorist or terrorism related
offenses.

Serious criminal offense is defined as an offense "punishable by a maximum
deprivation of liberty of more than one year or a more serious penalty" and
"for the United States, serious crimes shall be deemed also to include any
criminal offense that would render an individual inadmissible to or remov-
able from the United States under US federal law" (US and Estonia n.d.).
For reference, a first offense driving under the influence (DUI) carries a
maximum penalty of one-year imprisonment in many US states. The agreement also enables US and Estonian national contact points to conduct
anonymous automated searches of each other's fingerprint and DNA databases. If a submitted biometric produces a "hit," additional personal data
may be supplied according to rules governing mutual legal assistance. DHS
has also indicated that the Visa Waiver Program reforms instituted uniform
requirements for all Visa Waiver Program countries; that new members
would not become second-class members. As of May 2011, 34 of the then 36
VWP countries had signed agreements on sharing information on lost and
stolen passports; although all 36 VWP countries shared data according to
INTERPOL (GAO 2011, 21). As of January 2013, all of the then 36 VWP
countries signed agreements with the DHS pursuant to Homeland Security
Presidential Directive 6 (HSPD 6) on sharing information regarding known
and suspected terrorists (Siskin 2013, 14). DHS reports on a webpage that
states it was last updated November 1, 2013 that it "has completed Preventing
and Combating Serious Crime (PCSC) Agreements, or their equivalent with
35 Visa Waiver Program (VWP) countries and two additional countries to
share biographic and biometric information about potential terrorists and
serious criminals" (DHS n.d.).

EU response to Visa Waiver Program reforms

Visa Waiver Program applicants eagerly began to negotiate the Memoranda
of Understanding (MOUs) on counterterrorism cooperation required by
the 9/11 Act. However, the Polish Ambassador to the US was disappointed
with the reforms and called the visa refusal rate "an arbitrary and inflex-
ible standard" and suggested that it would be possible to lower the visa
refusal rate if "the rules that American consuls have to follow in granting
visas" were "rethought" (Reiter 2007). Poland's high visa refusal rate is also
caused by many visa applications (and rejections) from several poor rural
regions of the country and campaigns to discourage such fruitless applications were launched (Iglicka, 2008). The argument was also made that the
visa refusal rate is not an accurate proxy for actual visa overstay data but

US policymakers were unlikely to change program membership criteria until US-VISIT exit was implemented.

After the European Commission received draft DHS counterterrorism cooperation MOUs, it declared them unacceptable because they contained elements of EU responsibility such as ESTA and enhanced travel document standards. The US and EU then agreed to take a two-track approach to the agreements – EU and bilateral (European Commission 2008a, 8–9). Agreements on criminal data and terrorist information sharing would be handled bilaterally. Information sharing with respect to lost and stolen travel documents would be handled by having all states concerned contribute to INTERPOL's database.

This cleared the way for visa waiver status to go into effect on November 17, 2008 for the Czech Republic, Estonia, Latvia, Lithuania, Hungary, South Korea and Slovakia. The addition of Malta to the Visa Waiver Program became effective on December 30, 2008. Greece was the only country nominated by the State Department for membership and had a visa refusal rate below 3 percent but then subsequently gained membership in 2010. In any event, with the addition of seven new EU member states to the program, EU threats to impose temporary visa requirements on US nationals holding diplomatic and official passports did not materialize. Still, the European Commission must decide whether ESTA constitutes a visa. The European Commission made a preliminary determination that ESTA is not tantamount to a Schengen visa process (European Commission, 2008b) but a final determination is contingent on a final rule implementing ESTA, which has yet to be issued by DHS. After the DHS issued an interim final rule in 2010 that imposed a $14 fee, the European Commission sent written comments, to which the US did not reply as of November 2012 when the Commission issued its report on "visa requirements in breach of the principle of reciprocity" (European Commission 2011, 12). This report noted that the Commission continued to raise the issue of non-reciprocity in a series of EU–US meetings, lauded President Obama for saying in 2010 that he is "committed to make the accession of Member States to the VWP a priority, to be solved during his presidency" and welcomed the introduction of various pieces of legislation that would expand the VWP. The Commission also made clear that until the ESTA final rule is issued, maintaining transatlantic visa-free travel remains far from certain.

The reciprocity mechanism was revised with a December 2013 regulation (European Council 2013) that, as explained in the 2014 Commission report on visa reciprocity, "provides for a quicker and more efficient reaction in case a third country on the positive list introduces or maintains a visa requirement for one or more Member States" (European Commission 2014, 2). The 2014 report recapitulated the problem of non-reciprocity on the part of the US toward Bulgaria, Cyprus, Poland and Romania at the time of its previous report as well as toward new EU member state, Croatia. The report commented on the inflexibility of the 3 percent visa refusal rate and noted EU member state refusal rates for fiscal year 2013 being: "4% for Cyprus, 5.9%

for Croatia, 10.8% for Poland, 11.5% for Romania, and 19.9% for Bulgaria"
(European Commission 2014, 11). The report goes on to explain that the
complexity of the US visa system leads to many refused applications that
are subsequently refilled and approved and recounts a US State Department
agreement to provide better information about visa requirements and better
guidance on application processes so as to reduce the number of applications
filed that have little hope for being approved. At the end of the section on US
non-reciprocity, the report states

> the Commission has again requested information from the US authorities
> on the date of publication of the Final ESTA Rule, in view of completing
> its assessment of whether or not the ESTA system is equivalent to the
> Schengen visa application procedures. The US informed that it would be
> published in the coming months.

Given that the US has not issued a final rule on ESTA in over four years, it
appears that both sides seem to be content with operating under the existing interim rule so as to not provoke a confrontation over the ESTA fee
(Commission 2014, 12).

Future expansion of the Visa Waiver Program?

It is unlikely that there will be much more growth in Visa Waiver Program
membership. DHS authority to expand the Visa Waiver Program to states
with 3 to 10 percent visa refusal rates lapsed on July 1, 2009 when the biometric air exit capabilities of US-VISIT were not in place. Given that the DHS
has not issued a plan to collect exit data collection mandated by Visa Waiver
Program expansion legislation and once it does, system implementation will
probably take considerable time, it is unlikely that Poland, Romania and
Bulgaria will be included in the Visa Waiver Program in the near future, even
if they achieve a visa refusal rate below 10 percent.

This raises the question, if the systems mandated to make the Visa Waiver
Program secure are not in place, should those countries that were admitted
in anticipation of expected system deployment be removed? Although here is
a visa refusal rate it only includes those nationals of VWP countries whose
terms of travel or their particular status require them to apply. For example,
the refusal rate of UK nationals who applied for B visas was 16.9 percent in
2013.[2] Visa overstay statistics are not made public and their accuracy remains
doubtful. It would, therefore, be very difficult, politically speaking, to re-
impose visa requirements. In the end, those seven EU member states that
managed to get into the Visa Waiver Program in 2008 will remain in the
program, while those that were excluded will most likely remain outside.

Twenty-two former presidents, foreign and defense ministers argued in a
July 15, 2009 "Open Letter to the Obama Administration from Central and
Eastern Europe" that

It is absurd that Poland and Romania – arguably the two biggest and most pro-American states in the CEE region, which are making substantial contributions in Iraq and Afghanistan – have not yet been brought into the visa waiver program. It is incomprehensible that a critic like the French anti-globalization activist Jose Bove does not require a visa for the United States but former Solidarity activist and Nobel Peace prize-winner Lech Walesa does. This issue will be resolved only if it is made a political priority by the President of the United States.

(*Gazeta Wyborcza* 2009)

In line with these sentiments, Representative Mike Quigley (Democrat-Illinois) introduced legislation to extend DHS authority to waive the 3 percent visa refusal requirement for two more years as "part of a broader strategy to ultimately extend visa waiver privileges to Poland" (*Polish News* 2009). Comprehensive immigration reform legislation passed by the Senate in July 2013 subsequently included provisions loosening existing Visa Waiver Program member criteria along the lines of these bills, although not mentioning any country by name (US Senate 2013, Section 4506). The Senate bill or similar comprehensive immigration reform legislation failed to be introduced in the House of Representatives and by July 2014 Senate sponsors and President Obama essentially gave up on pushing for the Senate bill in the face of Speaker of House John Boehner's refusal to introduce the legislation without a majority of the Republican members in favor of it.

As reports of Europeans traveling to Syria to fight against the Assad regime and joining the Islamic State (also referred to as ISIS) increased over the course of 2014, members of Congress began to raise concerns about ISIS and terrorist groups exploiting the Visa Waiver Program and in January 2015 Rep. Candice Miller introduced the Visa Waiver Program Improvement Act of 2015, which if enacted, would allow the Secretary of the Department of Homeland Security to immediately suspend a country's participation in the VWP if the country fails to provide the United States with pertinent traveler information related to security threats. Other members of Congress also called for ending the Visa Waiver Program but no legislation has made significant progress toward enactment. Proponents of expanding the Visa Waiver Program to include Poland and other key allies made headway when provisions were included in Fiscal Year 2016 Homeland Security appropriations legislation passed by the Senate Appropriations Committee in June 2015; however, similar legislation has not yet been adopted by the House. It remains to be seen whether the 2008 expansion of the Visa Waiver Program will be enough 'progress' for the EU to maintain the political equilibrium in US–EU relations on this issue or if excluded EU member states, most notably Poland, press the issue of visa policy inequality in the EU to the extent that the EU reverts to threatening restrictions on visa-free travel for certain US nationals.

Political compromises and changing international norms

Congressional compromises that pitted border security and immigration law enforcement against foreign policy considerations and travel facilitation produced very complicated requirements for Visa Waiver Program membership. These complicated legislative requirements necessitated that generalist diplomats move aside for direct negotiations between interior ministries on increasingly technical matters such as the signing of MOUs on antiterrorism cooperation and then bilateral information sharing agreements between the DHS and interior ministries in Visa Waiver Program countries.

Information sharing agreements between the DHS and European interior ministries may be efficacious for screening travelers; however, they raise broader social questions with political and legal repercussions. The signing of agreements to share citizens' data with the DHS has become a concern of civil libertarians and data privacy advocates in EU member states. If the EU deploys its planned electronic travel authorization system, the same may increasingly happen in the US. US citizens traveling to Europe will then become increasingly aware of the information sharing agreements that give EU member states access to the personal data of each US citizen authorities believe "will commit or has committed a serious criminal offense …(or)…will commit or has committed terrorist or terrorism related offenses" (US and Estonia, n.d.).

While the negotiation of information sharing agreements between the DHS and counterpart interior ministries may seem peripheral within the broader scope of international relations, their consequences may be much greater than expected. Through such information sharing, states are de facto becoming obligated to determine whether their citizens are fit to travel internationally. This reverses the norms established ninety years ago when League of Nations member states decided that states receiving travelers are responsible for investigating credentials and making entry decisions. While traveler-sending states might not deny a passport to a citizen based on a past criminal conviction, they will now share information with receiving states that, in practice, may have the same effect. Such information sharing may facilitate terrorist screening by all states involved but it also means that governments might be able to make it very difficult for some of their citizens to travel abroad. This, in turn, increases the possibilities for misunderstandings, misuse and mischief.

Looking forward

It is very unlikely that Congress would vote to eliminate the Visa Waiver Program, unless, of course, someone who travels to the US visa-free successfully executes a major terrorist attack. It is also unlikely that the Visa Waiver Program will be expanded to countries regardless of high visa refusal rates given that President Bush failed to get Iraq War allies Poland and Romania in the program; such foreign policy considerations resonate even less in the Obama Administration, and skeptics of Visa Waiver Program expansion, like Senators Dianne Feinstein (Democrat-California) hold key committee leadership positions. If comprehensive immigration reform legislation such as that passed by the Senate with looser Visa Waiver Program criteria were to be enacted, then Poland and other excluded EU member states may have another chance at joining the program; however prospects for enactment have faded and it is unlikely that any immigration legislation will become law before the 2016 elections. As it stands, the Obama Administration and Congress will have little choice but to navigate within the policy parameters of the political compromise embodied in the Visa Waiver Program reforms of the 9/11 Act.

Although few realistic policy options remain, the Obama Administration can follow through on past legislation and fully implement ESTA and US-VISIT air exit as soon as possible. Rather than wasting any more time considering other approaches, the DHS should decide to collect travelers' biometric exit data at departure gates of the 80 airports with direct international flights rather than at the check-in counters or TSA checkpoints of over 400 airports from which travelers could take connecting flights out of the country. Exit data collection at departure gates may cost taxpayers more in terms of increasing Customs and Border Protection inspections staff and rebuilding gate areas but it minimizes disruption of domestic flight operations and provides more certainty that those individuals whose biometric data was entered into US-VISIT actually boarded the departing aircraft. Given the potential for significantly greater costs, Congress will need to considerably increase DHS appropriations for several years to implement the departure gate option. If members of Congress mean what they say about the necessity of collecting biometric exit data, then they must be willing to raise the revenue necessary to pay for it. Once this is accomplished, the administration could set a reasonable maximum overstay rate, expand membership to those countries that meet this criteria and publish the visa overstay rates of all countries to better justify the exclusion of those states that fail to meet the bar. Only then will the US be on solid ground in negotiating visa-free travel with the EU and explaining its visa policies to the rest of the world.

Notes

1 This chapter is a revised version of an article entitled "Visa Policy, Security and Transatlantic Relations," that was published in *Studia Diplomatica.* The research upon which this paper is based was made possible by a fellowship of the Transatlantic Academy in Washington, DC.
2 Source: FY 2013 table posted on US State Department (n.d.).

Bibliography

DHS n.d. *International Engagement Results*. Department of Homeland Security. Available from: www.dhs.gov/international-engagement-results.

European Commission 2008a. Fourth Report from the Commission to the Council and the European Parliament on Certain Third Countries' Maintenance of Visa Requirements. *Commission of the European Communities. COM(2008) 486 final/2*, Brussels, September 9.

European Commission 2008b. The U.S. Electronic System for Travel Authorization (ESTA). *Commission Staff Working Document*. SEC(2008) 2991 final, Brussels December 2. Available at: www.docser.com/file/commission-of-the-european-communities-statewatch-05d9ec4d.

European Commission 2011. Seventh Report on Certain Third Countries' Maintenance of Visa Requirements in Breach of the Principle of Reciprocity. *Report From the Commission to the European Parliament and the Council, COM(2012) 681 final*. Brussels, 26 November.

European Commission 2014. Report from the Commission Assessing the Situation of Non-Reciprocity With Certain Third Countries in the Area of Visa Policy. *C(2014) 7218 final*, Brussels, 10 October.

European Council 2005. Council Regulation (EC) No. 851/2005 of 2 June 2005 amending Regulation (EC) No. 539/2001. *Official Journal of the European Union* L 141/3. June 4.

European Council 2013. Council Regulation (EC) No. 1289/2013 of 11 December 2013 amending Council Regulation (EC) 539/2001 *Official Journal of the European Union* L 347, December 20.

Faull, J. 2004. Fortress America? The Implications of Homeland Security on Transatlantic Relations. *Panel discussion*, American Enterprise Institute, March 4.

GAO 2002. *Implications of Eliminating the Visa Waiver Program*. United States Government Accountability Office, GAO-3-38.

GAO 2011. *Visa Waiver Program: DHS has Implemented Electronic System for Travel Authorization, but Further Steps Needed to Address Potential Program Risks*, United States Government Accountability Office, GAO-11-335, May.

Gazeta Wyborcza 2009. Open Letter to the Obama Administration from Central and Eastern Europe. July 15.

Iglicka, K. 2008. *U.S. Visas: Myths, Facts, Recommendations*. Center for International Relations, Warsaw. May.

Kamen, A. 2004. Turning a Photo Op into a Lobbying Op. *Washington Post*, January 28.

Keohane R. O. and Nye, J. S. 1974. Transgovernmental Relations and International Organizations. *World Politics*, Vol. 27, No. 1, pp. 39–62.

Leiken, R. 2005. Europe's Angry Muslims. *Foreign Affairs*. Vol. 84, No. 4, pp. 120–135

Lloyd, M. 2003. *The Passport: The History of Man's Most Travelled Document*. Stroud, UK: Sutton.

McConnell, M. 2007. *Hearing on Confronting the Terrorist Threat to the Homeland Six Years after 9/11*. Senate Committee on Homeland Security and Governmental Affairs, September 10.

Passel, J., D'Vera Cohn, S. and Gonzalez-Barrera, A. 2013. *Population Decline of Unauthorized Immigrants Stalls, May Have Reversed*. Pew Research Center, September 23.

Polish News 2009. *Rep. Quigley Introduces Bill to Extend Visa Waiver Program.* June 23.

Reiter, J. 2007. The Visa Barrier. *Washington Post*, August 29.

Salter, M. B. 2003. *Rights of Passage: The Passport in International Relations.* Boulder CO: Lynne Rienner.

Sikorski, R. 2004. Fortress America? The Implications of Homeland Security on Transatlantic Relations. *Panel discussion.* American Enterprise Institute, March 4.

Siskin, A. 2013. *Visa Waiver Program*, CRS Report for Congress, Congressional Research Service, 7-5700, January 15.

Slaughter, A.-M. 2005. *A New World Order.* Princeton: Princeton University Press.

Stringer, K. D. 2004. The Visa Dimension of Diplomacy. *Discussion Papers In Diplomacy, No. 91.* Netherlands Institute of International Relations (Clingendael).

US and Estonia n.d. *Agreement between the Government of the United States of America and the Government of the Republic of Estonia on Enhancing Cooperation in Preventing Serious Crime.* Available from:www.dhs.gov/xlibrary/assets/agreement_usestonia_seriouscrime.pdf.

US Congress 2007. *Implementing Recommendations of the 9/11 Commission Act of 2007,* 110th Congress, Public Law 110-53.

US Senate 2006. *Comprehensive Immigration Reform Act of 2006,* S. 2611, Title IV, Section 413.

US Senate 2013. *Border Security, Economic Opportunity, and Immigration Modernization Act*, S. 744.

US State Department n.d. *Calculation of the Adjusted Visa Refusal Rate for Tourist and Business travelers under the Guidelines of the Visa Waiver Program.* Available from: http://travel.state.gov/content/dam/visas/Statistics/Non-Immigrant-Statistics/refusalratelanguage.pdf.

10 Judging borders

Expanding the Canada–United States border through legal decision making

Joshua Labove

Introduction

The law and borders are strange bedfellows. The law likes certainties and works to produce clear processes and decisions that promote legal clarity. Borders, on the other hand, particularly in the post-9/11 security environment, thrive in more ambiguous settings where jurisdiction and even location are not as obvious, but are open to manipulation and retooling. When scholars speak of borders, it is often that ambiguity which creates difficulty for the way we theorize how borders work and how they produce legal realities. It is helpful then to look at borders not as static but instead as a process, developed and retooled through legal decision making. This chapters aims to re-center our discussion of borders – and indeed, how far beyond the edges of territory those lines can be drawn – by identifying border-makers in judges and legal actors who hear cases at the border. While the Supreme Court of Canada has been largely silent on refugee appeals to the Safe Third Country Agreement (STCA), this silence, along with decisions in the Federal Court and Federal Court of Appeals, reveal bordering at work. This chapter intends to investigate that work – the drawing at times of new, larger borders. I suggest that much can be learned from the mechanics of the law itself, which privileges establishing clarity and setting up relational positions. Drawing upon research into bracketing and performativity, I argue here that understanding the way judges produce legal meaning at the border offers valuable insights toward the expansion and creation of new borders in North America.

This research extends previous work done to critically consider the expansion of the Canadian border by looking specifically at the role of Canadian courts and judges in the production and expansion of borders and bordering processes. Benefiting from scholars who have examined the formulation of border policy, here I consider how borders are worked into and out of the Canadian legal system, and the tools, notably the *Canadian Charter of Rights and Freedoms*, that make such legal framings possible. This is not to say that borders are uniformly spaces of rights, but that as the court system in Canada has been remade through *Charter* appeals, rights, and the vocabulary of rights jurisprudence become a useful mechanism through which borders can be expanded.

When one looks at the expanding use of remote detention, safe third country agreements, and strange manipulations of geography to particular legal ends, we quickly envision a vector moving in one direction – borders are becoming "more diffuse" (Côté-Boucher 2002). When we pull the law back in, however, we see spaces and practices in constant tension and reinvention operating at a range of scales. It produces "'detached geographies' through which detainees are spatially separated from the services that guarantee their rights" in detention centers where migrants and asylum seekers wait, and it constantly reimagines the legal limits of bordering (Martin and Mitchelson 2009, 466). What all of these re-workings of space tell us is that law matters. They become material expressions to change our relationship to state power, space, and to the law itself.

Rather, borders are constantly in a productive tension that defines how border work is carried out at the edge of the nation-state and far beyond. New securitization demands have reinforced the need for borders to be 'extralegal,' but they still need to be sites of state power. Rather than simply speak of borders as 'exceptional,' we should look at the way law – with its fixation on relational effects and producing clarity – is instrumental in bordering. Those creative tensions acknowledge the liminality, the 'inbetweeness' of bordering, and point us to how that liminality is constantly performed. For border studies, law can speak to the 'how' of bordering and unravel some of the ways borders function, particularly as the post-9/11 era frequently yields a confounding geography of new borders of security and economy.

I argue here that as border scholars we need to take law seriously and see the questions judges and lawyers ask as producing border spaces. To do this, I take from Blomley (2014) the suggestion that central to understanding the way the law works is to understand the law's ability to engage in bracketing. Bracketing "entails the attempt to stabilize and fix a boundary within which interactions take place more or less independently of their surrounding context" (Blomley 2014). Investigating what gets bracketed, from what else, and how, is important to understanding how the law is productive of bordering. Lastly, through these brackets, we become aware of scale and see how, through new legal tools and bracketing, bordering today occurs across a multiplicity of scales and no longer can be seen as strictly a practice at the level of the nation-state. Before we see bordering as increasingly remote, and expanding beyond the territorial limits of the nation-state, we can see through courts and judges how bordering is deeply embedded in more internal questions of law and jurisdiction and across a range of scales.

This chapter proceeds first by offering a case study, *Canada Council for Refugees v. Canada*, where interveners launched a legal appeal of the Safe Third Country Agreement on grounds the Agreement violated pre-existing international commitments such as the Refugee Convention as well as domestic law through the *Charter of Rights and Freedoms*. After a discussion of the case and the arguments undergirding the decision, this piece considers the way the *Charter* is deployed as a mechanism through which borders can be made and expanded. Finally, this chapter concludes by suggesting a research agenda for law at the border, bringing socio-legal studies in closer conversation with border studies in the hopes of mapping the material and real expansion of bordering practices through subtle steps in the courts.

Border expansion in the courts: the STCA and *Canada Council for Refugees v. Canada*

The legislative basis for the designation of a country as 'safe' for the purposes of refugee resettlement and adjudication began quietly with the 1988 amendments to the 1976 Immigration Act, then known as Bill C-55. As Bourbeau (2011) and others have noted, Bill C-55 begins a consistent trajectory of "securitization of migration" coupled with immigration strategies and approaches that challenge humanitarian access to resettlement (2011, 1). While no 'safe third country' was demarcated in 1988, Bill C-55 laid the groundwork for expanding the legal realm of 'Canada' for refugee applicants. A refugee who enters Canada via a safe third country, for example, would be "denied an opportunity to claim in Canada" automatically (Canada Council for Refugees v. Canada, 2007 FC 1262). Functionally, this serves to expand Canada's borders for specific, however some of the neediest migrants by foreclosing any asylum adjudication from a country deemed safe or on behalf of an individual transiting through a safe country. In the decade since Canada officially signed the Safe Third Country Agreement with the United States, academics and legal practitioners have called attention to the blind-spots of such policy, the potentially unintended but no less significant and real consequences of limiting access to the asylum adjudication process. Notably Arbel and Brenner (2013) exhaustively document the deficiencies of the STCA to protect those who seek refuge in Canada, advancing discourse of the border as a security nexus, conflating humanitarian migration with larger geopolitical worries of safety and economy in a post-9/11 world. Read alongside a range of new strategies to screen migrants further from ports of entry, the STCA is seen both as "pushing the border out," to a distinct legal end where "Canada seeks to avoid its legal obligations, and in so doing, weakens the legal protections available to asylum seekers under domestic and international legal instruments" (Arbel and Brenner 2013, 2).

Along with immigration practitioners, Arbel and Brenner have noted a range of international legal precedents that should govern Canada's commitments to refugees and should inform the way the border is administered for those seeking asylum. Here, however, I wish to focus on the specific and often tactical use of Canadian jurisprudence to refocus the shifting geographies of border work back within the jurisdiction of Canadian courts. The post-*Charter* era has seen courts willing to take on – and equipped with the tools to consider a broader range of Constitutional challenges – reframing the individual and, in turn, social groups beyond the state, as far more powerful than ever before. This has made for not only a far more substantial Supreme Court of Canada, as many have noted, but an equally empowered court system throughout as *Charter* challenges work their way through various provincial and federal venues. Following Sharpe and Roach (2005) the *Charter* provides the Court the latitude to consider issues "less constrained by strict legal principles, but also of greater significance to the average citizen than those relating to federalism" (2005, 25). This reimagining of the Court's reach has brought issues of bordering into Canadian jurisprudence and in so doing, re/performs the border through legal decision making that asserts Canadian sovereignty over an expansive (though not always transparent) constellation of spaces.

As Chief Justice Beverley McLachlin (1989) notes, "the Charter means that judges are called upon to answer questions they never dreamed they would have to face" and that answering means establishing practices of legal decision making that draw some material, relations, and facts in, while foreclosing others as beyond the legal process (1989, 579). As an objection to the Safe Country Agreement, *Canada Council for Refugees v. Canada* draws in a broad set of spaces and actors – Canada, the United States, our land borders, immigration policies in the United States, and that country's commitment to international pledges to *non-refoulement*. In order for legal decisions to be meaningful, however, the process through which decisions are rendered must still be narrow enough to specifically address the questions of the case. This balance is worked out through the construction of legal 'brackets,' productive of relationships between individuals, the state, and material space. These brackets reveal the way border work is pulled into the space(s) of Canadian courts and offer greater insight into the way bordering has expanded beyond the edges of the territory.

In *Canada Council for Refugees,* the deployment of *Charter* rights represents an opportunity to bring new spaces under the jurisdictional reach of Canadian courts. Interveners on behalf of John Doe, an asylum seeker from Colombia, seek redress from the Federal Court by calling into question the way the STCA is in conflict with pre-existing law and international obligations. John Doe entered the United States with his wife on a tourist visa in 2000. Removal proceedings began the following year, and at that time, Doe "submitted an application for asylum, and in the alternative, a withholding of removal based on fear of persecution" (Canada v. Canada Council for Refugees 2008, sec. 19). John Doe was informed not to approach the Canadian border and therefore did not make an application for refugee status within Canada.

Canada Council for Refugees is a case of standing, that is to say, who is entitled to bring issues before the Canadian courts, and whether a Colombian national who had not formally sought asylum from within Canada could bring an issue for redress before Canadian courts. If Mr. Doe does have standing, it would signal a wider interpretation of the limits of Canadian jurisprudence, itself a kind of bordering principle. If Mr. Doe would be seen to not have standing, he would have been caught in the interstices of the STCA and a newly fashioned North American security perimeter. While the 'expansion of the border' is often a complaint reserved for watchers of the post-9/11 security border, it is useful to note in the functioning and performance of law, border expansion goes both ways; the border can be seen as pushed out to already place Mr. Doe in safe territory, but it can also be read as a space of rights and so Mr. Doe's access to Canadian law from outside the country would constitute its own, however oppositional, border expansion.

Initially granting Mr. Doe and his interveners standing, the Federal Court said by taking up the case that the border can be extended, to bring into Canadian law those who are outside Canada. This allows us to see the expansion of the border to encompass Mr. Doe as built on a liberal, expansive understanding of rights. In the Federal Court decision, *Charter* rights could extend beyond the limits of the Canadian territory. Specifically, Mr. Doe and the intervening parties utilize Sections 7 and 15 of the *Charter,* in addition to a litany of international treaties and laws governing the rights of refugees. Section 7 provides for due process, both substantively and procedurally. Put simply, the court is equipped to protect 'life, liberty, and security' broadly conceived as facets of substantive due process and to provide equal protection under the law as a cornerstone of procedural due process. Section 15 provides for equality under the law. The court is charged then, with discerning how Mr. Doe's rights to due process and equal protection were violated and given that these are *Charter* rights not Constitutional rights from the United States, it is implied as well that these encroachments on Mr. Doe's rights happened within the jurisdiction of Canada and Canadian courts. At the same time, it is Mr. Doe's inability to physically access Canada that has been at the core of his rights grievance.

The case then hinges the way the STCA produces "a lack of discretion for a Canadian immigration officer" (2007, sec. 291). The STCA forecloses "fundamental principles of justice" by potentially returning asylum applicants to the United States and by not allowing applicants a chance to present a case for protection within Canada. The automatic nature of the STCA – whereby all but those who meet "enumerated exceptions" are to be returned to the United States – threatens an individual's security as well as protection under the law (Canada v. Canada Council for Refugees 2008, sec. 102.12). In launching a Section 15 *Charter* claim, Mr. Doe and his interveners note that his inability to access an asylum hearing is also discriminatory and that Colombian nationals would have been particularly adversely affected by such policy. Section 7 asks if there is equal and fair access to the law and Section 15 asks if that law disproportionately affects certain racial, ethnic, or religious groups.

While the Federal Court initially agreed with the appellant, the government appealed to the Federal Court of Appeal which objected to the idea that Mr. Doe had standing within Canadian courts. In this way, justices at the Appeals court roll the border back: "John Doe never presented himself at the Canadian border and therefore never requested a determination regarding his eligibility" (Canada v. Canada Council for Refugees 2008, sec. 102). In demanding that Mr. Doe reach the physical border, the justices roll up the border of legal protection, but extend the border as a security device, considering Mr. Doe within a 'safe' North American space.

Legal decision making as border work

This work of drawing lines and determining whether Mr. Doe and countless others fall in or out takes the border beyond its thing-ness and instead, as a series of legal determinations, constantly open to renegotiation and change. Considering how judges make decisions is helpful to understand how borders have grown and shifted over time. Decisions like *Canadian Council of Refugees* can be seen as pivot points through which the border is stretched, pulled, twisted, and moved.

How is a border to one judge static, but to another shifting? As Blomley (2014) notes, the law is invested in a practice of disentanglement, that is, separating, "carving off a distinctive realm unsullied by factionalism, that promises clarity and determinacy" (135). Cases heard in a court are inevitably entangled to a host of relations and knowledges. Blomley suggests that "for a legal transaction to occur, a space must be marked out within which a subject, object, and set of relations specified as legally consequential are bracketed, and detached from entanglements" (136). We have to bracket off the non-law in order to make decisions that are not merely descriptive of our world, but as Bourdieu (1986) notes, productive. Much as territorial boundaries represent the performance of geographic knowledge production, these brackets of what will be said, heard, and decided upon represent the performance of legal knowledge production, where the marking out of law and the rendering of legal opinion represents a productive "symbolic power" (1986, 868). The law proclaims, names, and, in its utterances and pronouncements, creates. Some relations and entanglements are severed and bracketed off, though not consistently. Bracketing demands particular readings of our relational position to the border and to state power. What are the grounds by which we seek legal action? Much as borders attempt a legal bracketing between the here and there, the brackets that undergird legal discourse have a kind of bordering effect – they do not simply describe but from their bracketing they derive the productive power to effectuate the border and determine who and what is bordered. In this way, the brackets that frame legal discourse become viewfinders – lenses through which borders are brought into practice. Brackets emerge all over legal proceedings: as admissible evidence, as case law that can be considered, as what constitutes a legal precedent. It is impossible to try to find all the brackets made to produce borders, but useful instead to consider how a discourse of rights informs the legal practice in making brackets. As routes to court, *Charter* claims are bracketed and become the venue through which substantive understanding and change occurs at the Canadian border.

The *Charter* becomes instrumental to conceptualizing the border as something that can be legally expanded, contracted, bent, or made diffuse. The *Charter* is the legal frame through which 'Canada' is conceptualized and legally articulated, and so cases at the border that invoke a *Charter* claim become opportunities to (re)consider the limits of the *Charter* as brackets to Canadian jurisprudence. If the law is making borders, the border is in part made out of the text of the *Charter of Rights and Freedoms* – where disagreements in the application of the *Charter* can serve to expand or contract the legal scope of Canada. Rights settle disagreements much as they are the basis for disagreements in liberal societies. We use rights speak as a way to coalesce state power around "certain issues worthy of special protection" (Blomley and Pratt 2001, 152). The mechanics of rights – the way rights work to arbitrate disagreements – can go a long way to helping us understand the way rights have an important, though at times quite complex, role in making space. The space of rights is not simply the space of Canadian state; in fact, the space of rights is continually expanding and contracting and evolving geography.

The way the law makes use of framing or bracketing techniques to make decisions and order within specific and often narrow constructions helps us see law as an effect, that is to say, that law is effectuated and made through a series of practices, approaches, vocabularies, and mechanisms that collectively reassert the existence of the law at the same time they invest in the law the power to arbitrate disputes. Borrowing from Mitchell's (1991) concept of the state as effect, I suggest that we can think about law similarly – as neither an autonomous institution nor a vague, amorphous quality. Instead, in thinking about the law as an effect, we accept that there are real brackets not simply within law but between law and society as well. As Blomley notes, the divide could be "itself produced as a bracket," a convenient partitioning that in turn produces law's productive power to create, organize, name, and make space (141). Blomley is clear to tell us, however, that 'effect' does not simply mean a "head trick we play on ourselves" (142). Rather, seeing the law as effect does not discount the law's real power, its ability to make in its declarations, and to call into being real political spaces.

Neither the brackets law makes, nor the decisions, are consistent, provoking fear and confusion from those caught in the borderlands and those who speak on their behalf. At one instance, a shifting border is utilized to pull the line of sovereign power back, to deny individuals access to due process. On other occasions, the same shifting border is deployed to hold broader sets of individuals outside the cartographic limits of the nation-state to increasingly punitive measures. To make matters more difficult, some cases reveal judges applying a static and shifting border at the same time. In *Charkaoui* (2007) the border undergoes seemingly unnatural "topological twists" to be read as a fixed line and a shifting concept (Allen 2011, 284). The border is at once shifting, culling non-citizen residents out from within the territory while at the same time a fixed, determinate line twinned with sovereign power.

Competing and new scales of border work

Coupled with static and shifting approaches to the border, new scales of bordering have opened up, further changing the way courts and border-crossers alike conceptualize the edge of the territory. For instance the STCA works on a transnational scale, in advance of and now alongside other measures in the *Beyond the Border Action Plan* that today speak of a 'security perimeter.' Shiprider agreements between British Columbia detachments of the Royal Canadian Mounted Police and United States authorities as well as more informal agreements between the Washington Department of Fish and Game and the Canada Border Services Agency (CBSA) reveal regional borders responding to unique issues and policy objectives for a particular geography and population. In the United States, for instance, the county has become an important new scale at which to locate immigration enforcement practices, through conscription of counties into a federal immigration-information sharing project entitled Secure Communities (Coleman 2009; Labove 2011). Somewhere, amidst all these new scales of border work, the stalwart national scale of bordering still endures. The physical symbol of the tension of border studies is, in fact, the border itself – on one hand tired and outmoded by new techniques and geographies that go within and beyond the border and "the devotion of unprecedented funds, energies, and technologies to border fortification on the other" (Brown 2010, 8).

These multiple scalings of border work matter, not the least because the law itself is so mired in its own scalar logic, where cases proceed up from the municipal to provincial to the federal, but as well because each scale brings with it unique demands of the border. Where a Washington–British Columbia border may be spoken of as a bridge, a corridor of economic activity, a federal border may well be thought of as a test site for new anti-terrorism legislation and labour market protection.

When border work is unhinged from the scale of the nation-state, we witness border expansion through added complexity. Rather than assume a whole excising of border work to remote locations, we should see these re-scalings of border work as instructive. When the border can emerge at new scales, it is devolved from the cartographic boundary through the complexity of trying to read a clear geography from multiple legal precedents. The expansion of post-9/11 border work, then, takes shape in an emergent geography of legal decisions and new policy architectures. The STCA, for instance, disrupts the static position of the border in refugee cases by dislocating American standards for the arbitration of refugee cases north to the Canadian border. Where before the STCA Canada and the United States would have made asylum determinations on their own grounds, the agreement demands Canada see United States' refugee laws as in legal practice the same as their own. While the *Immigration and Refugee Protection Act* (Government of Canada 2001) speaks of a Canadian border and the Anti-Terrorism Act expands a uniquely domestic idea of the border, bilateral agreements such as *Beyond the Border* create a transnational emergence of bordering. Each of these policy tools occurs simultaneously, and while judges may bracket off perceived external relational effects, most of us do hold such a privilege. We attempt to read order and jurisdiction out of at times competing and conflicting information – many laws operating to create a modern border that is unhinged from geography, able to go well within the territory and extend far beyond as well.

Conclusion: emergent geographies of bordering

A focus on legal decisions at the border produces a more complex geography of bordering. Rather than conceptualizing borders as lines on a map, we begin to see the unevenness through which border work is done and where border work is expanding. A new, emergent geography of borders is appearing as nation-states attempt to draw lines around certain kinds of persons and movement. Not all experience borders in the same way and indeed the expanding constellation of bordering points in and outside of the edges of a territory begin to reveal new borderings, new exclusions – from Canada among others.

While much has been made in the post-9/11 era of an expanding border – a practice by no means unique to Canada – far less has been said about the way the law effectuates such perceived expansions. By fixating our attention on law, we examine the unevenness through which borders emerge and expand beyond the cartographic boundary line. In particular, examining the way law functions at the border demands we 'disentangle' and de-familiarize the border. We have to, put simply, engage in 'verbing' the noun – where border-as-a-thing becomes border-as-a-process, "consequential condensation points where wider changes in state-making and the nature of citizenship are worked out on the ground" (Sparke 2006, 152). When we do this, we begin to recognize that bordering is not simply expanding or contracting, because bordering is less a thing that expands and more a bundle of practices that re-emerges across jurisdictions and scales. Making sense of the discretionary performative power of judges to make space and reconceive of borders through individual cases and decisions allows us to recognize the way borders do not simply expand, but are constantly open to reinterpretation and performance.

Bibliography

Allen, J. 2011. Topological Twists: Power's Shifting Geographies. *Dialogues in Human Geography.* Vol. 1, No.3, pp. 283–98.

Arbel, E., and Brenner, A. 2013. *Bordering on Failure: Canada-US Border Policy and the Politics of Refugee Exclusion. Harvard Immigration and Refugee Law Clinical Program.* Cambridge: Harvard Law School. http://ssrn.com/abstract=240854.

Blomley, N. K. 2014. Disentangling Law: The Practice of Bracketing. *Annual Review of Law and Social Science.* Vol. 10, No.1, pp. 133–48.

Blomley, N. K., and Pratt, G. 2001. Canada and the Political Geographies of Rights. *The Canadian Geographer/Le Géographe Canadien.* Vol. 45, No.1, pp. 151–66.

Bourbeau, P. 2011. *The Securitization of Migration. A Study of Movement and Order.* New York: Taylor & Francis.

Bourdieu, P. 1986. Force of Law: Toward a Sociology of the Juridical Field. *Hastings Law Journal* .Vol. 38, p. 868.

Brown, W. 2010. *Walled States, Waning Sovereignty.* Brooklyn, NY: Zone Books.

Canada Council for Refugees v. Canada. 2007 Federal Court. [2008] 3 FCR 606, 2007 FC 1262 (CanLII).

Canada v. Canada Council for Refugees. 2008 Federal Court of Appeal. 2008 FCA 40 (CanLII).

Charkaoui v. Canada. 2007 Supreme Court of Canada. [2007] 1 SCR 350, 2007 SCC 9 (CanLII).

Coleman, M. 2009. What Counts as the Politics and Practice of Security, and Where? Devolution and Immigrant Insecurity after 9/11. *Annals of the Association of American Geographers.* Vol. 99, No. 5, pp. 904–13.

Côté-Boucher, K. 2002. The Diffuse Border: Intelligence-Sharing, Control and Confinement along Canada's Smart Border. *Surveillance & Society.* Vol. 5, No. 2, pp. 143–4.

Government of Canada. 2001. *Immigration and Refugee Protection Act.* SC 2001, c 27. Ottawa: Parliament of Canada.

Labove, J. 2011. *On the Border: Discursive and Legal Foundations of Post-9/11 Immigration Enforcement.* MA Thesis, Dartmouth College, Hanover, NH.

Martin, L. L., and Mitchelson, M. Il. 2009. Geographies of Detention and Imprisonment: Interrogating Spatial Practices of Confinement, Discipline, Law, and State Power. *Geography Compas.* Vol. 3, No.1, pp. 459–77.

McLachlin, B. 1989. The Charter of Rights and Freedoms: A Judicial Perspective. *University of British Columbia Law Review.* Vol. 23, No 3, pp. 579–90.

Mitchell, T. 1991. The Limits of the State: Beyond Statist Approaches and Their Critics. *The American Political Science Review.* Vol. 85, No. 1, pp. 77–96.

Sharpe, R. J., and Roach, K. 2005. *The Charter of Rights and Freedoms.* Toronto: Irwin Law.

Sparke, M. 2006. A Neoliberal Nexus: Economy, Security and the Biopolitics of Citizenship on the Border. *Political Geography.* Vol. 85, No. 1, pp. 77–96.

11 Visas as technologies in the externalization of asylum management

The case of Canada's entry requirements for Mexican nationals

Liette Gilbert[1]

Introduction

A border, van Houtum (2010: 959) argues, "should first and foremost be understood as a process." Thus, van Houtum and Naerssen's (2002) dimensions of bordering, ordering and othering are particularly appropriate to understand the growing reliance of many liberal democracies on externalizing migration and asylum control and its tendencies to immobilize, illegalize, and dehumanize particular migrants. In recent years, the politics of bordering, ordering and othering have been central to the intense debate on immigration and refugee reform in Canada. Like other countries of the Global North, Canada has increasingly engaged in scare-mongering politics at the particular expense of migrants and refugees. Under the Conservative government, Canada's immigration and refugee regime has taken a clear turn towards restrictive legislation (e.g., imposing visas, restricting admission and citizenship), precarisation (expansion of temporary migrant workers programs, shrinking of social services), securitization (increased deportation and detention), and criminalization of migrants (through a politician-led public rhetoric of abuse and fraud).

Immigration and refugee reform has been at the center stage of recent legislative activities in Canada. The bordering, ordering and othering at play in the current migration reform climate is epitomized by former Canada's Minister of Citizenship, Immigration and Multiculturalism Jason Kenney's announcement of an entry visa requirement on Mexico and the Czech Republic in July 2009 – then two countries in the top sources of refugee applicants.

> The visa requirement I am announcing today will give us a greater ability to manage the flow of people into Canada and verify bona fides. By taking this important step towards reducing the burden on our refugee system, we will be better equipped to process genuine refugee claims faster.
>
> (Kenney in Citizenship and Immigration Canada, 2009a)

As the country's official gatekeeper, Kenney clearly reaffirms national sovereignty and border control management while normalizing an order against which othering is defined. Intended to quickly reduce the 'burden' of allegedly deceptive *mala fides* on the refugee system, the visa policy was literally imposed overnight. The urgency and the imperative obligation of the visa was rationalized by Conservative politicians through public allegations of 'fraud' and 'abuse' justifying the tightening national borders and reforming immigration and refugee laws. The inflammatory and polarizing public discourse particularly constructed Mexican nationals as a 'threat' to Canada's national borders and welfare system – despite 15 years of NAFTA partnership.

This chapter examines visa requirement as a technology used in the externalization of migration and asylum management in Canada. It argues that the imposition of a visa on Mexican nationals sought to deter refugee claimants through a set of practices reaffirming Canada's rebordering. As Casas et al. (2010: 76) state in the context of the European Union, if border policy

> is producing an intricate process of border externalization and a multiplication of 'bordering' instruments within the limits of territories, then the abyssal function of borders no longer occurs at the edge of a territorially defined social formation, but criss-crosses within and without that particular body politics generating a constant policing and reassertion of what is 'inside' and 'outside.'

The policing of Canada's visa requirement on Mexicans took multiple forms. First, the visa requirement was imposed in July 2009 as a direct strategy to remotely control the flow of Mexican arrivals in Canada. The visa policy enabled the control of the Canadian border some 3,500 kms away in Mexico City. Second, the visa restriction also revoked the exemption clause in the 2004 Canada–US Safe Third Country Agreement allowing a person who does not require a visa to travel to Canada but who required a visa to enter the US to claim refugee status at the land border ports of entry. The visa requirement removed this exemption previously used by Mexican nationals claiming refugee status in Canada via the United States more significantly in summer 2007 (Gilbert, 2013). These particular claimants were allegedly fleeing Homeland Security's immigration enforcement raids and the growing legislative attempts to control unauthorized migrants. Arriving directly from the US, Mexicans (who were then not required to have a visa to enter Canada) were allowed to file refugee claims under the exception of the 2004 Safe Third Country Agreement. Their claims were, however, rapidly constructed by media and Conservative politicians as using the refugee system to jump the immigration queue. The number of asylum claimants from Mexico had started to increase in the late 1990s but doubled from 2005 to 2007 (when 7,248 claims were made that year) and reached an all times high in 2008 (with 9,491 claims). Following the adoption of the visa policy, the number of claims from Mexico promptly dropped to 1,200 in 2010 and to as few as 84 in 2013 (Citizenship and Immigration Canada, 2014a).

Third, amidst the many immigration and refugee legislative changes, the Protecting Canada's Immigration System Act of 2012 gave the Minister of Citizenship and Immigration Jason Kenney the discretionary provision to create a list of so-called safe countries, officially known as designated country of origin, "that do not normally produce refugees, but do respect human rights and offer state protection" (Citizenship and Immigration Canada, 2012). From his office in Canada, the Minister's list serves as a "divisionary view of the world" preempting the refugee determination process (van Houtum 2010: 964). Mexico was placed on the list of designated countries of origin in February 2013 despite the deadly violence of a rampant narco-trade and forlorn 'war on drugs' and Canada's contradictory travel advisory generally recommending caution while traveling in Mexico due to high levels of criminality and violence in certain regions.

As part of a larger restrictive policy regime to control people from different locations, the visa, the resultant removal of exemption measure and the 'safe country' list serve to border, to order, and to other refugee claimants on an individual and collective/national basis through a re-spatialization of Canada's border and demarcating functions. This externalization of refugee control was implemented through the visa policy as well as through a public discourse constructing Mexican refugee claimants as 'fraudulent claimants' and 'system abusers,' which in turn was used to justify and normalize the need for the visa requirement.

The following section examines the externalization of border control and the tension between migration and the globalized economy, and the constitutive discourse of the 'illegality' attributed to Mexican migrants in North America. I then examine Canada's visa policy on Mexican travellers as an 'effective' policy and discursive instrument to stop the flow of refugee claimant following a border externalization logic. I also analyse how the conjuncture of visa policy, public discourse, and legislative reform erodes the rights of refugee claimants in Canada. The last section looks at how the current narco-violence in Mexico poses a particular challenge to the definition of refugee and Canada's designation of Mexico as a 'safe country.'

Externalization of migration and asylum control

The globalized economy has generated an unprecedented pool of migrants by favoring the internationalization of labor, yet unevenly enabling cross-border mobility. National immigration regimes have responded by favoring highly skilled migrants as permanent citizens and relying on temporary workers programs for many sectors of their economies. As a result, moving across borders has, for some, become a condition of their cosmopolitan lives and professional activities, expedited by the convenience of border preclearance registration programs. However, for others, moving across borders becomes an unsettled and even perilous journey through increasingly layered borders and policed territories.

According to United Nations population estimates, 232 million migrants were on the move worldwide in 2013. Of this number, 15.7 million were refugees, i.e., people seeking protection in another country (United Nations, 2013). Dauvergne (2008) estimates that, in 2005, as many as 50 million people worldwide were caught in a state of extralegality created by the contradictions of economic globalization, state sovereignty, and human rights. These are migrants who do not fit any categories of national immigration laws or international refugee agreements and yet are labeled 'illegal' rather than 'illegalized' through institutional and political processes (Bauder, 2014). Still, 'illegality' is associated with anyone entering a country in breach of immigration law or overstaying permitted stays in a country. Yet, as Dauvergne (2008: 16) argues, the term illegality "underscores a shift in perception regarding the moral worthiness of migrants" and creates a moral superiority for those perceiving immigration infringements as criminal activity. The divisive construction between 'legal' and 'illegal' has allowed governments, media, and neoconservative groups to reassert rhetorical and moral borders over migrants when national borders remain selectively porous to satisfy labor demands (Bigo, 2002; De Genova, 2004).

This perceived criminality is directly linked to migrants' unauthorized presence in a country and their assumed disregard for 'legal' immigration processes and queues. But as Carens (2010: 42) remarks,

> [t]here are almost no immigration lines for unskilled workers without close family ties to current citizens or residents. Most of those who settle as irregular migrants would have no possibility of getting in through any authorized channel. To say that they should stand in a line which does not exist or does not move is disingenuous.

As a result of the contentious relation between the unremitting flows of globalization and the intensified securitization of national borders by sovereign states, immigration and refugee laws have increasingly become the determination of those who must be turned away rather than those who are allowed to enter. Embedded in this logic of exclusion, a multitude of bordering technologies, actors, and operations have increasingly curtailed migrant mobilities.

The particular outsourcing of control practices outside of national territories represents an externalization of migration management (Casas et al., 2010; Triandafyllidou, 2014). Such arrangements range from visa policy, to extraterritorial detention centers, intensification of police and military presence (on land, air, and water), computerization of border monitoring (from biometrics to shared databases), intergovernmental collaborations and bilateral security agreements, and privatization of control sanctions (by transportation companies). The underlying goal of externalization has been to expand the policing of borders and various control mechanisms whether they be deterrence, interception, detention, or removal into new buffer zones. But contrary to externalizing technologies that seek to apprehend migrants entering a country or a neighboring territory, visa policy is a control strategy directly deployed in the country of origin.

Visas are required prior to departure in order to enter a country and as such they aimed at remotely filtering and controlling entries (Samers, 2004). The more restrictive the requirements are for granting visas, the more effective such a strategy will be at reducing arrivals. Visa requirements represent an explicit obstacle faced by refugees when having to flee their country of origin. Even if successfully granted, the visa expresses a conditional authorization for a limited duration and never constitutes a guarantee of entry into the granting country. Such determination rests at the port-of-entry with the border agency. Besides the visa granting only a temporary permission of entry into a host country, it is also an evidence (documented in some cases, unspoken in others) of intent to return to the home country. Overstaying a visa (whether exit strategies are in place or not) automatically result in being without status or so-called 'illegal' in the host country, thus exposed to removal orders.

The motivations for requiring a visa might differ across countries. As Samers (2004: 32) explains,

> [p]articular restrictive immigration policies are applied to specific countries on a case-by-case basis and are often the result of risk assessments based on the profile of both the socio-economic characteristics of their citizens (especially the stock of desirable labour market skills) and on whether there is likely to be a 'mass migration' from these countries.

The visa requirement did not stand alone in Canada's recent reorientation of his immigration regime to respond to economic efficiency and national security.

In 2008, the year preceding the imposition of the visa requirement for Mexican travelers and that of the highest number of Mexican refugee claimants, the number of temporary workers from Mexico in Canada was seven times higher than that of Mexican permanent residents. Temporary migrant worker programs have been extended from agriculture to construction and tourism. As a result, Canada has admitted more temporary foreign workers than permanent residents since 2006 (Citizenship and Immigration Canada, 2015).[2] These measures and trends result in diverting particular types of migrants away from the permanent status authorized by immigration law into temporary, precarious, and disposable working conditions. Canada has long been dependent on Mexican temporary agricultural workers. Numbers of Mexican workers in temporary agricultural programs has in fact doubled since 2005 to reach almost 42,000 in 2013 (Citizenship and Immigration Canada, 2015). Mexicans in Canada are thus preferred as temporary workers rather than as permanent citizens.

Indeed, temporary migrant/worker programs have been favored to mitigate the adverse effects of structural programs that federal and state domestic policies have failed to resolve both in Canada and Mexico (Hernández Velázquez et al., 2011).

Visa as technology in the externalization of asylum control

On July 13, 2009, Canada announced a visa requirement for Mexican travelers to Canada, effective the very next day. The Canadian government legitimized its decision to impose a visa condition on Mexicans given the rising number of refugee claims in recent years. The announcement of the visa requirement took many people by surprise but apparently Canada's Prime Minister Harper had discussed his policy intent with the Calderón administration in November 2008 (Studer Noguez, 2009). Canada's visa requirement for Mexican travellers was seen by many as a severe blow to fragile continental trilateralism, and more specifically to the "aspirations of Canada–Mexico strategic partnership" (Studer Noguez, 2009). The visa prerequisite significantly challenged the myth of seamless continental integration and accentuated the asymmetrical relationships among NAFTA partners. Until now, Mexico saw Canada as the more reasonable albeit less important partner. Bilateral Mexican–US trade relations have been successful but constantly strained by immigration issues specifically related to border enforcement and the significant unauthorized Mexican population living north of the border. Canada's sudden visa requirement for Mexicans generated a similar anxiety about Canadian border management politics, now extended from the border itself into the visa offices in Mexico. Literally overnight, the image of Canada in Mexico shifted from that of an open to an exclusionary society.

Initial reactions to the visa requirement from the Mexican government were restrained. Then President Felipe Calderón's response was rather subdued: "We deeply respect Canada's right to make decisions about its immigration system. However, I must share with you our regret for this series of events and decisions" (quoted in Valpy, 2009: A13). His timid response was accompanied by an acknowledgement of the surge in "illegitimate" refugee claims created by the operations of dubious "intermediary groups and organizations" (i.e., *polleros de cuello blanco* or white collar people smugglers) in Mexico exploiting Canada's slow refugee determination process (Valpy, 2009; Martínez, 2009). Although narco-violence was raging in parts of the countries, no mention of the correlation between the rise of refugee claims and the militarized drug war declared by the Calderón administration in 2006 was uttered.

Mexico has been the leading country presenting refugee claims from 2005 to 2009 – with a recognition rate varying from 28 percent (in 2006) to 8 percent (in 2009). By 2010, Mexico ranked 5th (with a recognition rate of 11 percent), dropping rank constantly to 7th in 2011 (17 percent), 19th in 2012 (19 percent), and 32nd in 2013 (18 percent) (Rehaag, 2014). Despite the public accusations of fraud and abuse, the acceptance or recognition rate for Mexicans nevertheless suggests a significant number of claims meet the definition of refugee. Yet, Canada insisted on the entry visa as the "sole method of control when we face a torrent of bogus applications" (Harper quoted in CampionSmith, 2010: 14).

Less than one month after the visa announcement, Canadian Prime Minister Harper participating in the annual North American Leaders' Summit in Guadalajara, or so-called "Three Amigos Summit," reiterated the need for the visa policy on the basis of the rise in "bogus claims" from Mexicans. Harper (quoted in Mayeda, 2009) stated,

> We are spending an enormous amount of money on bogus refugee claims ... This is not the fault of the government of Mexico. Let me be very clear about this. This is a problem with Canadian refugee law, which encourages bogus claims ... I hope our Parliament will take advantage of the attention that's been brought on this issue to deal with this problem.

Like his Minister Kenney, Harper adamantly maintained that claims and claimants from Mexico were illegitimate while recognizing the operational inefficiencies of the Canadian refugee system. In shifting the "fault" to the need for Canadian refugee law reform, the statement reads as a poor attempt to assuage stressed diplomatic relations between Canada and Mexico. Both Canadian and Mexican leaders avoided the issue of violence preferring instead to offer a paltry *mea-culpa* externalizing the need for action to institutions and organizations.

The new visa policy imposed on all travellers from Mexico meant that all Mexicans wanting to travel to Canada would first need to apply for a temporary resident visa before their departure. Canada's visa application requirements were particularly demanding. The visa requires a person to demonstrate that s/he would not overstay their time in Canada, have enough money to cover their stay, be in good health, have no criminal record, and not be considered a security risk to Canadians (Citizenship and Immigration Canada, 2009b). Applicants were asked to produce a panoply of documentation, including a passport, a letter from employer(s) granting a leave of absence (initially in English or French), recent pay slips, original bank documents showing financial history from the previous six months (eventually reduced to three months), evidence of assets in Mexico, and invitations where applicable, in addition to a processing fee (Canada Visa Application Centre, 2011). By comparison, the US visitor visa requires a valid passport, fee payment, travel information (purpose, dates), evidence of sufficient funds, and an interview in one of the 25 application centers (United States, 2014). For many applicants and observers, the documentation required by Canada was and is still judged as excessively onerous, and even simply impossible to assemble for many people. By refusal or attrition, the visa acts as a deterrence to travel to Canada (Ramírez Meda and Biderbost Moyano, 2011).

Moreover, there were serious geographical distances and related costs with the application. Visa applicants initially had to travel to Canada's Embassy in Mexico City until additional visa application centers opened in Monterrey and Guadalajara. Canada's visa application centers in Mexico are operated by London-based VSF Global (wholly-owned by Zurich based Kuoni Group, a "leading global travel and destination management service company" (Kuoni Destination Management, 2014)). VSF Global presents itself as "the world's largest outsourcing and technology services specialist for governments and diplomatic missions worldwide" (VSF Global, 2015). With 1,610 application centers operating in 120 countries for 45 'client governments,' the mobility management company has processed over 100 million visa and passport applications since 2001 (at the time of writing and the real-time counter on their website keeps tallying). Among other services, the company specializes in biometrics, data management, and security 'solutions' (VSF Global, 2015). VSF Global screen visa applications but decisions remain in the hands of Canadian visa officials. This externalization of visa processing exemplifies how border policy has become a terrain occupied by a myriad of new actors (Casas et al., 2010).

Many commentators considered the visa provision as more detrimental to the majority of legitimate Mexican travelers to Canada than to the smaller portion of alleged 'illegitimate' refugee claimants (Ramírez Meda and Biderbost Moyano, 2011; Gilbert, 2013). The undifferentiated construction of all Mexicans, whether refugee claimants or visitors, as a 'threat' to Canada's immigration and refugee system, social welfare, and national borders conveniently disregards the dominant economic discourse of Mexico as a close political friend and strategic trading partner. Since the visa imposition in 2009, Mexican officials have continually reiterated (with growing exasperation) their hopes for the removal of the visa requirement – to which Canada continually defended (with staunch assertiveness) that the visa requirement would stay as long as necessary for the country to reform its refugee system.

Reform to the Canadian refugee system came in December 2012 with the contested passing of the Protecting Canada's Immigration System Act (Bill C-31)[3] amending the not even fully in force Balanced Refugee Reform Act (Bill C-11). Among many other provisions, legislative changes sought to improve the refugee status determination system by establishing shorter timeframes for application and decision processes. Faster decisions meant alleviating a backlog of 60,000 applications and 18-month wait, removing unsuccessful claimants more rapidly, and therefore deterring applicants (Showler and Maytree, 2009). The legislation also gave full power to the Minister of Citizenship, Immigration, and Multiculturalism to compile a list of 'safe' countries, i.e. countries where human rights are respected, state protection is offered and which do not normally produce refugees. Claims from these 'designated countries of origin' would be considered but expedited and without appeal. This contested reform created a de-facto two-track asylum processing system.

Jason Kenney, the longest serving Minister of Citizenship, Immigration, and Multiculturalism from 2008 to 2013, initially identified 27 countries (the US, Croatia, and 25 EU members at the time)[4] on his first list of 'designated country of origin' in December 2012. In February 2013, Mexico was added to the list along with Australia, Iceland, Israel (excluding Gaza and the West Bank), Japan, New Zealand, Norway, South Korea, and Switzerland. Chile was added later in May 2013. The new Minister of Citizenship and Immigration, Chris Alexander (appointed in July 2013), added Romania and European city-states to the 'designated country of origin' list in October 2014 (Citizenship and Immigration Canada, 2014c).

Adding Mexico to a 'safe countries' list along with countries like Australia and New Zealand, Israel or Switzerland (to name only a few) raises significant question on the criteria used by the Minister to create such a list. In fact, criteria used to produce such list of 'safe countries' have never been transparent – another casualty of discretionary power. The full list includes other top source countries of asylum claims in recent years (e.g., Hungary, Croatia, and Czech Republic) which do not require an entry visa. Of the eight countries designated in February 2013 at the same time as Mexico, none appears in the top 50 sending countries of refugee claimants (Citizenship and Immigration Canada, 2014a). Economic considerations and political relations seem to influence the presence of some countries with whom Canada has, or was negotiating at the time, free trade agreements. It would be, after all, somewhat embarrassing to accept refugees from trading partners.

In designating Mexico as a 'safe country,' it was suggested and expected that Canada would eliminate the need for the visa. Over the years politicians have repeatedly indicated that the visa requirement for Mexican travelers could eventually be lifted with reforming immigration and refugee policy, yet they simultaneously dampened down expectations as they saw appropriate. It is now evident that the addition to Mexico on the 'safe country' list was not enough to reconsider the visa requirement. Recognizing Mexico as a designate country of origin has been, at best, vaguely understood as "a positive indicator" of an eventual elimination (Kenney quoted in Shane, 2013a). In the words of a government representative (cited in Shane, 2013b), "[t]he government of Canada will closely monitor the impact of the DCO [designated country of origin] list on claims, and will continue to consider all established criteria before making any changes to its visa policy."

Canada is postponing any decision on the visa until the implementation (and 'closed monitoring') of its imminent electronic travel authorization program for screening visa-exempt travellers as part of the security perimeter pact with the United States – which could be extended to Mexicans (Shane, 2013a). Meanwhile, Citizenship and Immigration Minister Chris Alexander announced, in May 2014, a new program to facilitate trade and travel with Mexico. Under the Can+ program, Mexican nationals who have travelled to Canada or the United States within the last 10 years would become eligible for expedited visa processing. The Citizenship and Immigration Canada website ironically presents the Can+ program as "yet another option to make it easier for Mexican travellers to come to Canada" (Citizenship and Immigration Canada, 2014b). Can+ appears to be the online version of the same visa application process, expedited if applications show recent travelling in Canada and in the United States.

Amnesty International (2009) has denounced the use of the visa requirement for Mexico as a tool to explicitly restrict access to refugee determination procedures. Amnesty International considers that it is not the personal prerogative of Conservative politicians but it is rather the role of the Immigration and Refugee Board of Canada to determine eligibility. Such view is echoed by the United Nations High Commissioner for Refugees. The High Commissioner, via his top representative in Canada, has expressed serious concerns about the preemptive use of a visa to reduce refugee claims and insisted on Canada's duty and responsibility to review individual claims in order to determine refugee status (quoted in Davis, 2009). The Canadian Council for Refugees (2009a) has also condemned the closing of the door on refugees from Mexico as an unfair attempt to undermine the legitimacy of claimants. The Council further argues that the 'designated country of origin' list actually violates the UN refugee Conventions as denying a right to appeal is an issue of access to justice. Moreover, the addition of Mexico to the list of designated countries of origin certainly is contradicted by abundant and constantly growing evidence of the lack of safety for many Mexicans. Mexico as a 'designated country of origin' denies the human rights crisis created by the violent drug wars and the erosion of the power of the state to protect and to govern (Estévez, 2012a; Wright, 2012).

The erosion of refugee rights

The externalization of migration and asylum policies privileges control and efficiency over the respect of fundamental rights notably the right to asylum. With the increasing trend to govern mobility as a matter of national security, marginalized migrants and refugees are targeted similarly even though their respective modalities of mobility and exclusion are different. Targeted by punitive policy and populist xenophobic rhetoric, many migrants and asylum seekers are constructed as criminals, leaving them especially vulnerable to precarity and scapegoating. For refugee or asylum claimants, vulnerability is played out right at their arrival in their demand for protection and their obligation to prove they would be persecuted if they returned home. However, in the nexus of criminality/security given priority by many governments, the presumption of fraud and abuse prevails until proof of persecution is officially sanctioned (Pratt, 2005; Dauvergne, 2008; Mountz, 2010).

Canada's imposition of visas as a response to the arrival of Mexican refugee claimants is an integral part of the externalizing practice of control and security because it preemptively controls mobility. It also extends a public discourse of criminality (i.e., 'bogus' and 'fraudulent') to all Mexicans whether they have submitted an asylum claim or not. Mexican refugee claimants have been perceived as a problem in Canada because the Conservative government has increasingly claimed that 'real' refugees are those waiting destitute in camps overseas. Mexican refugee claimants do not fit this profile and therefore they are seen by Canadian authorities as coming to take advantage of its refugee and welfare system. When talking to the Mexican president in Guadalajara in August 2009, Harper insisted that the visa requirement, although necessary to stop the flow of 'bogus' refugee claimants, had nothing to do with the Mexican government, but instead rested with problematic Canadian legislation. Harper and Kenney have both used this kind of double-speak when talking about the refugee system, switching from blaming the *mala fides* who damage the prospects of the *bona fides* to blaming the broken Canadian refugee determination system rendered inefficient by legislation intended to deter entry. Over the years, the Immigration and Refugee Board has often accumulated backlogs, but none as long lasting as the estimated 60,000 to 65,000 pending claims restraining the refugee determination process in 2009. The problem resides not solely at the entrance of the system, but rather in the rest of it because the Conservative government failed to fill vacancies on the board over time (Office of the Auditor General of Canada, 2009). Delays and inefficiencies have caused a massive backlog and a window of opportunity for the growing industry of ill-intended immigration/refugee consultants promising a way into Canada's system.

While Canada has, since its signing of the 1951 UN Refugee Convention in 1969, acquired an international reputation for taking refugees into its overseas sponsorship program selecting mainly from refugee camps, its inland process for individuals making a claim at a portofentry or at Citizenship and Immigration Canada offices has been more contentious because of structural inefficiencies. The opportunity created by the backlog seems, however, to disguise the Conservatives' contempt for the refugee law's consistency with the 1982 Canadian Charter of Rights and Freedoms and the subsequent 1985 *Singh v. Ministry of Employment and Immigration* landmark Supreme Court decision. Based on the Charter's Section 7, the Singh ruling held that foreign nationals on Canadian soil are protected by the provisions of the Charter, even if costs and delays required in providing a hearing to claimants and guaranteeing the provisions of the Charter are deemed administratively inefficient by governmental agencies (Kelley and Trebilock, 1998). The Singh decision actually led to the creation of the Immigration and Refugee Board in 1989, and entitles refugee claimants to an oral hearing in accordance with the principles of fundamental justice and international law. In other words, the Singh decision ensures the right of refugee claimants to liberty and security of their persons during the determination process of their claims by the Immigration and Refugee Board. Some Conservative politicians have consistently and openly expressed their dislike of the 1985 decision (Köhler, 2009). For them, the extension of the protection of Charter of Rights and Freedoms to refugee claimants results in excessive provisions, expenditures, and delays.

Even before being appointed Minister of Citizenship, Immigration, and Multiculturalism in 2006, Kenney had long condemned inland claims as 'bogus' refugee claims. From United States war resisters to Mexican refugee claimants, Kenney publicly accused alleged 'bogus' refugee claimants of threatening the integrity of the refugee and welfare systems in Canada and to be the reason for the chronic administrative backlog (CBC News, 2009). In a letter to the daily *National Post* shortly after the events of September 2001 in the United States, Kenney argued for the "overriding of the Singh decision" and the "detention of all undocumented arrivals until their identity is verified" (quoted in Köhler, 2009). Even then, Kenney defended the need for a refugee system that assists 'legitimate' refugees "rather than lawbreakers and queuejumpers" (quoted in Köhler, 2009). Kenney claimed that he

> cannot tolerate a situation where they see people getting a plane ticket, arriving here, saying the magic word 'refugee,' getting quasilanded status, getting a work permit and/or welfare benefits. That is an insult to the millions of people who aspire to come to Canada legally.
>
> (quoted in Köhler, 2009)

For Kenney, the suggestion of systematic abuse has long been central in the promotion of a climate of suspicion, distrust and fear.

Zimmerman (2011: 336) observes that "the state's processes of granting or refusing asylum are central to how meaning is assigned," and "abusive applicants are often framed as being social or economic migrants" and therefore as criminals. The problem of 'bogus' or 'false' claims and the perception that refugee claimants as economic migrants in disguise rather than 'real' refugees fleeing persecution (a dichotomy also described as fraudulent and genuine, deserving and undeserving, desirable and undesirable) is based on political and other forms of violence and does not recognize socioeconomic issues. These categorizations are based on a moralistic discourse that considers both the conduct and motivation of refugee claimants wrong and suspect.

This othering discourse operates on the assumption that genuinely deserving refugees are languishing in camps overseas and therefore claimants who actually arrive in Canada are less in need of protection and consequently less likely to be 'genuine' refugees (Pratt, 2005). Adding to this problematic assumption of un/deserving claimants is the neoliberal credo that values the entrepreneurial ethos and stigmatizes the person in need as immediately suspected of not contributing or abusing the state's largesse. Mexican refugee claimants in Canada, as explicitly stated by Conservative politicians, were widely assumed to be fraudulently trying to take unfair advantage of the system. As Pratt (2005: 216) demonstrates,

> [t]he emergence and conflation of the problems of the 'bogus refugee' and the 'welfare cheat' and the way in which these linked threats [are]

framed by the crimesecurity nexus and the specter of fraud ... produced a powerful, new, hybrid threat – the fraudulent criminal refugee.

Through this process, the 'at risk' refugee is constructed and recast as the 'risky' refugee (Pratt, 2005: 18). In the case of Mexican refugee claimants, the visa requirement served first and foremost as a deterrence and risk management technology, where risk is constituted by moral considerations shaping the deserving and undeserving.

In maintaining the apparent distinction between 'false' and 'real' refugee claimants in their numerous public addresses, Harper and Kenney have been seen as preempting decisions of the Immigration and Refugee Board – the independent administrative tribunal solely responsible for determining the admissibility of refugee claims in Canada – by concluding that the majority of Mexican claimants are 'bogus' refugee claimants and by insisting that 'real' refugees only exist overseas in desperate camps (Collins, 2009a). While the Minister of Citizenship and Immigration has discretionary powers to admit claimants on particular humanitarian grounds, explicit position on alleged 'bogus' claimants has been increasingly perceived as political interference in due process by creating systematic discrimination and prejudice against a targeted group of claimants. Such discriminatory political discourses legitimize a favorable situation for both visa enforcement and legislative reform.

While imposing a visa on a particular country is within the power of the Canadian government, linking the visa requirement to the need to curtail so-called 'bogus' refugee claims from a specific national group stands against international refugee law which stipulates that claims must be examined individually. Questioning the legitimacy of refugee claimants and deterring claims have been ways to impair and erode refugee applicants' human rights by reasserting border control. Many expert organizations, such as Amnesty International (2009), the Canadian Council for Refugees (2009b), and the Refugee Lawyers' Association of Ontario (quoted in Collins, 2009b), agree that it is inappropriate for politicians to comment on who is or is not a legitimate refugee and conclude that an individual does not have a wellfounded fear of persecution based only on his/her national origin. Commenting on the numerous verbal gaffes of Minister Jason Kenney, former Immigration and Refugee Board chairman Peter Showler said he was unaware "of a single previous minister of immigration who has made such remarks, who has intruded on the judicial process in this way" (quoted in Collins, 2009b). Such comments were echoed by the Refugee Lawyer's Association of Toronto, which contends that the "Canadian public should be shocked that a minister would interfere in the work of an independent body" (quoted in Collins, 2009b). Such prejudicial bias is seen as particularly problematic given that, under the recent refugee law changes, the minister will have full discretion over determining Canada's list of designated countries of origin or the socalled list of 'safe countries.' As Dauvergne (2008) reminds us, refugee law should not be about protecting sovereignty but rather the human rights of those needing protection.

Mexico, a 'safe country'?

Harper and Kenney have publicly condemned the alleged 'bogus' refugee claimants without any attempt to situate the phenomenon within the larger context of Mexico's drug war and impunity. The Mexican government's unhappiness about the visa measure and lack of power to change the situation also hide a complicit silence about Mexico's structural problems, violence, and corruption (Hernández Velázquez et al., 2011; Estévez, 2012a; Wright, 2012).

Under the Calderón administration and its 'war on drugs,' more than 50,000 military personnel have been deployed on the streets, mostly in the states of Baja California, Chihuahua, Nuevo León, Veracruz, and Tamaulipas (Estévez, 2012a). Military assaults have focused on the capture and killing of drug cartel leaders, leaving cartels to compete in the strategic restructuring of relations and territories. As a result of intensified violence, citizens have been caught in "shootings, crossfire, beheadings and car bombings directed at the authorities, as well as kidnappings and executions" (Estévez, 2012a: 22).

The recent 'disappearance' and killing of 42 students of Ayotzinapa – and the way investigations were conducted – is a sad reminder of the grave human rights crisis in Mexico. But sadly, these were not isolated cases. As Neve (2015) states, "more than 22,000 people have disappeared or gone missing since December 2006. Almost half of them disappeared between 2012 and 2014, under the current government of President Peña Nieto". By some estimates more than 150,00 have been killed since the beginning of the so-called drug war in the late 1990s, and more than 27,000 disappeared – but estimates are as high as 63,000 (Paulson, 2014). An estimated 230,000 others have left their homes (Estévez, 2012a: 24; Luthnow, 2012).

Mexican authorities have defended that the drug war has been limited to certain states and cities where criminals are killing each other. But as Wright (2012) remarks, while the brutality of the narco-violence and conflicts created by both criminal activities and military harassment might be concentrated in certain areas, it is nevertheless part of the consciousness of all Mexicans (inside and outside Mexico).

Moreover, the drug war has led to a serious human rights crisis. According to Amnesty International (2014), drug cartels and other criminal gangs, sometimes acting in collusion with police and public officials, are responsible for human rights abuses including kidnapping, rape, and murder. Migrants, journalists, women, and defenders of human rights have been the victims of killings or numerous threats that have never been investigated (Estévez, 2012a; Wright, 2012; Amnesty International, 2012, 2014; Human Rights Watch, 2014). Such human rights violations, along with others such as intimidation, torture, forced disappearances, extrajudicial killings, and the use of excessive force, are widely reported, but the large majority of these abuses go unpunished. In fact, the Mexican government considers such violations "exceptional" and remains without effective measures to prevent, investigate, or punish serious human rights violations committed by criminal organizations and police forces (Estévez, 2012b; Amnesty International, 2012, 2014; Human Rights Watch, 2014) – as the recent Ayotzinapa events demonstrates.

It is in the absence of state protection that violence against citizens becomes a human rights violation. García Clark (2011: 13) argues the

> human rights crisis in Mexico is a crisis of distrust of the Mexican state's capacity to safeguard its citizens' fundamental rights. A determining factor in this distrust is institutional inefficacy, manifested particularly in a highcost public security policy with negligible results, as well as the partisan bias and inconsistency in the administration of justice.

The human rights crisis in Mexico has therefore been exacerbated by a high level of impunity, and, as Estévez (2012a: 23) argues, "[i]mpunity in Mexico is not simply the result of incompetence or the inability to investigate, but the consequence of the high levels of corruption and the penetration of the criminal justice system and the police forces." In its incapacity to protect its citizens, the state becomes involved in perpetuating human rights violations.

The nature of narco-violence and impunity in human rights violations pose a particular challenge to the dominant understanding of asylum seekers or refugee claimants as persons suffering persecution and being unable to secure any form of protection. By insisting that human rights violations are exceptional or letting them go unanswered (due to a lack of judicial investigation) and affirming that the violence generated by the drug war is concentrated only in certain areas (and therefore persons displaced by violence can actually relocate to another area), the Mexican state has constructed citizens needing and seeking protection outside of the conventional definition of the 'asylum seeker.' The government's discourse of drug violence therefore normalizes a 'blame the victim story,' where drug violence is presented as strictly internal to the drug trade, i.e., perpetrated by criminals against criminals and therefore not affecting lawabiding citizens (Wright, 2012). As Wright (2012: 572) explains,

> A key component of this discourse is that the murderers target people for specific reasons; the violence is not random. Since these reasons are internal to the drug trade, the general public cannot know them, but as the violence reflects the killing of criminals by criminals, the general public, which is largely innocent of criminal activity, need not worry.

Still, since 2006, many Mexicans motivated by the fear of violence and persecution sought protection and asylum in other countries, in the United States and in Canada, as well as in many European countries (Estévez, 2012a: 29). In Canada, the majority of asylum applications are being rejected on the basis that the Mexican state is able to protect its citizens even though such claims about the Mexican government's ability to control and protect its citizens against the ongoing narco-violence are increasingly contested. According to Amnesty International (2014), violence keeps spiraling in Mexico and torture and other ill-treatment has grown 600 percent higher in 2013 compared to 2003.

For Canada to classify Mexico as a 'safe country' is a denial that safety and security concerns in Mexico might contribute and explain a rise in refugee claims. In the wake of the recent mass killing of 43 students from Guerrero and the wave of outrage and protests, the lack of understanding of the violent reality and humanitarian crisis in Mexico by the Canadian government seems more problematic than ever. It is particularly so when recognizing that drug "cartels pose a serious threat to the security of Mexico," and when Canada and the US provided technical assistance to train federal police in Mexico in 2009 through an Anti-Crime Capacity Building Program (Prime Minister of Canada Stephen Harper, 2012). Of course, the most glaring hypocrisy of the Canadian government is that it frequently advises its own citizens to not travel to some regions of Mexico.

Conclusion: ongoing violence

Borders and their control functions are becoming radically re-spatialized and externalized. This chapter shows that Canada's visa policy for Mexicans serves as an externalizing technology to preempt the entry of refugee claimants. This is part of a shift in Canada's immigration and refugee politics. In the recent years, Canada has stepped up migration and security enforcement by restricting programs, reforming immigration and refugee laws, and introducing stricter penalties for migration infringements. For example, recent legislative changes introduced by the Canadian government give greater discretionary powers to the immigration minister and officers to select, limit, or fast-track new immigration and refugee applications for those they deem desirable, and accordingly to refuse or conditionally authorize migrants to Canada.

In 2009 alone, the Harper government imposed new visa requirements, lifted a moratorium on removal, and carried out workplace raids. The Conservatives also defended policies justifying rendition to torture and security certificates. Temporary migrant worker programs have been extended from agriculture to construction and tourism. As a result, Canada has admitted more temporary foreign workers than permanent residents since 2006 (Citizenship and Immigration Canada, 2011). These measures and trends result in diverting particular types of migrants away from the permanent status authorized by immigration law into temporary, precarious, and disposable working conditions.

Canada's restrictionism is part of a larger trend. In recent years, many governments (the United States and Australia, among others) have created new immigrant related crimes, increasing penalties and prosecutions at the border and the workplace, expanding the grounds for being outside the law and therefore expanding the grounds for illegality (Kanstroom, 2004; Dauvergne, 2008). Migrants are thus rendered 'illegals' by a normalized criminalizing process in immigration laws, public discourses, and neoconservative politics.

While reporting on a multitude of abuses ranging from death, disappearances, torture, corruption, collusion, and impunity to the curtail of freedom of expression for human rights defenders, activists, and journalists, a Human Rights Watch (2014) report identified Australia, the US, and Canada for increasingly undercutting migrant and refugee protections. Many European Union members are cited for human rights violations within and externally to its borders and for the lack of efforts to counter extreme forms of intolerance such as racism, anti-semitism, xenophobia, homophobia, and islamophobia.

There is an ongoing violence against migrants and refugee claimants. Canada has played a particular role in sustaining such violence against Mexican refugee claimants despite 20 years of NAFTA partnerships and the 70th anniversary of bilateral relations. The visa and 'safe country' list adds onerous procedural obstacles making it more difficult for people to move, claim refugee status, and get a fair hearing. But as the extension of the visa, denied and rejected claims can of course be used as evidence fulfilling the prophecy of 'bogus' claims. Through this construct of alleged abuse, politicians have reduced a complex migration condition to 'bogus' claims, trivialized refugee and human rights, and failed to recognize Mexico's complex political reality.

Murders, disappearances, reports of torture, levels of corruption and collusion, absence of prosecution, military abuses and impunity, biased criminal system, and the fact that such issues are rarely brought to justice does not sum up into a 'safe' climate. In designating Mexico as a 'safe country,' it was expected that Canada would eliminate the need for the visa, but only by denying the human rights crisis created by the violent drug wars and the erosion of the power of the state to protect and to govern (Estévez, 2012a; Wright, 2012). Yet the visa requirement – imposed pre-emptively in the summer of 2009 after a surge in Mexican refugee claimants – remains in place even though the Harper government has reformed its refugee system. The visa imposed on Mexicans citizens thus proves to curb and to restrict access to the basic rights of mobility, interdicting them to cross and reach borders. But it also proves to be a process of ordering and othering.

Notes

1 An earlier version of this article was published in NORTEAMÉRICA, Year 8, Number 1, January–June, 2013: 139–161.
2 In 2005, there were 12,266 foreign temporary workers from Mexico, compared to 2,837 permanent residents. The number of accepted refugees from Mexico for the same year was 714. Five years later, the numbers were 18,152 temporary workers compared to 3,865 permanent residents and 655 refugee claimants. The latest figures available for 2013 show a consistent growth of 21,842 temporary

workers compared to 3996 permanent residents and 182 refugees (Citizenship and Immigration Canada, 2015).
3   Bill C-31's full title is An Act to Amend the Immigration and Refugee Protection Act, the Balanced Refugee Reform Act, the Marine Transportation Security Act and the Department of Citizenship and Immigration Act. Omnibus bills have been favored by the Conservative government but are criticized for limited opportunities for debate and scrutiny.
4   Countries included in December 2012 are Austria, Belgium, Cyprus, Czech Republic, Denmark, Estonia, Finland, France, Germany, Greece, Hungary, Ireland, Italy, Latvia, Lithuania, Luxembourg, Malta, Netherlands, Poland, Portugal, Slovak Republic, Slovenia, Spain, Sweden, and the United Kingdom. City-states designated in October 2014 are Andorra, Liechtenstein, Monaco, and San Marino.

Bibliography

Amnesty International. 2009. *Letter to the Honourable Jason Kenney*. www.amnesty.ca/resource_centre/news/view.php?load=arcview&article=4855&c=Resource+Centre+News.

Amnesty International. 2012. *Informe anual 2012: El estado de los derechos humanos en el mundo. México*. www.amnesty.org/es/region/mexico/report-2012.

Amnesty International. 2014. *Out of Control: Torture and Other Ill-Treatment in Mexico*. London: Amnesty International Ltd.

Bauder, H. 2014. Why Should We Use the Term 'Illegalized' Refugee or Immigrant: A Commentary. *International Journal of Refugee Law*. Vol. 26, No. 3, pp. 327–332.

Bigo, D. 2002. Security and Immigration: Toward a Critique of the Governmentality of Unease. *Alternatives*. Vol. 27, No. 1, pp. 63-92.

Campion-Smith, B. 2010. Ottawa's Visa Rule Is Costing Canada Visitors, Mexican President Says. *Toronto Star*. May 27. www.thestar.com/news/canada/article/815228–ottawasvisaruleiscostingcanadavisitorsmexicanpresidentsays.

Canada Visa Application Centre. 2011. *Required Documents*. http://cicmex.com.mex/enMx/selfservice/CAN_required_ documents.

Canadian Council of Refugees. 2009a. *Lives in the Balance: Understanding Current Challenges to the Refugee Claim Process*. June. www.ccrweb.ca/documents/refugeeclaimsFAQ.pdf.

Canadian Council of Refugees. 2009b. *The Challenge of Fair and Effective Refugee Determination*. www.ccrweb.ca/documents/fairdeterminationsummary.pdf.

Carens, J. H. 2010. *Immigrants and the Right to Stay*. Cambridge: MIT Press.

Casas, M., Cobarrubias, S., and Pickles, J. 2010. Stretching Borders beyond Sovereign Territories? Mapping EU and Spain's Border Externalization Policies. *Geopolitica(s)*. Vol. 2, No. 1, pp. 71–90.

CBC News. 2009. Kenney's Comments Prejudice Hearing for War Resisters, Critic Says. *CBC News*. January 9. www.cbc.ca/canada/story/2009/01/09/refugeewar.html.

Citizenship and Immigration Canada. 2015. *Facts and Figures 2013 – Immigration Overview: Permanent and Temporary Residents*. January 26. www.cic.gc.ca/english/resources/statistics/menu-fact.asp.

Citizenship and Immigration Canada. 2014a. *Facts and Figures: 10.2. Refugee Claimants by Top 50 Countries of Citizenship, 2004 to 2013*. www.cic.gc.ca/english/resources/statistics/facts2013/temporary/10-2.asp?_ga=1.6158195.1938659596.1422768832.

Citizenship and Immigration Canada. 2014b. *Can+ Program to Facilitate Trade and Travel with Mexico.* http://news.gc.ca/web/article-en.do?nid=847149.

Citizenship and Immigration Canada. 2014c. *Designated Countries of Origin.* www.cic.gc.ca/english/refugees/reform-safe.asp.

Citizenship and Immigration Canada. 2012. *Facts and Figures 2011: Immigration Overview–Permanent and Temporary Residents.* www.cic.gc.ca/english/pdf/researchstats/facts2011.pdf.

Citizenship and Immigration Canada. 2011. *Facts and Figures 2010: Immigration Overview–Permanent and Temporary Residents.* www.cic.gc.ca/english/pdf/researchstats/facts2010.pdf.

Citizenship and Immigration Canada. 2009a. *Canada Imposes a Visa on Mexico. Press Release.* July 13. www.cic.gc.ca/english/DEPARTMENT/MEDIA/releases/2009/20090713.asp.

Citizenship and Immigration Canada. 2009b. *Application for Temporary Resident Visa (TRV).* www.cic.gc.ca/english/information/applications/visa.asp.

Collins, M. 2009a. Cabinet Pulls the Plug on Mexican and Czech Visafree Travel. *Embassy.* July 15. www.embassymag.ca/ page/view/adios_canada-7152009.

Collins, M. 2009b. Political Interference Crippling Refugee Board: Former Chair. *Embassy.* July 22. www.embassymag.ca/page/view/political_interference7222009.

Dauvergne, C. 2008. *Making People Illegal: What Globalization Means for Migration and Law.* New York: Cambridge University Press.

Davis, J. 2009. UN Refugee Agency Cries Foul on Mexican, Czech Visas. *Embassy.* September 19. www.embassymag.ca/page/view/un_refugee_agency8192009.

De Genova, N. 2004. The Legal Production of Mexican/Migrant 'Illegality.' *Latino Studies.* 2: 16085.

Estévez, A. 2012a. La violencia en México como crisis de derechos humanos: las dinámicas violatorias de un conflicto inédito. *Contemporánea.* Vol. 2, No. 1, pp. 21–44.

Estévez, A. 2012b. Asilo y derechos humanos en Estados Unidos y Canadá. Cuestionamientos a Giorgio Agamben. *Norteamérica.* Vol. 7, No. 1, pp. 183–205.

García Clark, R. 2011. Human Rights Crisis in Mexico. *Voices of Mexico.* Vol. 89, No. 1, pp. 10–13.

Gilbert, L. 2013. The Discursive Production of a Mexican Refugee Crisis in Canadian Media and Policy. *Journal of Ethnic and Migration Studies.* Vol. 39, No. 5, pp. 827–43.

Government of Canada. 1982. *Canadian Charter of Rights and Freedoms.* www.efc.ca/pages/law/charter/charter.text.html.

Hernández Velázquez, J. A., Leco Tomás, C., and Aguilar Armendáriz, L.. 2011. México y Canadá: complementariedades desatendidas en materia migratoria. *Escenarios XXI.* Vol. 7, No. 1, pp. 48–64.

Human Rights Watch. 2014. *World Report 2014: Events of 2103.* New York: Seven Stories Press. www.hrw.org.

Kanstroom, D. 2004. Criminalizing the Undocumented: Ironic Boundaries of the postSeptember 11th 'Pale of Law.' *North Carolina Journal of International Law and Commercial Regulation.* Vol. 29, pp. 639–70.

Kelley, N. and Trebilock, M.J. 1998. *The Making of the Mosaic: A History of Canadian Immigration Policy.* Toronto: University of Toronto Press.

Köhler, N. 2009. A Crackdown on Queue Jumpers. *Maclean's.* July 28. www2.macleans.ca/2009/07/28/acrackdownonqueuejumpers/.

Kuoni Destination Management. 2014. *About Kuoni Group.* http://kuoni-dmceurope. com.

Luthnow, D. 2012. Mexico Drug Violence Shows Decline. *The Wall Street Journal.* June 12. http://online.wsj.com.

Martínez, M. 2009. Justifica Canadá la visa por el exceso de falsos refugiados mexicanos. *La Prensa.* August 10. www.oem.com.mx/laprensa/notas/n1279988.htm.

Mayeda, A. 2009. Harper Blames Canada's Refugee System for Mexican Visa Uproar. *National Post.* August 10. www.nationalpost.com/news/ story.html?id=1877551.

Mountz, A. 2010. *Seeking Asylum: Human Smuggling and Bureaucracy at the Border.* Minneapolis: University of Minnesota Press.

Neve, A. 2015. Ottawa must take action on Mexico's growing human rights crisis. *The Toronto Star.* January 22. www.thestar.com/opinion/commentary/2015/01/21/ ottawa-must-take-action-on-mexicos-growing-human-rights-crisis.html.

Office of the Auditor General of Canada. 2009. *Status Report of the Auditor General of Canada to the House of Commons.* Ottawa: Ministry of Public Works and Government Services Canada. www.oagbvg.gc.ca/internet/docs/parl_ oag_200903_02_e.pdf.

Paulson, J. 2014. 43 Students and the Future of Mexico. *The Bullet: Socialist Project.* E-Bulletin No. 1059. November 26.

Pratt, A. 2005. *Securing Borders: Detention and Deportation in Canada.* Vancouver: University of British Columbia Press.

Prime Minister of Canada Stephen Harper. 2012. *Canada Provides Basic Training for Federal Recruits.* August 9. http://pm.gc.ca.

Ramírez Meda, K. M. and Biderbost Moyano, P. N. 2011 Acción y reacción: Los motivos de la modificación política migratoria canadiense y la respuesta de la clase política Mexicana. *Revista Mexicana de Estudios Canadienses.* Vol. 21, pp. 23–34.

Rehaag, S. 2014. *2013 Refugee Claim Data and IRB Member Recognition Rates.* 14 April. http://ccrweb.ca/en/2013-refugee-claim-data.

Samers, M. 2004. An Emerging Geopolitics of 'Illegal' Immigration in the European Union. *European Journal of Migration and Law.* Vol. 6, pp. 27–45.

Singh v. Minister of Employment and Immigration. 1985. 1 S.C.R. 177. http://csc. lexum.umontreal.ca/en/1985/1985rcs1177/1985rcs1177.html.

Shane, K. 2013a. Mexico's 'Safe Country' Label Gets Mixed Reviews. *Embassy.* July 17. *Embassy.* www.embassymag.ca/ news/2013/02/26/ mexico's-'safe-country'-label-gets-mixed-reviews/43364.

Shane, K. 2013b. EU, Canada Far Apart on Visa Expectations: Documents. July 27. *Embassy.* www.embassymag.ca/ news/2013/07/16/ eu-canada-far-apart-on-visa-expectations-documents/44194.

Showler, P. and Maytree. 2009. *Fast, Fair and Final: Reforming Canada's Refugee System.* September. The Maytree Foundation. www.maytree.com/policy.

Studer Noguez, I. 2009. CanadáMéxico: adiós a la 'relación estratégica (I). *Offnews. info.* July 4. www.offnews.info/verArticulo.php?contenidoID=15900.

Triandafyllidou, A. 2014. Multi-levelling and Externalizing Migration and Asylum: Lessons From the Southern European Islands. *Island Studies Journal.* Vol. 9, No. 1, pp. 7–22.

United Nations, Department of Economic and Social Affairs, Population Division. 2013. *International Migration Report 2013.* New York: United Nations.

United Nations High Commissioner for Refugees. 1951. *Convention and Protocol Related to the Status of Refugees.* www.unhcr.org/ protect/PROTECTION/3b66c2aa10.pdf.

United States. 2014. *United States Diplomatic Mission to Mexico: Non-Immigrant Visas*. http://mexico.usembassy.gov/visas/non-immigrant-visas.html.

van Houtum, H. 2010. Human blacklisting: the global apartheid of the EU's external border regime. *Environment and Planning D: Society and Space*. Vol.28, pp. 957–76.

van Houtum, H., and van Naerssen, T. 2002. Bordering, Ordering and Othering. *Tijdschrift voor Economische en Sociale Geografie*. Vol. 93, pp. 125–36.

Valpy, M. 2009. Visa Controls on Mexico 'Humiliating,' Senator Says. *The Globe and Mail*, October 24, p. A13.

VSF Global. 2015. *About Us*. http://vsfglobal.com.

Wright, M. W. 2012. Wars of Interpretations. *Antipode*. Vol. 44, No. 3, pp. 564–80.

Zimmerman, S. E. 2011. Reconsidering the Problem of 'Bogus' Refugees with 'Socio-economic Motivations' for Seeking Asylum. *Mobilities*. Vol. 6, No. 3, pp. 335–52.

12 The south/north axis of border management in Mexico

Roberto Dominguez and Martín Iñiguez Ramos

Introduction

This chapter examines the main transformations in the border policies in Mexico over the past decade. Due to its geopolitical position as a gateway between the United States (US) and the rest of Latin America, the management of borders in Mexico has traditionally followed a twofold logic of reactions to the developments of the US border policies and disinterest to control of the southern border. While this differentiated attention devoted to northern and southern borders essentially remains as a defining feature of Mexican border policies, important events are occurring in the North and Central America around the management of borders. In the north, the full implementation of the North American Free Trade Agreement (NAFTA) has produced a selective redefinition of the US–Mexico border by opening the free circulation of goods and keeping restrictions on movement of people. At the same time, the number of Mexicans migrating to the US has decreased while the number of actions and resources of the US border to control the southern border has increased. In the south, Mexican border controls face limits to deal with the multiple problems derived from the spread of power of organized crime in Mexico and Central America affecting transit migrants.

In the light of these transformations, this chapter argues that a twofold process of externalization of borders has driven the redefinition of border practices in Mexico in the past decade. On the one hand, the externalization of US borders has influenced the design and implementation of border policies in Mexico with regard to the monitoring of the wide range of transborder networks of migrants. On the other, Mexico has also attempted to externalize its borders toward Central America in order to control networks of transit migrants, albeit with limited effect, and focused on mechanisms of control in its southern border. The combination of the US and Mexican externalization of borders has produced an incipient process of informal regionalization of borders. The chapter also suggests that, even though with limited success, Mexico has focused on improving the rates of detention of Central American migrants, respecting their human rights, and combating migration associated with criminality. In order to develop the argument of this chapter, the analysis is divided into four sections. The first defines some concepts of the US externalization of borders and the border management in Mexico. The second section looks at the changes in the border policies in Mexico, particularly around migration and cooperation with the US. The third observes the dynamic of Mexico as a source of migration, and the fourth describes the role of Mexico as a transit country.

The US externalization of borders in Mexico and Central America

The management of borders in Mexico is associated with the evolution of three main events. The first is the transformation of the border policies in the US. More or less restrictive US policies affect the flows of Mexican migrants in many different ways including, but not limited to, the levels of repatriation, the new routes of undocumented migrants, and the transformations in the strategies of organized crime groups to move illegal drugs as well as people in US territory. The second is the weaknesses and strengths of the Mexican state to improve the living standards of Mexicans, and to effectively implement the rule of law, particularly about transit migrants from Central America. The third is the changes in the emigration patterns of Central American populations to the US.

These three broad elements are closely intertwined and produce multiple collective challenges. For instance, increasing and decreasing levels of violence in Mexico and Central America affect migration patterns of people, border controls and immigration policy in the US. Two emblematic examples are the special immigration status for Salvadorians or Nicaraguans in the context of their civil wars in the 1980s and the increasing deportation of Central Americans as a result of the larger number of felonies included in the Illegal Immigration Reform and Immigrant Responsibility Act (IIRIRA) (Blanchard et al. 2011).

In other words, the challenges posed to border policies in the US and Mexico form part of a regional collective phenomenon that is largely addressed in a fragmented fashion based on national premises. Against this background, the perspective of the US is very significant because it is the magnet for migrants pursuing better living standards and also for organized crime organizations seeking profits derived from illegal activities. Keeping at bay these and other perceived threats to the US is an objective of the border policies, which are implemented through two main actions. The first is through actions within the borders ranging from policing the borders to naturalization of immigrants. The main agency in charge of the implementation of such policies is the Department of Homeland Security (DHS), which was endowed with a budget of $60.1 billion and 240,000 employees in 2015. Within the DHS, three areas are focused on security and immigration: Customs and Border Protection (62,552 employees); Immigration and Customs Enforcement

(19,374); and US Citizenship and Immigration Services (13,196) (Department of Homeland Security 2014).

The second way to manage borders is through instruments implemented outside of the territorial borders to deter the flow of people and illegal activities in Mexico and Central America, namely, the externalization of borders. From the regional perspective, in North America the development of border management coordination policies is practically non-existent or is in very incipient stages. In spite of the increasing role of economic deepening sparked by NAFTA and the Central American Free Trade Agreement (CAFTA), the free movement of people remains off the agenda. In contrast, a process of re-bordering has taken place in Europe with the enlargements through the opening of internal borders to new members and the creation of new external ones to non-European Union (EU) states (Follis 2012). Rather than re-bordering, a double, subtle process of US externalization of border policies is occurring in Mexico.

The concept of externalization or remote control of borders is formally associated with actions, particularly those beyond the territory of the state, such as preventive information campaigns, developmental aid, and visa restrictions (Kimball 2007). From an informal and broader perspective, this chapter argues that the externalization of borders can be amplified or even be more effective when there is a coincidence of interests between the country implementing remote controls and the recipient state. This is the case of US externalization of border controls in Mexico in the recent years, which entails a difficult and sensitive exercise of self-limited sovereignty, as Joppke (1998) argues. The scholarly debate agrees on the influential role of the US, but it remains unclear to what extent and how receptive Mexico is or is not to the pressure from the US. On the one hand, Valdes Ugalde argues that the geopolitical conception of the US includes Mexico as a border member that plays the role of buffer zone to face external threats and Mexico has responded partially to the security demands from the US, focusing on the northern border between both countries and leaving as a second priority the southern border with Central America (Valdes Ugalde 2012). On the other hand, Kimball argues that Mexico has opted to restrict transit migration for geopolitical reasons, even in a draconian and restrictive way; but more than a mere instrument of US interests, Mexico also is using such restrictions as a chip in the bargaining game with the US (Kimball 2007). Beyond the scholarly debate, the Mexican government launched in July 2014, the Southern Border Plan, which reiterates that transit migration is a national security problem, but excludes concrete strategies to combat organized crime. After a year of implementation, the Southern Border Plan has not produced the expected improvements and social leaders such as the priest Fray Tomás González in Tenosique, Tabasco, and Heyman Vázquez in Arriaga, Chiapas, have denounced that corruption remains a common practice of Mexican authorities and the Plan has been uneffective (Robles Maloof 2015). All in all, further cooperation between the US and Mexico in border management coordination remains incipient and limited due to the lack of incentives of the leader state, in this case the US, to develop long term and deeper regional arrangements (Koslowski 2011).

Border policies in Mexico

Border policies in Mexico face the paradox of coping with opposite challenges and planning different strategies for its northern and southern neighbors. Numerous voices in Mexican society advocate immigration reforms and respect for the human rights of migrants in the US, and condemn US actions to stop the flow of migrants mostly focused on the use of walls and agents displaced along the border. However, when one looks at Mexican border policies to Central America, it seems that Mexico emulates the US border policies albeit with two major weaknesses: organized crime has eroded the capacity of institutions to provide security for migrants crossing Mexican territory, and the anemic implementation of the rule of law has paved the way for human rights violations. The southern border of Mexico has become more significant in the past few years because Mexico's incentive to control the flow of transit migration from Central America for security reasons coincides with the interest of the US to use Mexico as a buffer zone for Central American immigration. This coincidence produces a contradiction because, as Kimball argues, while Mexico advocates treating its migrants in the US as economic or human rights migrants, the Mexican political rhetoric has increasingly focused on immigration from Central America as an issue of national security or crime (Kimball 2007).

The evolution of border management policies in Mexico has been reactive. Since the mid-1990s, the Mexico/Guatemala border became the site of a plan targeted to control regional transmigration and arrest transit migrants, drug traffickers, and human smugglers. The National Immigration Institute (INM) developed the Integral Migratory Policy Proposal for the Mexican Southern Border (IMPPMSB) in 2005, which called for increasing border authorities at the border and a technological revolution for securing it (Kimball 2007), recommendations that resemble US strategies to secure its southern border. The results of the IMPPMSB were far from tangible and drug trafficking organizations have extended their area of influence in Central America since the second half of the 2000s.

In order to address these challenges, the Mexican government implemented several legal changes and policies. In 2006, the INM developed new detention procedures that require citizens of Central American countries to sign a repatriation form that may expedite their removal from Mexico. In August 2010, after 72 migrants were executed at the hands of one of the drug cartels in Tamaulipas state (100 miles from Brownsville, Texas), the Mexican Congress unanimously approved the 2011 Migration Law. This new law replaced the 1974 General Population Law, which restricted the flow of migrants and addressed potential risks to national security based on the premises of the Cold War. The 2011 Migration Law introduced several innovations such as simplification and clarification of administrative rules (temporary and permanent residence) to all countries. It also complied with the International Convention on the Protection of the Rights of All Migrant Workers and Members of Their Families (ICMW) and is more explicit to respond to the deteriorating circumstances for migrants in transit, recognizing its obligation to ensure humane conditions for migrants through Mexico. According to Alba (2014), the Mexican government's adoption of a new migration law in 2011 was a ground-breaking recognition, not only of the scale of transit migration, but also of government accountability to ensure migrants' rights, including those of trans-migrants. Some other instruments dealing with the migration lead the actions of the Mexican Government. The 2013–2018 Strategic Plan of the INM focused modernizing and improving the performance of the governments in areas such as bettering control of incoming foreigners, training of public officials, consolidating a career service, and using high technology in the implementation of policies (Secretaria de Gobernacion 2014).

Nevertheless, the effective implementation of these laws has been quite challenging in areas such as detention conditions, duration of detention, and access to translators, beefing up border personnel, and modernizing control infrastructure. Human rights and civil society groups, including Mexico's National Human Rights Commission, have issued reports pointing out that many detention facilities are overcrowded and unhygienic and that detainees receive poor treatment. They also point out that corruption among government officials has not been eliminated and that organized criminal groups prey on many migrants. Indicative of the magnitude of the problem are the declarations of INM's commissioner, Ardelio Vargas Fosado, who has on several occasions acknowledged the deep-rooted corrupt practices to the point that under his watch he fired more than one thousand INM workers associated with different forms of corruption (Ballinas and Becerril 2013).

Central Americans in transit and potential Mexican emigrants have been victims of criminal activity for decades. In 1990, a pilot program was created in Tijuana, in the state of Baja California, which consisted of 45 non-uniformed armed police members who aimed to identify individuals robbing migrants. In light of the increasing power of organized crime, this pilot group revisited its goals and was transformed into the Beta Group. The model expanded to other regions in the country and became the humanitarian branch of the INM composed of unarmed groups focused on assisting migrants in cases of human rights violations or rescuing them rather than combating crime. The Beta Group, Nogales, Sonora, was formed in 1994, another in Tecate, Baja California in 1995, and one more in Matamoros, Tamaulipas. As of 2015 there were 22 Beta Groups, which are made up of members of the three levels of government and work in nine states: Baja California, Sonora, Chihuahua, Coahuila, Tamaulipas, Veracruz, Tabasco, Chiapas, and Oaxaca. While the Beta groups contribute to addressing human rights violations, they are inherently weak and the performance of the 22 groups vary in relation to the personal commitment of the members of each group (Wolf 2013).

Mexico has also received US aid to improve the capacity to manage borders and deal more effectively with migrant groups. The funding is usually designed as part of the programs that include Central America. President Obama's request for foreign assistance to Mexico, Guatemala, Honduras, and El Salvador totaled more than $280 million for the 2015 fiscal year ($137 million to Mexico). After the 2014 unaccompanied children crisis, President Obama urged Congress to quickly provide almost $4 billion to address the surge of young migrants from Central America crossing the border into Texas. However, the US Congress approved only $694 million in funding for federal agencies involved in border control and the housing and care of unaccompanied migrant children.

Mexico as a source of emigration and border policies

The near-zero net migration from Mexico to the US represents a new trend in the bilateral relationship. There is no one single factor accounting for this change, rather the confluence of events such as the decrease in outmigration, the combination of economic factors in the US and Mexico – unemployment in the former, paired with Mexico's improving economic stability – and the increase in the costs associated with crossing the border. A recent trend is the 115,000 Mexicans who have arrived in the border cities between 2006 and 2011 out of fear of drug-related violence and extortion, producing a new demand for new housing developments in McAllen and Brownville (Rios Contreras 2014). Likewise, the immigration enforcement at the US–Mexico border and within the US has also played a significant role, as indicated in the steady growth in the pace of deportations that began during the Bush administration and have continued under President Obama (Alba 2014).

From the Mexican border perspective, the general trend of the deportation of Mexicans from the US is in decline. From 601,356 Mexicans repatriated in 2009, the number declined to 332,865 in 2013. Based on the repatriation trend in the first half of 2014, it is possible to argue that the tendency will be a continued decline (as of mid-2014, 137,429). In the case of repatriated children, in the first half of 2014, 8,049 Mexican children were repatriated, of which 80 percent (6,453) crossed over the border unaccompanied by a family member. The number of children is almost equivalent to 6 percent of all countrymen who were repatriated in the first half of 2014 (Martínez 2014). In order to address the flow of repatriated Mexicans, in July 2013 the Mexican Government created the Repatriation Process to Inside Mexico (*Procedimiento de Repatriación al Interior de México*-PRIM). This program consists of two government sponsored weekly flights from the US to repatriate Mexicans to Mexico City rather than to the border areas. During the PRIM's first year of operation, 13,597 men and four women were repatriated on the 103 PRIM flights (Lopez 2014).

Once Mexicans are repatriated, the Mexican government has implemented patterns to help former migrants to stay in their local communities. The Minister of Interior (Secretaria de Gobernacion-SG), Miguel Angel Osorio Chong, has indicated that the Mexican federal government is offering employment to deported adults in their original local communities. In the case of migrant children, the national agency for children – Desarrollo Integral de la Familia (DIF) – has developed programs to help them to remain in their communities (Milenio 2014). At the state level, several institutions are involved with assisting repatriated Mexicans including the Tamaulipas Institute for Migrants (ITM) and the Institute of Migrants Oaxaca (IMO). During the summer and vacation periods, the INM sets up information stands to assist Mexicans returning to Mexico. Other programs also try to produce better conditions to ameliorate push factors; particularly significant is the 3x1 program, funded by four sources: remittances, local, state, and federal government. Among other benefits, the program opens opportunities for migrants seeking to work and support children, and young people of school age (Rodríguez 2014).

Two additional elements influence the flow of migration and the effectiveness of the border policies. One is the number of visas the US approves to Mexican applicants. The US Embassy Mexico City's Non-immigrant Visa Section is one of the largest visa processing units in the world, processing over 450,000 applications in more than 20 different visa categories each year. The second element is related to the role of criminal organizations in the patterns of migration. In the view of Goddard (2012), the problem is not the border crossers, but the criminal organizations that make their crossing possible. The arrest and deportation of those who make it across simply gives the cartels more customers and whether arrests are up or down is inconsequential as long as the cartels are in operation. A key element to disrupt these organizations is to follow up their financial transactions and plan strategies to freeze their assets after thorough investigations. However, the lack of bi-national coordination provides cartel leaders a sanctuary south of the border. Most importantly, it is estimated that the illegal flow of funds across the border into cartel pockets is around $40 billion annually (Goddard 2012).

Mexico as a transit country

While Mexican border policies to the north are reactive to events in the US, policies related to the southern border present a completely different perspective. The effective implementation of policies is challenging because the weakness of the rule of law and the practices of corruption undermine the spirit of new laws and regulations. In the view of Isacson et al. (2014: 3), the way to fix it is not by applying the US prescription of walls, patrols, soldiers, and technology. In a situation of low skills, poor coordination, and high impunity, beefing up existing border security will increase abuses and trigger more violence without actually reducing migrant flows or trafficking.

Mexico has not been the only country that changed its immigration and border control policies in recent years. In June 2006, El Salvador, Honduras, Guatemala, and Nicaragua signed the Central America-4 (CA-4) Border Control Agreement that created a common passport and eliminated border controls and movement restrictions between the four states. The removal of political barriers to movement has decreased the cost of migrating northward toward the US (Nowrasteh 2014). Now that Central Americans are able to circulate more freely within their own region, the road to the US has become saturated with obstacles. As the CA-4 countries need a visa to travel to Mexico, undocumented migrants avoid border controls and face two main logistical transit obstacles on their way to the US.[1] The first is the transportation by foot or car between Ciudad Hidalgo and Arriaga, one of the most dangerous areas in Mexico. The second is dealing with organized crime on board 'the Beast.'

In October 2005, Hurricane Stan had a disastrous effect on the state of Chiapas. In addition to close to 100 deaths, the hurricane destroyed 70 bridges and several railways, creating 280 kilometers of inoperable land between Ciudad Hidalgo and Arriaga. The drive between these two cities is around three and half hours, but for undocumented migrants travelling on foot, the journey – including avoiding checkpoints and roadblocks – takes from five to six days. It is very common that migrants are assaulted, women are raped, and children become victims of organ trafficking and child prostitution. Hurricane Stan not only changed the migratory route of undocumented immigrants, particularly Central Americans, it also extended the criminal operations of Mara Salvatrucha (MS13) from Soconusco to Arriaga.

Once migrants are able to reach Arriaga, they literally jump on 'the Beast,' also known as the train of death because of the multiple accidents that cause deaths or loss of limbs to migrants as well as the numerous violent acts that are committed within cars. 'The Beast' travels the states of Yucatan, Campeche, Tabasco, Chiapas, and Veracruz. In 2014, Ferrocarriles Chiapas-Mayab (FCCM), owner of 'the Beast,' announced that it had put into operation a project that sought to end undocumented immigrant traffic in the region. This consists of two elements: a) rehabilitate 100 kilometers of the railroads in order to increase the speed of the train from 10 to 30 miles per hour; and b) install center-traffic control, a satellite system to monitor both the railroad path and the loading of the cars of the train (Xicoténcatl 2014).

Despite the ordeal undocumented Central American migrants are facing in their transit to the US, there are no indications that the push factors will discourage migration through Mexico in the coming years. In 2013, for the first time, more than one third of migrants apprehended by US Border Patrol were not Mexican. The overwhelming majority of these 153,055 "other than Mexican" apprehended migrants came from Central America, mainly El Salvador, Guatemala, and Honduras. The number of "other than Mexican" migrants apprehended has nearly tripled in two years reaching 54,098 in 2011. The wave of Central American migrants has intensified; nine months into the US government's 2014 fiscal year, (which began in October 2013), the Border Patrol had already apprehended 202,951 (Isacson et al. 2014: 3–5).

In the case of Central American migrants in Mexico, according to statistics from the Ministry of the Interior, Mexico detained 86,298 foreign individuals in 2013. The INM returned or deported 93 percent of those detained (80,079, a slight increase over the 79,416 people deported in 2012). Of those deported in 2013, nearly all came from Central America: Honduras (32,800), Guatemala (30,005), or El Salvador (14,427). Data from the first four months of 2014 indicates an approximate 9 percent one-year jump in Mexico's deportations of migrants from these three countries (Isacson et al. 2014: 3–5).

Unlike formal deportations, the number of migration events is larger and has declined over time (an event is a migration action in which the same person can be returned more than once). From 1995 to 2005, the annual number of events increased from 200,000 to 433,000. The trend decreased by 70 percent to 140,000 events from 2007 to 2011. The main Central American migrants crossing through Mexico toward the US come from Honduras, Guatemala, El Salvador, and Nicaragua, representing between 92 and 95 percent of the total (Chávez et al. 2011). In 2013 and calculations as of mid-2014 indicate that the migration trend is stable at an average of 150,000 per year.

The outflow of Central American migration has been affected by the economic crisis in the US, the strengthening of its borders, and the insecurity in Mexico, making transit extremely expensive and dangerous through Mexico. Since the mid-2000s, the number of cases of Central American migrants kidnaped has increased not only by the so-called 'Maras' on the southern border, but by Mexican organized crime groups as well. The 2011 Comisión Nacional de Derechos Humanos (CNDH) report on undocumented migrants in Mexico indicated that kidnaping, including sexual assaults on women, has been highly profitable because the victims are forced to contact their families in the US to pay the ransom, and when released they do not report to the authorities out of fear of being deported back to their country of origin. The report estimates that between 2008 and 2009, 9,758 undocumented migrants in Mexico were kidnaped: 55 percent in the southern border, 11.2 percent in the north, 1.2 percent in the center, and 32 percent around the country.

Another element linked to the borders is asylum policies. While the numbers of asylum claimants from Central America and Mexico have increased, United States Citizenship and Immigration Services (USCIS) shows low numbers of asylum grants to Salvadorans, Guatemalans, Hondurans, and

Mexicans as a general trend from the fiscal year 2003 to 2012. In 2012, for example, immigration courts granted asylum at rates of 6 percent to Salvadorian applicants, 7 percent to Guatemalan, 7 percent to Honduran, and 1 percent to Mexican applications. These figures contrast with asylum granted rates of more than 80 percent to applicants from Egypt, Iran, and Somalia for the same period (Campos and Friedland 2014: 13). In the case of Central Americans requesting refugee status in Mexico, statistics show 1,164 requests for refugee status in 2013. Applicants' top four countries of origin were Honduras (455), El Salvador (285), Cuba (92), and India (88). Of these claims, only 245 were granted refugee status and 35 received complementary protection. Within this group, the top four countries of origin were Honduras (99), El Salvador (84), Nigeria (14), and Syria (11) (Isacson et al. 2014: 16). According to figures of the Mexican Commission for Aid to Refugees (COMAR) in the period 2013–mid-2014, 56 children were recognized as refugees in Mexico (Rivera 2014).

The case of children crossing borders is quite sensitive for legal and ethical reasons. Unlike children from Mexico or Canada, who can be turned back at their borders unless they can prove a credible fear of persecution – which they rarely achieve – the flood of children from El Salvador, Honduras, and Guatemala cannot be swiftly deported. Federal agencies have to take care of the children in line with a US Supreme Court decision and two laws passed during the George W. Bush administration, aimed at finding more humanitarian ways to deal with unaccompanied immigrant children. In 1997, the *Flores vs. Reno Settlement Agreement*, a Supreme Court decision, set a precedent that immigration officials should treat children differently than adults and provide basic services like food, water, toilets, medical care, and supervision. The 2002 Homeland Security Act codified standards set by *Flores vs. Reno* and transferred the care of unaccompanied immigrant children to Department of Health and Human Services (DHHS). The Trafficking Victims Protection Reauthorization Act of 2008 reinforced the previous laws and specified that the children should be turned over to the DHHS within 72 hours and be held in the least restrictive setting that is in the best interests of the child, which has come to mean reuniting them with parents in the country when possible (Ortega 2014).

The assistance and respect for human rights has been another area of concern in the implementation of border policies. The capacity of the Mexican government is surpassed and the role of non-governmental organizations has been quite significant. The Director of the Human Rights Center *Fray Francisco de Vitoria*, Miguel Concha Malo, has argued that Mexico should stop defining the immigration issue as a national security issue and consider it an issue of human solidarity. Other important actors assisting Central Americans on the way to the US are Alejandro Solalinde, director of the Global Village (GV) shelter *Hermanos en el Camino, Sin Fronteras* and *el Centro de Derechos Humanos Miguel Agustín Pro Juárez* (Centro Pro), which have played an important role in activism around the kidnapping and extortion of migrants in Mexico.[2] The CNDH has been another key actor in the protection of human rights of migrants. It began to take a particular interest in migration in 2005 when it created its Fifth General Visitation focusing on the issue of human rights violations toward migrants. Amnesty International has also followed the human rights of migrants with greater determination, including statements about it in its annual reports.

Conclusion

The precedent ideas aimed at drawing some of the main elements involved in the transformation of border policies in Mexico. Current border management practices in Mexico are the result of regional and domestic transformations paving the way to two different practices of borders. In the case of the northern border, the Mexican government is reactive to the effects of US border policies and its externalization in areas such as deportation or combating organized crime. The southern border is the opposite case; infrastructure is insufficient, and impunity of criminal groups have produced extremely dangerous areas for the migrants in transit. Rather than concentrating on controlling its southern border, Mexico has chosen also to increase interior enforcement. For example, it has set up checkpoints along major highways and expanded detention facilities, doubling the number of such facilities from 22 to 48 between 2000 and 2008 (Alba 2014). In this regard, border policies need to be recalibrated on a regular basis because push factors are moving targets and in fact have produced two different trends depending on whether Mexico is a source of migration or transit country. In the former, Mexican migration to the US has been around zero in the past few years. In the latter, the push factors are still strong in Central American societies.

Notes

1 Citizens of Costa Rica, Panama, and Belize do not required a visa.
2 Other prominent shelters working with migrants are the following: el Albergue Belén (Tapachula, Chiapas), el Albergue Belén Posada del Migrante (Saltillo, Coahuila), el Albergue Hogar de la Misericordia (Arriaga, Chiapas), el Albergue Hermanos del Camino (Ixtepec, Oaxaca), el Albergue Parroquial Guadalupano (Tierra Blanca, Veracruz), la Casa Betania (Mexicali, Baja California), la Casa de la Caridad Cristiana, Cáritas (San Luis Potosí, San Luis Potosí), la Casa del Migrante (Tijuana, Baja California), Nazaret Casa del Migrante (Nuevo Laredo, Tamaulipas) and la Parroquia de Cristo Crucificado (Tenosique, Tabasco).

Bibliography

Alba, F. 2014. Mexico: The New Migration Narrative. *Migration Information Source.* Vol. 24, April.
Ballinas, V. and Becerril, A. 2013. Corrupción generalizada en el INM, admite Vargas Fosado. *La Jornada*, 8 November.

Blanchard, S., Hamilton, E., Rodriguez, N., and Yoshioka, H. 2011. Shifting Trends in Central Amerian Migration: A Demographic Examination of Increasing Honduran–US Immigration and Deportation. *The Latin Americanist* (December), pp. 61–84.

Campos, S. and Friedland, J. 2014. *Mexican and Central American Asylum and Credible Fear Claims Background and Context.* May. Washington DC: American Immigration Council.

Chávez, E., Berumen, S., and Ramos, L. F. 2011. *Apuntes sobre migración. Migración centroamericana de tránsito irregular en México. Estimaciones y características generales.* Mexico: Instituto Nacional de Migracion.

Department of Homeland Security 2014. *Budget-in-Brief. Fiscal Year 2015.* Available: www.dhs.gov/sites/default/files/publications/FY15-BIB.pdf (17 December 2014).

Follis, K. 2012. *Democracy, Citizenship, and Constitutionalism. Building Fortress Europe.* Philadelphia: University of Pennsylvania Press.

Goddard, T. 2012. *How To Fix A Broken Border: Disrupting Smuggling At Its Source.* February. Washington DC: Immigration Policy Center.

Isacson, A., Meyer, M., and Morales, G. 2014. *Mexico's Other Border Security, Migration, and the Humanitarian Crisis at the Line with Central America.* August. Washington DC: Washington Office on Latin America.

Joppke, C. 1998. Why Liberal States Accept Unwanted Immigration. *World Politics.* Vol. 50, No. 2, pp. 266–293.

Kimball, A. 2007. The Transit State: A Comparative Analysis of Mexican and Moroccan Immigration Policies. *Center for Iberian and Latin American Studies and Center for Comparative Immigration Studies. Working Paper 150.* University of California, San Diego, June. San Diego: The Center for Comparative Immigration Studies.

Koslowski, R. 2011. Global Mobility Regimes: A Conceptual Framework. In Koslowski, R. (ed.) *Global Mobility Regimes.* Gordonsvile, VA: Palgrave Macmillan, pp. 1–24.

Lopez, L. 2014. Recibe Osorio Chong a 135 migrantes en el AICM. *Milenio,* 24 July.

Martínez, F. 2014. Este año, EU ha deportado a 6,453 menores que iban solos. *La Jornada,* 13 July.

Milenio 2014. Osorio: deportará EU a 60 mil mexicanos más en 2014. *Milenio,* 28 July.

Nowrasteh, A. 2014. Mexican Immigration Policy Lowers the Cost of Central American Migration to the US. *CATO at Liberty,* 17 July.

Ortega, B. 2014. Why are kids from Central America treated differently than Mexican kids? *The Arizona Republic,* 13 July.

Rios Contreras, V. 2014. The Role of Drug-Related Violence and Extortion in Pormoting Mexican Migration. *Latin American Research Review.* Vol. 49, No.3, pp. 199–217.

Rivera, C. 2014. Meade: no new policy will be shelter for migrant children. *Milenio,* 10 July.

Robles Maloof, J. 2015. El Plan Frontera Sur. *Sin Embargo,* 15 March. Rodríguez, O. 2014. In three months 680 children repatriated to Oaxaca. *Milenio,* 10 July.

Secretaria de Gobernacion 2014 *Plan Estratégico del Instituto Nacional de Migración 2013-2018.* Instituto Nacional de Migración, Mexico DF.

Valdes Ugalde, J. L. 2012. Gobernanza fromteriza, cooperacion y crimen organizado en la relacion Mexico-Estados Unidos: Hacia un nuevo paradigma?. In Rodriguez Sumano, A. (ed.) *Agendas comunes y diferencias en la seguridad de America del Norte. De donde venimos? Donde estamos? y A donde queremos ir?* Guadalajara: Centro de Estudios Superiores Navales de la Armada de Mexico y Universidad de Guadalajara, pp. 269–283.

Wolf, S. 2013. *Los Grupos Beta: ¿El Rostro Benévolo del INM?* INM-Dirección de Migración y Derechos Humanos, Mexico DF.

Xicoténcatl, F. 2014. Modernizan a 'La Bestia' para acabar con tráfico de migrantes. *Excelsior*, 24 July.

13 Secluding North America's labor migrants

Notes on the International Organization for Migration's compassionate mercenary business

Bruno Dupeyron

Introduction

The re(b)ordering efforts made by states over the last three decades, for instance the securitization of some border areas and harsher visa policies, may denote an evolution of the international migration regime. The increase of migrant and refugee flows in the 1970s and 1980s (Hatton 2012), coupled with demographic and security challenges in developing countries (Geddes 2005), started to significantly alter an international migration regime that was essentially based on the notion of 'control' (Pécoud 2010; Georgi 2010). In the 1990s, the collapse of the Soviet bloc and the wars in Iraq and the former Yugoslavia added further policy makers' concerns about the regulation of permanent and temporary migrations and refugee flows. A new regime, based on a global policy agenda relying particularly on the concept of 'migration management,' was originally formulated by Bimal Ghosh, in 1993. Ghosh further developed this concept of 'migration management' in the 1996 project known as the New International Regime for Orderly Movements of People (NIROMP), funded by the Swedish, Dutch and Swiss governments (Ghosh 2000). Ghosh proposed a comprehensive international migration regime, designed to tackle what was perceived as current and future migration policy crises, and focusing on both migrants and refugees (Geiger and Pécoud 2010). Yet, Sassen argues that these two categories, migrants and refugees, cannot be merged: "there are separate regimes for refugees in all these countries and an international regime as well, something that can hardly be said for immigration" (1996, 64). Nonetheless, Ghosh's 'migration management' approach was welcomed and later borrowed by the International Organization for Migration (IOM). This notion of 'migration management' became a mantra of the IOM, "committed to the principle that humane and orderly migration benefits migrants and society" (IOM 2015e).

The IOM is an intergovernmental organization (IGO), created in 1951 as an operational organization (as opposed to a full-fledged migration organization now), known as the Provisional Intergovernmental Committee for the Movement of Migrants from Europe (PICMME). The context of the origins of the PICMME deserves to be examined briefly. After the Second World War, approximately 11 million displaced people brought about serious socio-economic concerns for European governments. The problem was generally framed as an economic issue of surplus populations in Europe, surplus populations that required resettlement to certain countries with manpower needs, e.g., the United States (US), Canada, Australia, and so on. Yet, there was no consensus over an appropriate solution, as the United States disagreed with multilateral international organizations, e.g., the United Nations (UN). In the midst of the cold war, the United States insisted on limiting international interferences over migrant and refugee policy issues, and thus proposed to create a basic institution with specific functions, based on an intergovernmental structure driven by nation states' economic agenda. Conversely, the UN supported the idea of an international organization that would lead international cooperation on migration and refugee issues, based on humanitarian purposes, i.e., the United Nations High Commission for Refugees (UNHCR) (Karatani 2005). In 1951, the conferences of Naples and Brussels ratified the US approach, which led to the creation of the PICMME, then re-christened several times before acquiring in 1989 the acronym IOM we know today. In short, Düvell argues that the IOM "was always intended to offer an economic counter-agency to the humanitarian UNHCR, set up the year before" (2003).

Georgi analyzes the historical development of the IOM in five major phases: during the cold war, the organization is a modest "anti-communist logistics agency" that supports some of western countries' migration policies. Secondly, from the 1980s to 1993, the organization experiences an opportunistic phase, in the context of globalization and the end of the Soviet bloc; from 1994 to 2000, an expansion by solving the 'migration control' problem with a 'migration management' solution, although Pécoud notes that the difference between control and management is extremely tenuous (2010, 194); from 2000 to 2008, an exceptional growth of the IOM under Brunson McKinley's leadership; and from 2008 on, the IOM was in a post-neoliberal era (Georgi 2010, 49–61). We could also summarize this shift by suggesting that the IOM, originally a transport agency, shifted opportunistically to a multi-service agency for states.

The IOM, which is headquartered in Geneva, describes itself as a growing organization, from 67 member states in 1998 to 156 in 2014; from a total expenditure of US dollars (USD) 242.2 million in 1998 to USD 1.3 billion in 2013; IOM offices are located in more than 150 countries; operational staff was about 1,100 in 1998 and reaches at present 8,400, "almost entirely in the field" (IOM 2015g; Migreurop 2009). As opposed to numerous UN agencies that are going through significant organizational downsizing (Hammerstad 2014), the IOM provides an interesting example of international organizations that have the wind astern, symptomatic of nation states' shifting international priorities in the 21st century.

The expansion of the IOM, along with an increasing focus on border and migration security policy issues in diverse social science disciplines, may explain the emerging scholarly interest in this organization in the 2000s and 2010s. However, what explains the fact that it was and is still a relative research blind spot? First, the IOM is not an organization that belongs to the UN system. The absence of the IOM in an organizational framework that is over-scrutinized by international relations students may thus explain in part this research myopia. Second, and this is related to the first point, due to its narrow mandate and limited resources and activities until the early 1990s, the IOM has been a marginal organization for decades, which strived to reproduce itself, especially in the 1960s, when the 'iron curtain' threatened its existence (Georgi 2010, 51). Third, the IOM's lack of transparency regarding the services provided to its member states is another obstacle. Most of the contracts concluded by the IOM to offer these services are kept confidential. Besides, field research seems problematic: it is certainly possible to interview some IOM employees (see for instance Geiger 2010), and pretty challenging to converse with asylum seekers who are detained on the island of Lombok, east of Bali (Ashutosh and Mountz 2011), but it is extremely difficult to interview temporary foreign workers who have signed an IOM contract and understandably fear reprisals for voicing concerns (see for instance Vargas-Foronda 2010a; Ancheita Pagaza and Bonnici 2013).

At least two topics seem to draw consensus among researchers who study the IOM. First, many scholars agree that the IOM fosters neoliberal values and policies in a globalized context (Andrijasevic and Walters 2010; Geiger and Pécoud 2010; Vargas-Foronda 2010a). Similarly, Kalm uses the expression of 'neoliberal governmentality,' borrowed from Foucault, which allows us to think about migration management in relation to the government of populations, where the maximization of human capital is seen as a key objective for governments and individuals (2010). Not far from Kalm's analysis, Andrijasevic and Walters contend that the concepts of 'international liberalism' and 'global governmentality' allow us to grasp how the IOM works in a global field of power at the service of its member states; also, they argue that IOM policies should be analyzed through its practices, starting with specific ones defined by the IOM, for instance 'assisted voluntary return' (2010). Georgi identifies the IOM's political project as a form of "neoliberal global migration governance" (Georgi 2010, 67). Ashutosh and Mountz show how the IOM works "at the intersection of nation states, international human rights regimes, and neoliberal governance" (2010, 22).

Second, several scholars emphasize the fact that IOM activities also represent a significant business, relying for instance on EU funds, as the organization was able to secure European Commission funding to offer policy services to Albania (Geiger 2010, 157), on a vast repertoire of national ministries and agencies that consider that the IOM knows and defends migrants and asylum seekers' human rights (Pécoud 2010, 195; Geiger 2010), or on temporary foreign workers' remittances (Vargas-Foronda 2010b).

In the context of this literature, the externalization and internationalization of migration management seems to be one of the core features of the current regime. Yet, this international migration management regime is increasingly questioned by scholars (Geiger and Pécoud 2010; Kunz et al. 2011) who have worked on the IOM. Although the literature proposes stimulating approaches to migration governance, it fails to propose a robust coherent framework that would make sense of the role of the IOM as a key component in the neoliberalizing processes we witness at multiple levels – local, national, transnational and international. In the first section of this chapter, we suggest an alternative model to analyze how neoliberal migration management requires focusing not only on the notions of workfare and prisonfare, but also on the 'theory of fields' theoretical framework. In the second section, we will analyze the role of the IOM in a labor contract for Guatemalan temporary foreign workers who are employed in farms in Canada.

Methodologically, this study relies on primary and secondary sources: primary sources include website pages from the IOM's main website, as well as contracts, appendices to contracts and documentation in Spanish provided by Jacobo Vargas-Foronda; secondary sources comprise scholarly, government, IGO and non-government organization (NGO) documents.

The role of the IOM in the field of 'borderfare'

Wacquant argues that, as opposed to neoliberalism analyzed from three main perspectives – i.e., one that can relate to an economist ideology, a mode of governance rooted in the notion of governmentality, and a repertoire of policy instruments – its analysis is restricted to a polarized approach: on the one hand, a monolithic economic approach that diverges into neoclassical and neo-Marxist declensions, and on the other, an unsteady use of the Foucaultian concept of governmentality. Nonetheless, Wacquant contends that both views of neoliberalism fail to make sense of the reshaping and unfolding of the state that anchors the norm of market rule into individuals' representations and social practices. In short, he argues that both conceptions are too "thin" for conceiving an "anthropology of neoliberalism": the economist approach is too restricted to the fantasy of markets, and does not take notice of broader non-economic dynamics; the governmentality perspective expands far beyond the economic domain, but does not clearly explain how technologies of conduct and norms are specifically neoliberal, how they flow, and how they structure the society (2012a).

Instead, Wacquant proposes to conceive neoliberalism as "market-conforming state crafting" (2012a, 71). By this, he means that the fundamental feature of neoliberalism is "an articulation of state, market and citizenship" (2012a, 71), and therefore proposes a "thick" sociological conception of neoliberalism that has three dimensions. The first one contends that neoliberalism, far from being an economic project, is a political one that is implemented, not by shortening sail, but by "reengineering the state." The second refers to the argument that neoliberalism shifts the "bureaucratic field" (Bourdieu 1994) — generally fitted with two wings, one that is both economic and penal, and one that is essentially social and protective, that struggle over the definition and distribution of public goods — toward the economic and penal one. This shift contributes to structure the state around two sets of policies, the first one analyzed as "workfare" policies by Peck (2001), the second that builds on this work by proposing the related notion of "prisonfare" (Wacquant 2010, 2012b). The third dimension refers to the expansion and praise of the penal wing of the state. Wacquant suggests that the penal apparatus is one of the core features of the neoliberal state, as the neoliberal state must deal with the consequences of neoliberal policies that generate social inequality, work instability and ethno-racial anxiety.

This redefinition of the neoliberal state from Wacquant will allow us to analyze how the IOM supports states that have implemented neoliberal reforms in several forms and degrees of intensity. To do so, though, we will suggest two alterations to this foundational model: first, we will argue that the IOM undoubtedly serves states that implement workfare and prisonfare policies by using similar policy repertoires, but the added value of the IOM resides precisely in the fact that it offers a complementary set of externalized policies, entitled "border and migration management" by the organization, and that we call with a critical view "borderfare." The second alteration to this model is the use, not of the notion of the bureaucratic state that is limited to its two wings, but of the theory of fields (Fligstein and McAdam 2011, 2012). The theory of fields draws partly on Bourdieu's work, solves some of Bourdieu's theoretical problems, and is at the same time more flexible in the sense that it can be used at the intersection of several fields – intergovernmental, transnational, national, local – and finally allows us to add a third wing to Wacquant's model.

Thus, this section will be divided into two sub-sections. The first one will focus on the field of borderfare. The second will analyze the IOM as an IGO that contributes to shape borderfare states.

The field of borderfare: neoliberal migration control as a partial externalization of states' migration control

Analogous with the notions of workfare and prisonfare (Wacquant 2012b, 208), the concept of borderfare refers to the multilevel policy regime that addresses migration control problems by deploying militarized border patrols, offshore and domestic detention centers, domestic police forces and specialized courts, bilateral labor migration channels, along with their appendices – for instance other ministries – private organizations, and IGOs such as the IOM. In other words, the borderfare regime offers a set of global solutions to block unwanted migrants and asylum seekers, and accept strictly what labor markets need in terms of temporary and permanent workers.

In order to explore this concept of borderfare further, we will use the theory of fields, the theoretical framework developed by Fligstein and McAdam, which they note is "an integrated theory that explains how stability and change are achieved by social actors in circumscribed social arenas" (2012, 3). They suggest first to circumscribe the "strategic action field" (or field) that is studied, with the understanding that this field is not isolated, but closely linked with other fields. Fields are defined as "constructed social orders that define an arena within which a set of consensually defined and mutually attuned actors vie for advantage" (2012, 64). In this sense, the field of borderfare allows us to analyze the ways in which workfare and prisonfare take place, not just in domestic spaces, but also in multilevel and transnational spaces.

Fields are populated by three main categories of actors: first, *incumbents* "are those actors who wield disproportionate influence within a field and whose interests and views tend to be heavily reflected in the dominant organization of the strategic action field" (Fligstein and McAdam 2012, 13). In the field of borderfare, incumbents are nation states, although it might be more pertinent to suggest that only some nation states are incumbents in practice, namely the United States, that have considerable interests and resources regarding migration flows, along with other western countries. On the contrary, *challengers*

> occupy less privileged niches within the field and ordinarily wield little influence over its operation. While they recognize the nature of the field and the dominant logic of incumbent actors, they can usually articulate an alternative vision of the field and their position in it.
>
> (Fligstein and McAdam 2012, 13)

Challengers in the field of borderfare may be identified as the United Nations and its extensions, such as the UNHCR, the ILO and the United Nations Development Program (UNDP), as well as NGOs and trade unions that share an international human rights-based view on migrations. Finally, *internal governance units* (IGUs) "are charged with overseeing compliance with field rules and, in general, facilitating the overall smooth functioning and reproduction of the system" (Fligstein and McAdam 2012, 13–14). IGUs in the borderfare field are the IOM that constructs neoliberal policies, practices and representations, as well as regional intergovernmental organizations, such as the International Centre for Migration Policy Development (ICMPD) and Frontex in Europe.

Fields are constantly in flux, and to understand the dynamics of strategic action, three stages of the field are identified: the *field formation or emergence*, which is one of the most challenging of the three states; the *stable field* that requires institutionalization and reproduction; and the *field crisis* that usually entails a *resettlement* of the field (Fligstein and McAdam 2012, 165–167). In the case of the borderfare field, the focus may be restricted to the field formation, and to the status quo. The emergence of the field of borderfare goes beyond the scope of this chapter, so we would only situate its stability around the 1990s–2000s, when a new migration regime is established to externalize the control of migrants, asylum seekers and migrant workers. Instead of controlling them only at the doors of each country, which appears to be increasingly difficult and entails several international and domestic legal obligations for some countries, new strategies are suggested to anticipate this control and elude those legal requirements: controlling them in their own country, which is equivalent to limiting the free movement of people and secluding categories of populations in their own country; controlling them when they are in transit, for instance by transport companies and by transit countries; and controlling them where they work, for instance in western countries. In this context, we will preferably use the notion of 'control' rather than the IOM's concept of 'management,' as the latter is borrowed from a managerial semantic field meant to depoliticize and technicize migration control. This management newspeak is a key aspect of the neoliberal discourse, exemplified by Geiger who interviewed an IOM expert: "We have to move away from a control-oriented approach towards a proper, comprehensive and more managed approach to migration. A proactive way to look on migration not as a threat but as a benefit" (expert interview, in Geiger 2010, 155).

The externalization of migration control follows this logic of internationalizing and transnationalizing such control. It consists in observing how states select migrants and asylum seekers, not just through national apparatuses, but increasingly through international or a combination of national and international, policy and administrative tools. A vast repertoire of multilevel administrative procedures is thus made available for the selection of permanent and temporary migrants and refugees, and based on several factors, essentially class, age, education/profession, race and gender, e.g., higher skilled labor migrants, lower-skilled labor migrants, seasonal workers, family migrants and asylum seekers (Geddes 2005). At the same time, the multiplicity of administrative tools tends to blur where states are and where they are not. In fact, the IOM is one of those international institutions that serve the nation states' neoliberal migration objectives, i.e., proposing tailored programs, communicating the hegemonic discourse on global migration management, eluding policy issues of human rights and being opportunistically absent or present in the national and the international realms.

The IOM, a neoliberal migration agency that crafts 'borderfare' states

The IOM is a neoliberal migration agency that crafts borderfare states in two main directions: first, by moving the focus of a sociology of neoliberal states from the diptyque 'workfare-prisonfare' to the triptyque 'workfare-prisonfare-borderfare,' it is possible to add an analytical layer that makes sense of the role of the IOM with its member states, hegemonic ones and subalterns. Secondly, the IOM is a 'centaur' organization that helps select migrants, refugees and workers.

From the Diptyque 'workfare-prisonfare' to the Triptyque 'workfare-prisonfare-borderfare'

The move from the diptyque 'workfare-prisonfare' to the triptyque 'work-fare-prisonfare-borderfare' with the theory of fields allows us to consider the role of the IOM, mainly in the borderfare field, but also where the bor-derfare field intersects with the fields of workfare and prisonfare. In some instances, only two fields will intersect. For instance, although scholars criti-cally analyze border and migration security practices (Dureau and Hily 2009; Ferrer-Gallardo 2008; Ferrer-Gallardo and Albet-Mas 2013), including the role of the IOM in their design and implementation (Ashutosh and Mountz 2011), the IOM takes a slightly different stance: in a dedicated website enti-tled "Missing Migrants" (IOM 2015b), the IOM also denounces migrants' deaths, not to expose questionable security policies, but rather to distill fear and guilt in sending countries, as well as to keep delivering the narrative of an iron hand in a velvet glove to receiving countries. The IOM is here positioned at the intersection of the borderfare and the prisonfare fields. Namely, four stories illustrate migrants' tragedies: the lesson of one of those 'migrant expe-riences' is offered by Kessah, a young Ethiopian who sought to find better job opportunities in Saudi Arabia and, after encountering inexplicit but trau-matic obstacles in his journey, was 'repatriated' to Ethiopia. Now, Kessah calls to sovereign powers to stop migrants; he emphatically passes on to oth-ers the (IOM) lesson he learnt, i.e., living in poverty in one's country should be celebrated:

> My only dream now is that somebody may stop these 'trips' that are full of pain and suffering for poor people. My only truth now is to tell my friends about what happened to me and warn them not to go through what I went.
>
> (IOM 2015c)

This case, that could very well be fictional, illustrates how the IOM contributes to shape a rhetoric that seeks to use two types of registers, at the intersection of borderfare and prisonfare: the first one relates to the notion of 'symbolic violence' (Bourdieu 1998), used to internalize the prohibition of migrating illegally; in other words, would-be migrants would ideally refrain from using illegal migration channels, would try to use (inexistent or microscopic) legal channels or would migrate to other countries. However, in parallel, this web-site is first and foremost accessible by first world countries that may take a humanitarian stance after reading these stories. This leads us to the second type of register mobilized by the IOM: first countries' citizens may consider that the IOM has a compassionate approach to migration. On this subject, Žižek contends that "compassion is the way to maintain the proper distance towards a neighbour in trouble" (Žižek 1994, 211). IOM's compassionate gaze towards migrants, a gaze that mainly originates from western countries, allows the avoidance of ethical concerns about those spaces of violence. This is precisely this compassionate gaze, according to Žižek, that reinforces the spaces and cycles of violence against migrants, through the quasi-*blanc-seing* given to the IOM to control migrations and consequently contributes to form borderfare states.

The IOM, a centaur organization to help manage centaur states' migrations

The IOM as a centaur organization helps in the selection of migrants, asylum seekers and temporary foreign workers (Wacquant 2012b). Namely, it is very liberal, laissez-faire and pleasant at the top of the hierarchy of the field, with employers and member states, and is conversely short-sighted, paternalistic and rude with those who are at the bottom: migrants, migrant workers and refugees.

The liberal approach of the IOM can be illustrated by Georgi (2010), who argues that the paradox of the IOM, using a positive, human rights-based discourse of migration that contradicts actions, policies and projects aimed at harsh migration control, is not the mere result of borrowing the international human rights narrative, as it was suggested by Human Rights Watch (HRW) (2003, 2). Instead, the gap that exists between IOM's actions and rhetoric can be found in IOM's funding sources that tend to impose a migration control agenda, as opposed to the relative autonomy that the IOM can have as an international organization and growing, divided bureaucracy with a slightly different, human rights-oriented agenda (IOM 2010, 62). However, Pécoud contends that the use of a human rights rhetoric allows the IOM to be funded by a larger repertoire of ministries and agencies, not just home affairs and interior ministries, which tends to explain why the type of control suggested by the IOM is fairly ambivalent, "in-between control and humanitarian agendas" (2010, 195). Besides, the borderfare field involves many different actors, and therefore must create a consensual policy agenda.

Conversely, the iron hand approach toward migrants and refugees can be observed through a complementary strategy of border security and migration management that consists of deploying an apparatus capable of intercepting migrants before they reach the territory of migration destination. There are also tools within receiving countries that detect illegal migrants through the collection of data by the welfare state, shared with border and migration departments. In addition, data can be collected by other actors, for instance actors from the private sector. Pécoud argues that "they rely on tools that have little to do with 'law and order,' but rather with information technologies, communication and the media (newspapers, video clips, etc.)" (2010, 195). Yet, we would argue that this data collection process has a lot to do with the remaking of 'law and order,' as it is a constitutive component of the silent techniques that allow pre-emptive and continuous targeted controls.

In addition, the IOM as a centaur organization helps manage specific centaur states' migrations. The IOM has been instrumental in several aspects of the quantitative success of the Programa de Trabajo Agrícolas Temporal en Cánada (PTAT-C), which allows temporary labor migration between Guatemala and Canada, for instance in the selection of workers in Guatemala, and in the definition of the best route of the trip. First, the selection of workers is extremely discriminatory: the IOM, following the employers' desiderata, recruits essentially men who are on average 25 years old, married, along with other criteria (Mantsch 2009):

> It is not required that they know how to write. We have categories. If they need to work in the fields, they need to know how to add up, and even better if they know how to multiply. With the Spanish, at least they can be understood. Depending on the type of crop, they need to have an understanding and experience with the type of work, and they do not need to be tall. We have different criteria for each type of job. Sometimes, they ask us that people be at least 1.65 meter tall, and in this case, we go to Oriente, Santa Rosa, Jutiapa, Progreso. Sometimes, they indicate to us that height does not matter, 1.40 or 1.50 meter. As for the gender, it is easier to get men than women. With them [women], a huge proportion cannot travel, due to a lack of identification documents, due to the fact that they are pregnant, they have kids, it is difficult to travel. As for the age, men must be already married, and even better if they have children. The Canadians consider that the average must be 25 years old, because people are more mature, they have plans, and for this reason they are more responsible, and thus have more opportunities [to be hired]. We have the experience that people between 19 and 20 years old, all of a sudden they realize that they earn a lot, they do not know what to do with this money, they do not have a vision, no plan, and for this reason, they want to return without fulfilling [their obligation]. People who are older have had better success.
>
> (Mantsch 2009, 3 – translation by the author)

The experience of the IOM and the support of Guatemala's Ministerio de Relaciones Exteriores (MRE) are also useful in mapping out the best route of the trip between Guatemala and Canada, taking into account heightened border and migration security measures, promoted and implemented by other branches of the IOM and its most powerful member states. Initially, the trip to Canada passed in transit through the United States, but due to mistreatments suffered by Guatemalan workers, the route was changed to

Mexico. Vargas-Foronda argues that, with MRE's assistance, it is fairly easy to get the Mexican transit visa. In Mexico it is possible to get a direct flight to Montreal (Vargas-Foronda 2010a, 25–26).

Secluding labor migrants: an example of IOM's compassionate mercenary business

This section examines how the IOM, when dealing with the business of sending Guatemalan temporary migrants to Quebec's farms from 2003 on, offers a compelling example of a compassionate mercenary at the service of states and employers. The PTAT-C was initiated in October 2000, when the Embassy of Guatemala in Ottawa contacted the Guatemalan Ministerio del Trabajo y Previsión Social (MTyPS) and the IOM to follow up regarding the possibility of creating and managing a program to send Guatemalan temporary foreign workers to Quebec's farms (Comunicado del MRE número 100-2003, June 12, 2003, in Coto Pineda 2010, 2–3). It is assumed that the Embassy was either contacted by Canadian employers or acted based on its own evaluation of the Canadian labor market. The first groups of workers went to Quebec in 2003, 36 male workers in June and 29 female workers in July. In total, 215 workers went to Quebec in 2003 for a "two-year pilot project," with 10 employers (IOM 2010). In 2010, the IOM claims that it was "assisting 1,000 migrants to travel to Canada" only in the month of June. For the whole year of 2010, the PTAT-C involved 14,000 migrants and 500 employers (IOM 2010). It has been expanding to other sectors, industry, services and tourism (Mantsch 2009). Nonetheless, the IOM has been increasingly excluded from the market of temporary foreign workers, due to the emergence of private firms, ironically made up by former IOM chiefs of mission and employees.

This PTAT-C program is a stimulating case study for several motives: first, Guatemala and Canada never signed a bilateral agreement; second, private actors are involved in the foreground (private firms, Quebec employers, Guatemalan workers), but many public actors are present in the background (IOM, MRE, MTyPS, the former Human Resources and Skills Development Canada (HRSDC), Citizenship and Immigration Canada (CIC)); fourth, this case study exemplifies how the triptyque workfare-prisonfare-borderfare works in practice; finally, it demonstrates how the IOM is a centaur organization, which is particularly liberal and laissez-faire with Canadian employers, and very paternalistic and brutal with Guatemalan workers.

In order to analyze the role of the IOM in the PTAT-C, we will first examine how the IOM gained a leading role in this program, and then lost it at the end of the 2000s; we will then scrutinize a 2009 labor contract prepared by the IOM and another one prepared by a private firm in 2011, Amigo Laboral, managed by former IOM employees. Those contracts were signed between Guatemalan temporary workers and an association of Quebecois agricultural employers, la Fondation des Entreprises en Recrutement de Main-d'œuvre agricole Étrangère (FERME).

The leading and fading role of the IOM in the PTAT-C program

It must be first noted that Canada and Guatemala have not signed a bilateral agreement in order to clarify the conditions under which the Guatemalan workers would be hired temporarily by Quebec agricultural employers. This is surprising for two reasons: on the one hand, a bilateral agreement exists between Canada and Mexico on the exact same matter, the Seasonal Agricultural Workers Program, established in 1974 with Mexico. This program has included Caribbean countries since 1966 on a bilateral basis as well, which means that extending it to Guatemala should not be difficult. On the other, the initiative of the program emanated from the MRE, which means that the MRE had the opportunity to construct and drive the program. However, the MRE sought to involve other actors, the MTyPS and the IOM. The MTyPS was brought to this initiative, as the Guatemalan labor legislation requires that Guatemalan workers, who sign a labor contract in Guatemala but work in another country, must be authorized by the MTyPS. At the same time, some actors involved in the process argue that the MTyPS had not the sufficient resources to pilot the program, e.g., recruiting (Coto Pineda 2010). However, the IOM had them, i.e., human, technical and financial resources (for instance, to cover the first workers' airline tickets in 2003, then reimbursed by FERME), to develop and implement the program. It was even argued that the IOM was especially interested in these issues, as it is in charge of them in Guatemala (Coto Pineda 2010, 2). A "Memorandum of Understanding for Technical Cooperation on Temporary Migrant Workers between the MTyPS and the IOM" was signed, according to the 2006 evaluation of the PTAT-C by the IOM, but the date and the text have not been made public yet (MRE, MTyPS and OIM 2006, 15). Finally, the construction of the program took place in IOM facilities in Guatemala, where MRE and MTyPS were invited. We can then observe here that the Guatemalan public administration had determined very early on that the IOM, which had already been a recipient of the externalization of these issues in Guatemala, should logically be the entity in charge of the PTAT-C. This project was created through an agreement between the IOM and FERME, under the supervision of HRSDC (MRE, MTyPS and OIM 2006, 13), specifically an "MOU between FERME and the IOM in the Guatemalan Office of the IOM, for the implementation of the project of migrant workers to Canada," signed on July 10, 2003 (MRE, MTyPS and OIM 2006, 14–16).

The failed strategy from the Guatemalan administration, or perhaps the posterior justification to legitimate the lack of leadership role in the PTAT-C, is summarized by a public servant from the MTyPS who was the assessor of the program in the 2000s: due to a basic lack of resources, the MTyPS decided to let the IOM take the reins, even if this implied to turn a blind eye to certain practices and provisions, in the perspective of taking the leadership later on and improving the conditions of the program for Guatemalan workers (Coto Pineda 2010). This did not happen, as the IOM sought to demonstrate that it managed a successful and growing program over the years (curiously, the 2006 evaluation was apparently carried out by the IOM itself, as well as by the MRE and the MTyPS), and as the MTyPS public servants eventually did not get the resources they expected. In addition, the IOM cannot assume a public administration role and lead the PTAT-C, since public administration cannot be delegated by the Guatemalan government, which would represent a breach of the Constitution of Guatemala, especially articles 154, 193 and 194 (Vargas-Foronda 2010a, 5). This may also reflect the influence of the IOM over the Guatemalan government and administration, and the fact that the IOM is a de facto governance entity in Guatemala, in charge of labor migration.

The lack of bilateral agreement between Guatemala and Canada, coupled with the lack of oversight from MRE, MTyPS, HRSDC and CIC, leads to a growing PTAT-C that allows extreme labor flexibility and abuse, as we will see below, in the examination of the contracts signed by Guatemalan workers. The growth of the program also emphasizes the fact that managing temporary migrant workers is a lucrative business, befitting relatively all actors involved: IOM Guatemala is congratulated for its good work by the IOM's hierarchy, Canadian employers are able to hire workers who are flexible and cheap; Guatemalan workers can earn several times what they can get in Guatemala (Coto Pineda 2010, 11), Canada recovers numerous taxes from temporary migrant workers and has very limited welfare expenses; remittances received by Guatemala are substantial, as they account for nearly 10 per cent of the GDP in 2013 (Pew Research Center 2013). Vargas-Foronda also shows that the IOM may benefit directly from remittances, sent by temporary foreign workers (TFWs) in Canada to their relatives: remittances in Canadian dollars go first to a BANRURAL account managed by the IOM, which are then transferred to BANRURAL accounts of migrants' relatives in Guatemalan Quetzales (2010a, 11). Choosing BANRURAL is not contested, as the financial entity has numerous branches in rural areas in Guatemala. However, it is surprising that migrants' relatives are not allowed to manage their funds in Canadian dollars.

In 2011, this lucrative business led to the Chief of Mission of the IOM, a German, to establish a private business, the firm Amigo Laboral, managed by one of his relatives, and populated by former IOM employees. Amigo Laboral takes the Quebec market of migrant workers (the most important one), and leaves the Anglophone provinces to the next Chief of Mission, an American, who also soon created his own business to send Guatemalan workers to Anglophone provinces. Ironically, this leaves the IOM, silent about those issues, without any market. However, the former Chiefs of Missions' mutual appetite leads them to compete. It is uncertain what the new IOM Chief of Mission for El Salvador, Guatemala and Honduras, who is from Latin America and studied in Europe (LinkedIn 2015), is going to do with the PTAT-C (Vargas-Foronda 2015, emails no. 1 and no. 2).

In practice, the PTAT-C, which is supposed to be mostly free for its applicants, involves a series of processes, detailed in the 2006 evaluation: a)

Recruitment; b) Recruitment form process; c) Reception of required documents from worker and payment of fees; d) Visa application sent to embassy; e) Medical visit; f) How an application from a Guatemalan worker is processed; g) Allocation of application; h) Application to HRSDC (work permit); i) Visa granted; j) Conversation prior to the trip; k) Day trip; l) Worker's protection in Canada; m) Workers' return (MRE, MTyPS and OIM 2006, 17–20). The omission of the signature of the contract in this series of steps is not excessively surprising, given the condescending and paternalistic tone that is used throughout; in other words, the IOM does not specify when contracts are signed by migrant workers, and in fact we are not supposed to know that migrant workers sign a contract at any point, but we are led to understand that the contract is tacit. Another omission is the deposit of 4,000 Quetzales that each worker must make. Moreover, this series of processes is frequently challenged by former and current migrant workers (Vargas-Foronda 2010a). We will not be exhaustive here, and will only focus on a few: recruitment, mandatory deposit and worker's protection in Canada.

First, recruitment is conducted by the IOM which uses databases and mappings of agricultural communities in Guatemala. Preliminary contacts and interviews are not announced by the IOM, as it claims that a public announcement can lead to abuses by IOM impersonators who charge a fee for nothing (Mantsch 2009). As we have seen above, the recruitment follows several criteria, for instance gender, age and height. But it also involves other requirements, for example physical condition (tests are conducted, for instance carrying a certain weight on a certain distance), limited literacy skills (illiteracy is rampant in rural areas where migrant workers are recruited) and professional skills (almost 70 percent of those who are interviewed are already working as small farmers for their own use, so that transferable skills can be detected easily) (Mantsch 2009; Vargas-Foronda 2010a, 7). Those criteria, requested by Canadian employers, and followed by the IOM, are clearly discriminatory and a breach of the Charte des droits et libertés de la personne du Québec, and the Canadian Human Rights Act.

Second, as opposed to what is described in the 2006 evaluation, migrant workers claim that a deposit is required by the IOM. The amount of 4,000 Quetzales must be deposited in one of IOM's accounts. Stefan Mantsch, in charge of the PTAT-C for the IOM, confirms that this deposit is required. If the worker does not have the entire amount, it can be paid in instalments, but it means that the bank has to lend this amount with a 15 to 30 per cent interest rate, which occurs for people who go to Canada for the first time. Next, Mantsch explains the reason why a deposit is required by the IOM:

> The origin of this deposit is if the individual does not fulfil the contract. There are several conditions that specify when it is given back, and when it is not, it depends on the circumstances. [If they do not fulfil the contract] for motives like "I do not want to live here," "fellow workers bother me," "I do not like the food," and if they want to return because

of this, with this deposit it is possible to send the worker who is going to replace him/her in Canada. This is the idea of the deposit.

(Mantsch 2009, 4 – translation by the author)

Besides the fact that this deposit appears to be purely and simply illegal, and does not constitute a potential compensation for the early termination of the contract, this deposit is never mentioned in the contract signed by migrant workers. Fortunately, if the contract is fulfilled, workers get their money back with a gain of 1 to 3 per cent interest (Mantsch 2009, 4).

Third, according to the 2006 evaluation, the worker's protection in Canada is supposedly held by the personnel of the "Consulate of Guatemala in Canada" (which is incorrect, and most likely refers to the Consulate General of Guatemala in Montreal, Quebec). This is surprising, as consulate personnel are not supposed to check the labor conditions of the country where they are posted. The protection of labor rights in Canada is generally the domain of federal and provincial governments, as well as trade unions. It must be noted that the Consulate welcomes migrant workers at the airport, and explains their rights and obligations in Canada. However, the Consulate does much more, as we discover in the 2006 evaluation:

> In Canada, unions have contacted [Guatemalan workers] so that they join. They deceive them by saying that, if they join them, they will get better benefits from the employer. The work of the Consulate is to warn workers since the very moment they land in Canada, so that they know the existence of this labor organization, which will court them to offer services that the Consulate can offer for free. The constant communication and advising of the Consulate to the workers is the best way to avoid that these organizations trick them.
>
> (Vargas-Foronda 2010a, 56)

This anti-union rhetoric, found in a report co-written by the MRE, MTyPS and IOM, and in breach of several international and national conventions that recognize the right to unionize, is not so surprising. After all, the personnel of the Consulate seem to protect first and foremost the interests of the sending state, which is also one of the provisions of the same Vienna Convention, and certainly a superior injunction than the protection of its nationals. Thus, one of the migrant workers explains:

> Those from the Consulate. They lead us to the boss, they talk about the wages, tell us that if the boss does not treat us well, we have to talk to them. They always come to the farm to see us, they ask 'How are you doing?', 'Are you working?', 'One must go to work, not waste time, not go for a walk'.
>
> (Vargas-Foronda 2010a, 26)

The *sentence arbitrale* (arbitration award) that we will examine below shows that the Consulate was the organization the employer sought to contact in order to take disciplinary action against workers, and lay them off, which meant, for the Consulate, to send them back to Guatemala with zeal. The Arbitration Tribunal rejected this curious reading of the contract.

Labor contract between IOM/Workers-FERME

The copies of the contract we are going to examine have been gathered by Jacobo Vargas-Foronda. The first one carries the logo of FERME. It is entitled "Contrato de trabajo de trabajadores agrícolas temporales de Guatemala en Quebec 2009," and has been signed by workers who were recruited by the IOM in order to work for FERME members (farm employers). The parties are the employer (with his exhaustive contact information) and the workers (listed in a separate appendix, "Lista de contractos de la finca," which includes the names, dates of contract and signature of four workers). It is five pages long, and contains 24 articles. The second contract has been used in 2011 between the Guatemalan firm Amigo Laboral and FERME. The version we possess is not complete (two pages long and until article 16), but it is clearly a carbon copy of the contract designed by the IOM: the same logo, wording and structure, with only a couple of noticeable differences. For instance Appendix I has been simplified to include only the first name, last name and signature of each worker (the template allows up to six workers). Due to the fact that the second contract cannot be studied satisfactorily, we will limit our analysis to the first one, negotiated in 2009 by the IOM and FERME. Before examining the contract, we would like to emphasize the striking absence of any contract between the IOM and the Guatemalan workers who are recruited in the framework of the PTAT-C, which shows that the IOM acts as a quasi-MTyPS.

The IOM, followed from afar by the MRE and the MTyPS, took the lead in the wording of this contract which has been used as a template by the IOM and FERME (Coto Pineda 2010); for instance, the balance between the contracting parties does not exist, but rather it is in the benefit of the employer (contracting parties, art. 22, signature section, Appendix I); the Consulate General of Guatemala in Quebec is wrongfully mandated "in all situations related to employment and stay" (art. 23); safety measures are barely mentioned (art. 3); information about holidays, conditions for being absent, psychological abuse is missing (art. 4–7).

The arbitration award we have mentioned earlier, between the union *Travailleurs et travailleuses unis de l'alimentation et du commerce, section locale 501* (TUAC-FTQ) and an employer, seeks to solve allegations of lay off and discrimination against two Guatemalan workers, by scrutinizing the provisions of the contract and additional evidence. The allegations were received on December 21, 2009, and the arbitration award is dated December 11, 2014. The Arbitration Tribunal concluded that the employer laid off the complainants without just and sufficient cause, that the employer breached articles 10 and 16 of the *Charte des droits et libertés de la personne* following the complainants' discriminatory treatment based on their ethnic or national origin and language, and that the Tribunal is competent to determine compensation other than reinstatement. In this arbitration award, the role of the Consulate is stigmatized, as the Guatemalan diplomatic personnel sought to use their symbolic power in order to return Guatemalan workers promptly to Central America, although they have been discriminated against by their employer. In this context, the Consulate General of Guatemala is not just another actor in charge of the social control of migrant workers, at the intersection of workfare and borderfare: the Consulate, and more generally the MRE and the Guatemalan state, indicate that they follow IOM's neoliberal representations, policies and practices, even when this superior order negates international, Guatemalan and Canadian legislation.

But the contract is only one facet of the workers' obligations, as they receive additional instructions verbally and in writing: for example, Vargas-Foronda has gathered documents that have been used in 2010 by the IOM with Guatemalan workers sent to Alberta, one of them entitled "Informaciones para el día del viaje a Alberta, Canada," the other "Reglamento de normas y conductas en Alberta, Canada" (Vargas-Foronda 2010a). The second document includes some reminders set in bold typeface:

> Beware of relations with women, 2 Guatemalans have been detained for sexual abuse.
> Important notice: when you arrive at the farm, the employer/manager will keep your passport during your stay in Canada. In this way, you can be assured that it won't get lost.
> Remember: you cannot compare yourself with Mexican workers, since they have other types of deductions.
>
> (Vargas-Foronda 2010a)

It is not possible to know whether these documents have also been handed out to Guatemalan workers in Quebec, but since they indicate clearly the same program (PTAT-C) administered by the IOM, it is likely that similar advice, rules and threats have been provided to all participants in the PTAT-C, in addition to the contract.

Altogether, the contract and these additional documents provide a context for considering another condition of workers' social control, this one lies at the intersection of the triptyque 'prisonfare-workfare-borderfare': workers are secluded in a farm/circumscribed area, where they are supposed to work as much as the employer wishes, under illegal labor conditions, and are threatened to return to Guatemala in disgrace and bearing the financial burden.

Conclusion

The field of borderfare allows connecting at the international and transnational levels the domestic fields of workfare and prisonfare. Within the field of borderfare, the IOM occupies the role of IGU, though its functions often overlap with the two other fields: for instance, when the IOM drafts the labor contract template of the PTAT-C, the IOM operates in the borderfare field to send temporary workers from Guatemala to Canada; also, the IOM acts in the workfare one, as workers fill unrewarding and underpaid jobs, though working in Canada is presented by the IOM as a privilege; finally, a euphemised prisonfare field intersects with the two other, as Guatemalan workers are confined to their workplace for several months, and face dire repercussions for not accepting it.

After this brief overview, several research agendas should certainly be explored:

- How do Canadian trade unions interact with temporary foreign workers in Quebec and in other provinces?
- How do they defend their rights? How is the judicial system used by temporary foreign workers?
- How is the IOM going to reassert its influence in Guatemala?
- Is it going to confront the migration services companies it generated at the end of 2000s, beginning of 2010s?
- What are the Guatemalan workers' paths during, after and between seasonal works in Canada?

These questions lead us to go back to what the IOM states on its websites: "[it] is dedicated to promoting humane and orderly migration for the benefit of all" (IOM 2015a). In this short sentence, the IOM is showing its contribution to the neoliberal project of its member states. Here, three brief comments should be made to scrutinize it. First, the assumption that this dual objective benefits all may be related to the notion that supporting a rational, progressive, modern neoliberal agenda is always justified, even though this agenda, backed by coercive means, creates spaces of violence, spaces of postcoloniality (Fanon 2004). Second, the apparent dilemma between "humane migration" and its opposing view, "orderly migration," might be solved by Althusser's idea of "ideological interpellation" or "hailing": an individual who walks on the street is suddenly hailed by a policeman; when this individual turns his head and looks at the policeman, this interpellation transforms the individual into a subject (Althusser 1971, 163). Similarly, the IOM's humane intervention works in such a way that migrants suddenly become subjects — subjects of an order that is beneficial to everyone, migrants and non-migrants. Naturally, the implicit fantasy is that migrants may not only become subjects of a broader order, but may be on a path to become superior subjects of this order, i.e., non-migrants. In other words, the IOM may be seen as the guardian of this order and the protector of migrants-as-subjects.

Finally, if we follow Žižek, this IOM's regime might well be an additional illustration of "how ideology works" (Žižek 1989); by adopting a certain formulation of values and justice that is supported by human rights NGOs, IGOs, public opinion, as well as several categories of employees within the IOM, the organization and its member states are protecting us, comforting us, reassuring us. Simultaneously, they prohibit or limit the construction of alternative models for regulating migrations, thus contributing to maintaining the neoliberal status quo in the field it safeguards.

Bibliography

Althusser, Louis. 1971. Ideology and Ideological State Apparatuses. In *Lenin and Philosophy and other Essays*. New York: Monthly Review Press, pp. 121–127.

Ancheita Pagaza, A.C. and Bonnici, G.L. 2013. ¿Quo vadis? Reclutamiento y contratación de trabajadores migrantes y su acceso a la seguridad social: dinámicas de los sistemas de trabajo temporal migratorio en Norte y Centroamérica. *Documento de Trabajo No. 4*, Febrero, Serie de Documentos de Trabajo INEDIM.

Andrijasevic, R. and Walters, W. 2010. The International Organization for Migration and the International Government of Borders. *Environment and Planning D: Society and Space*. Vol. 28, No. 6, pp. 977–99.

Ashutosh, I. and Mountz, A. 2011. Migration Management for the Benefit of Whom? Interrogating the Work of the International Organization for Migration. *Citizenship Studies* Vol. 15, No. 1, pp. 21–38.

Betts, A. 2014. The Global Governance of Crisis Migration. In Martin, S.F., Weerasinghe, S. and Taylor, A. (eds.) *Migration and Humanitarian Crisis: Causes, Consequences and Responses*. New York: Routledge.

Bourdieu, P. 1994 [1993]. Rethinking the State: On the Genesis and Structure of the Bureaucratic Field. *Sociological Theory*. Vol. 12, pp. 1–19.

Bourdieu, P. (ed.). 1998. *La Misère du monde*. Paris: Points-Seuil.

Coto Pineda, Ó. R. 2010. *Entrevistas directas abiertas sobre Canadá y Guatemala: Sus relaciones bilaterales y la Cooperación Canadiense en Guatemala*. Interview by Jacobo Vargas-Foronda, March 8.

Delgado Wise, R., Márquez Covarrubias, H. and Puentes, R. 2013. Reframing the Debate on Migration, Development and Human Rights. *Population, Space and Place*. Vol. 19, No. 4, pp. 430–43.

Dureau, F. and Hily, M.-A. (ed.) 2009. *Les mondes de la mobilité*. Rennes: Presses Universitaires de Rennes.

Düvell, F. 2003. The Globalisation of Migration Control. In *Open Democracy – Free Thinking for the World*. London: www.opendemocracy.net/people-migrationeurope/article_1274.jsp.

Fanon, F. 2004. *The Wretched of the Earth*, trans. Richard Philcox with commentary by Jean-Paul Sartre and Homi K. Bhabha. New York: Grove Press.

Ferrer-Gallardo, X. 2008. The Spanish–Moroccan Border Complex: Processes of Geopolitical, Functional and Symbolic Rebordering. *Political Geography*. Vol. 27, No. 3, pp. 301–21.

Ferrer-Gallardo, X. and Albet-Mas, A. 2013. EU-Limboscapes: Ceuta and the Proliferation of Migrant Detention Spaces across the European Union. *European Urban and Regional Studies*, November 11 2013. doi:10.1177/0969776413508766.

Fligstein, N. and McAdam, D. 2011. Toward a General Theory of Strategic Action Fields. *Sociological Theory*. Vol. 29, No. 1, pp. 1–26.

Fligstein, N. and McAdam, D. 2012. *A Theory of Fields*. New York: Oxford University Press.

Foucault, M. 1976. *Histoire de la folie à l'âge classique*. Collection Tel 9. Paris: Gallimard.

Foucault, M. 1978. *The History of Sexuality Vol. 1: The Will to Knowledge*. London: Penguin.

Foucault, M. 2002. '*Society Must Be Defende': Lectures at the Collège de France, 1975–76*, trans. David Macey. New York: Picador.

Geddes, A. 2005. Europe's Border Relationships and International Migration Relations. *JCMS: Journal of Common Market Studies*. Vol. 43, No. 4, pp. 787–806.

Geiger, M. 2010. Mobility, Development, Protection, EU-Integration! The IOM's National Migration Strategy for Albania. In Geiger, M. and Pécoud, A. (eds.) *The Politics of International Migration Management*. Basingstoke: Palgrave Macmillan, pp. 141–59.

Geiger, M. and Pécoud, A. 2010. *The Politics of International Migration Management*. Palgrave Macmillan.

Georgi, F. 2010. For the Benefit of Some: The International Organization for Migration and Its Global Migration Management. In Geiger, M. and Pécoud, A. (eds.) *The Politics of International Migration Management*. Basingstoke: Palgrave Macmillan, pp. 45–72.

Ghosh, B. (ed.). 2000. *Managing Migration: Time for a New International Regime?* Oxford: Oxford University Press.

Hammerstad, A. 2014. *The Rise and Decline of a Global Security Actor: UNHCR, Refugee Protection, and Security*. First edition. Oxford: Oxford University Press.

Hatton, T. 2012. Refugee and Asylum Migration to the OECD: A Short Overview. IZA Discussion Paper No. 7004. Bonn: Forschungsinstitut zur Zukunft der Arbeit (IZA).

HRW (Human Rights Watch). 2003. *The International Organization for Migration (IOM) and Human Rights Protection in the Field: Current Concerns* (Submitted to the IOM Governing Council Meeting 86th Session 18–21 November 2003, Geneva), http://hrw.org/backgrounder/migrants/iom-submission-1103.pdf

IOM. 2010. *IOM Assists 1,000 Guatemalan Labour Migrants in June*. Press Release 06/21/10. www.iom.int/news/iom-assists-1000-guatemalan-labour-migrants-june

IOM. 2014. C/105/4 – *Financial Report for the year ended 31 December 2013*, https:// governingbodies.iom.int/system/files/migrated_files/about-iom/governing-bodies/ en/council/105/C-105-4.pdf

IOM. 2015a. *Our Work*. https://www.iom.int/our-work

IOM. 2015b. *Missing Migrants Project*. http://missingmigrants.iom.int

IOM. 2015c. *Mission Migrants Project – Stories*. http://missingmigrants.iom.int/stories

IOM. 2015d. Assisted Voluntary Return and Reintegration. www.iom.int/ assisted-voluntary-return-and-reintegration

IOM. 2015e. *IOM Mission*. www.iom.int/mission

IOM. 2015f. *Governing Bodies – Essential Documents*. https://governingbodies.iom. int/essential-documents

IOM. 2015g. *IOM Publications*. http://publications.iom.int

Kalm, S. 2010. Informing Migrants to Manage Migration? An Analysis of IOM's Information Campaigns. In Geiger, M. and Pécoud, A. (eds.) *The Politics of International Migration Management*. Basingstoke: Palgrave Macmillan, pp. 21–44.

Karatani, R. 2005. How History Separated Refugee and Migrant Regimes: In Search of Their Institutional Origins. *International Journal of Refugee Law.* Vol. 17, No. 3, pp. 517–41.

Kunz, R., Lavenex, S. and Panizzon, M. (eds.) 2011. *Multilayered Migration Governance: The Promise of Partnership.* Routledge Advances in International Relations and Global Politics 89. Abingdon, Oxon; New York: Routledge.

LinkedIn. 2015. *LinkedIn Profile of Jorge Peraza Breedy.* https://ca.linkedin.com/pub/jorge-peraza-breedy/40/a2a/74b

Mantsch, S. 2009. *Entrevistas directas abiertas sobre Canadá y Guatemala: Sus relaciones bilaterales y la Cooperación Canadiense en Guatemala.* Interview by Jacobo Vargas-Foronda, August 5.

Migreurop. 2009. *Atlas des migrants en Europe : géographie critique des politiques migratoires européennes.* Paris: Armand Colin.

MRE, MTyPS and OIM. 2006. "Evaluación Proyecto Trabajadores(as) Agrícolas Temporales a Canadá." *Cuadernos de Trabajo sobre Migración* 22.

Newland, K. 2010. The Governance of International Migration: Mechanisms, Processes, and Institutions. *Global Governance: A Review of Multilateralism and International Organizations.* Vol. 16, No. 3, pp. 331–43.

Peck, Jamie. 2001. *Workfare States.* New York: Guilford Press.

Pécoud, A. 2010. Informing Migrants to Manage Migration? An Analysis of IOM's Information Campaigns. In Geiger, M. and Pécoud, A. (eds.) *The Politics of International Migration Management.* Basingstoke: Palgrave Macmillan, pp. 184–201.

Pew Research Center. 2013. Remittances Received by Guatemala, 2000–2013. *Pew Research Center's Hispanic Trends Project.* www.pewhispanic.org/2013/11/15/remittances-to-latin-america-recover-but-not-to-mexico/ph-remittances-11-2013-a-10/.

Salter, M. B. 2003. *Rights of Passage, The Passport in International Relations.* Boulder, CO: Lynne Rienner.

Sassen, S. 1996. *Losing Control? Sovereignty in an Age of Globalization.* New York: Columbia University Press.

Vargas-Foronda, J. 2010a. *El Programa de Trabajo Agrícola Temporal en Canadá en su VII Aniversario 2003–2010. Una hipócrita negociación: Exportamos Mano de Obra barata con enormes rendimientos y altos lucros. Su cruda perversión y magnificada degradación.* Manuscript.

Vargas-Foronda, J. 2010b. "El Programa de Trabajo Agrícola Temporal en Canadá (PTAT-C). Mano de obra barata de exportación." *Diálogo.* Vol. 16, pp. 2–7.

Vargas-Foronda, J. 2015. *E-mails no. 1 and no. 2.* Electronic exchanges between Vargas-Foronda and the author.

Wacquant, L. 2010. Crafting the Neoliberal State: Workfare, Prisonfare, and Social Insecurity. *Sociological Forum.* Vol. 25, No. 2, pp. 197–220.

Wacquant, L. 2012a. Three Steps to a Historical Anthropology of Actually Existing Neoliberalism. *Social Anthropology.* Vol. 20, No. 1, pp. 66–79.

Wacquant, L. 2012b. The Wedding of Workfare and Prisonfare in the 21st Century. *Journal of Poverty.* Vol. 16, No. 3, pp. 236–49.

Žižek, S. 1989. *The Sublime Object of Ideology.* London: Verso.

Žižek, S. 1994. *The Metastases of Enjoyment: Six Essays on Women and Causality.* London: Verso.

Part IV
Conclusions

14 Policy outsourcing and remote management

The present and future of border and migration politics

Martin Geiger

Introduction

Externalizing Migration provides a timely and compelling discussion on the current state of affairs in North American and European border and migration politics. *How did we get here?*, *What is currently happening?* and *What might be next?* are important questions which are addressed by the numerous contributions to this volume. Outsourcing, and the extension of "flow control" into the territory of "source" countries and regions, have indeed become the global norm, given that this can also be witnessed in other parts of the world, for example in Australia, which is expanding its long-standing practice of outsourcing and extraterritorialization as far as Cambodia (Taylor 2005b; Grewcock; 2013; Ahmed 2014). *Externalizing Migration* is a much-needed publication, especially since practices of outsourcing and remote management are in most cases almost completely hidden from the public. Rarely are they presented and discussed in a transparent and truly democratic manner; even experts, academics and important stakeholders such as parliamentarians lack complete insight. Little is known about these practices, specifically regarding the mechanisms and implications of outsourced and spatially-shifted control and management.

The book dissects the multiple practices applied by states and international entities such as the European Union (EU) or the group of EU Schengen States (see also Zaiotti 2011, 2012). Certain "best practices" and "tools" have emerged against the backdrop of a so-called "global migration crisis" (c.f. Weiner 1995) and "war on terror" (Amoore 2006). Some of these practices still include more or less "conventional" approaches at the physical border line, such as fences or walls, which have seen a further expansion in recent years (Jones 2012). Border control and migration management, however, have also been enhanced by modern electronic technology – combining physical fences with virtual fences, complex radar systems or heartbeat sensors, for example (e.g., Ackleson 2005; Amoore 2006; Feldman 2011; Magnet 2011; Muller 2010; Salter 2008). Other forms of control and management include more indirect techniques which aim to "discipline" migrant flows, whole country populations, and even the body and behaviour of individuals

(Geiger 2013), using, for example, information campaigns or health checks (e.g., Heller 2014; Barron 2014). At first sight, these types of "technologies" (in a Foucauldian sense) seem to (re-)affirm the agency of migrants. However, on closer examination, most of these strategies are rather attempts at moulding agency, by giving out messages such as "you are at risk, you should stay at home" in order to deter migration, for example, or, in other, quite opposite cases, by supporting the selection, mobilization and "welcoming" of wanted migrant types and flows, and making migrants available for the global labour market (Rodriguez 2010; Rodriguez and Schwenken 2013).

How are states able to remotely control and remotely manage mobility and migration? Explanations are arguably to be found in the fundamental transformation and expansion of border and migration politics since the late 19th century (e.g., Torpey 2000; Taylor 2005a; Geiger 2013; Amoore 2006), further intensified in the post-Cold war era, and again following 9/11. This transformation was accompanied by the emergence and strengthening of actors from outside state apparatuses; today, a plethora of non-state policy actors, including specialized international organizations, non-governmental organizations, think tanks and private corporations, are contributing not only to the design of policies but also the practical implementation of border and migration policies (Geiger and Pécoud 2010). Rhetoric and, in general, the discursive framing and justification of certain new approaches to migration and mobility matter greatly. These "new" policy-makers seem to be increasingly foregoing such terms as *politics*, *policies*, *rules*, *regulations* or *govern/governance*. Instead they speak more and more nebulously about "managing" migration, about *initiatives* or *programs* and their piloting/testing in reality, often while simultaneously avoiding regular and due process requirements to discuss, adopt and evaluate in an open and transparent manner policies, rules, laws or norms. The introduction of so-called *migration management*, a new "ideology" and approach, was quite significant in this respect. As will be argued in the following section, the quoted disappearance or evasion of politics can be traced back, to a considerable extent, to the advent of this new term and framework in the mid-1990s, and to the implications this had for the practical workings of border and migration politics.

All that management talk ... the disappearance of 'politics' from border/migration politics

The term migration management has become frequently used not only in political discussions (e.g., European Commission 2015; House of Commons 2009; US Department of State 2015) but also in an increasing number of academic publications (e.g., Loyd and Mountz 2014; Martin et al. 2006; Mosneaga 2015). There are different issues arising and existing with regards to this term, all dependant on how the "*m*-term" or "*m*-talk" is employed, by whom, and to what (political) end:

1. Migration management is often used simply as a synonym for migration politics, employed without much, if any, knowledge of the implications of talking about management versus politics, rules, laws etc., and not necessarily with any direct political intentions or ulterior motives involved.
2. Some scholars and experts intentionally use the term management, however, as an umbrella word because they wish to express a political statement: that migration politics are today indeed plural and not the prerogative of (single) states anymore.
3. Migration management, both as a term and as a concept, is also used strategically by certain actors to create and maintain a climate and political venue for potential stakeholders to participate in discussions and activities together. In fact, the origins of the term lie in exactly such an orientation (Ghosh 2012) that aims to facilitate overarching discussions and cooperation with the help of the management term and related "international migration narratives" (c.f. Pécoud 2014). These narratives are often deliberately left vague, so as to make it possible for everyone to relate to and "work" with them.
4. This strategic invention, character and instrumentalization of "management" is acknowledged and critically interrogated by a growing number of scholars, activists and experts in the field. These critics share a certain curiosity and uneasiness about the "*m*-term" and its proliferation over other, more specific terms, as well as about the evasion of "politics" in migration management deliberately promoted by this conceptual framework or "ideology," its managers and "believers." Migration management, according to these critical observers (e.g., Georgi 2010; Kalm 2010; Geiger and Pécoud 2010; Hess and Kasparek 2010; Overbeek 2002), needs to be interpreted as a genuinely different and particular new political project that is based on the intentions of policy stakeholders to pragmatize and commercialize border and migration policies and, with these goals in mind, to deliberately avoid the "politics" – the political debate, discussion and legitimization – of border control. Migration management, according to this reading, involves a problematic redistribution of political sovereignty and security activities, and necessarily results in a blurring of responsibilities between state, non-state and private actors. Neo-liberal practices of outsourcing control and management tasks are thus unmasked as the integral parts and key features of migration management.
5. Finally, there is also the problem of language communities and the usage of the "*m*-word" in languages other than English. In wake of a more evasive, frequent and general usage of the term management, even when talking about people, human resources, skills and other abilities, emergencies or aspects of normal daily life, triggered notably by the school of 'new public management' (e.g., Pollit 1990; Hood 1991) and its particular ideology, the "*m*-word" and "*m*-talk" are hardly avoidable in any discussion, be it at universities, in research, government, administration,

the private sector, media or the general public. In the Anglo-Saxon language community "migration management" is used interchangeably with "migration politics," with hardly any discussion regarding the real meaning of this term and why it matters to talk less and less about politics, rules and laws.

6.   However, the term migration management is less common in other languages, including German, French, Spanish and Italian. It could be argued that the term migration management therefore directly invites and provokes critique whenever it is used in English, for example in policy documents (e.g., European Commission 2008; European Commission 2015; IOM/FOM 2005). Particularly in the German-language context, the term management provokes some unease. Partly because of the country's problematic historical past, Germans are very reluctant to talk and think about "managing people," "managing migrants" or "managing refugees," and as a result, criticism of migration management is, by international comparison, particularly strong in Germany. It is also interesting to see that even EU bureaucrats seem to sense these language differences, the lack of clarity and the existing or potential uneasiness about the term of "management" in certain countries/language communities. Some EU documents in their German, French and other European language versions will instead of "management" employ other, albeit often still too technocratic or nebulous bureaucratic or legalistic terms (see, e.g., the case of European Commission 2004 in different European languages).[1]

In the context of the decay of communist regimes in Central and Eastern Europe, as well as other parts of the world, politicians and scholars began arguing that the world is challenged by a "global refugee crisis" or "global migration crisis" (Loescher and Loescher 1994; Weiner 1995) and the onset of a new "age of migration" (c.f., Castles and Miller 1993). A debate emerged in- and outside political and academic circles as to whether the nation-state was about to "lose control" (Sassen 1996) of migration and border policy due to increased cross-border flows of people – migrants and refugees, as well as travellers. The question arose of whether states should cooperate more closely, and if there was a need for a "world migration regime" (e.g., Bergsten 1994; Bhagwati 2003; Ghosh 2000a, 2000b). The Commission on Global Governance in the early 1990s consequently entrusted a senior UN expert on migration – Bimal Ghosh – with the task of exploring the possibilities of a globally "more comprehensive institutionalized co-operation regarding migration," cooperation that would include both "formal institutions and regimes empowered to enforce compliance, as well as informal arrangements that people and institutions either have agreed to or perceive to be in their interest" (c.f., The Commission on Global Governance 1995: 2, 206 and 207; Ghosh 1993). In essence, the Commission's aim was to supplement the previous, more state-driven and unilateral "government" of population flows with a form of better governance, assisted by existing intergovernmental organizations such as the International Organization for Migration (IOM), the International Labour Organization (ILO) or the UN High Commissioner for Refugees (UNHCR); other non-governmental entities; migrants and their organizations; and the private sector.

Representatives of states and other governmental actors, however, remained quite reluctant to use the term "governance" in their discussions of new, state-transcending and more cooperative, less hierarchical forms of coordination and collaboration. Ghosh and other experts therefore (re-) invented and strategically employed the term of "management of migration" as it did not challenge national sovereignty and "government" over foreign cross-border movements and state borders, while at the same time evading precise definition, allowing for a flexible interpretation in order to bring (and keep) everyone at the table (Ghosh 2012). The term became very popular, and there was more and more "*m*-talk," especially at the level of international organizations and specific global and regional consultative processes (Geiger and Pécoud 2010; Kalm 2010). This happened to a great extent due to the fact that the IOM, founded in 1951 and already an important organization in the 1990s, had taken the lead on Ghosh's project for a "New International Regime for Orderly Movements of People." While working on this project the IOM strongly promoted the use of "migration management" and developed their own concept and portfolio of activities under this umbrella terminology.

Despite the flexibility which the "*m*-word" allowed, migration management from its birth as a "magic umbrella" concept thus entailed a powerful rationality and a certain strategic calculation and preciseness. Referring to management instead of speaking about actual policy-making and policy-implementation was arguably based on the outright desire to take the "politics" – problematization, politicization, political debate etc. – almost completely out of the equation, in order to find and lobby for *ad hoc*, pragmatic and quick-fix solutions unhindered by excessive debate or public and institutional scrutiny (Geiger and Pécoud 2010: 11). Ghosh had deliberately lobbied for a more 'balanced approach' to avoid the over-problematization of migration which was so prevalent in the 1990s, especially in Europe. He and other proponents of migration management had also hoped that, on the basis of a decreased problematization and a more pragmatic approach, nation-states or other important stakeholders like corporations would be more likely to successfully lobby for greater opportunities for persons to cross borders and to migrate legally, with the result of more "regulated openness" towards migration than before. Another aim was that these more "balanced" and "open" attitudes would facilitate the triple-win objective of making migration more beneficial for sending and receiving states as well as migrants and their families (Ghosh 2000a, 2000b, 2012; Geiger and Pécoud 2010).

It could be argued that these good intentions were in the following years hijacked and exaggerated. Problematically, soon after the introduction of the term management a specialized "toolbox" of migration management emerged. The expansion of a certain "business mentality" towards migration greatly facilitated the evasion of open political discussion, transparency, oversight and accountability, by moving necessary discussions towards exclusionary and secretive panels of mid- and high-ranking bureaucrats and experts of international organizations, interior ministries and security-concerned corporations and think tanks. The increasing "toolization" and commercialization of border and migration politics, and the consequent availability of tailored services (or tools), advertised and offered by specialized providers, encourages states and other actors like the EU to not only "upload" migration management towards supranational entities such as intergovernmental institutions, but also to spatially shift important policy tasks and thus "remotely control" and "manage" migration already in transit states or directly at its "source" in origin countries/regions.

The "toolization" and commercialization of border and migration politics

Migration management and the arrival of advanced technology mark important hallmarks in the noted transformation of migration and border politics. In the context of this transformation, actors beyond the state, such as inter-governmental organizations, and non-governmental actors – including diaspora associations, partisan think tanks and private businesses such as security industries – have acquired new stakes or significantly expanded their previous already strong involvement in the field of border and migration policy. What has emerged is a global 'migration industry' (Hernández-León 2013): stakeholders of different origins and with typically quite diverse orientations and agendas, including vested economic interests, who assist states worldwide in their activities to manage borders and migration. The rise of these 'industry' stakeholders, with the plethora of programs and 'tools' they had begun to offer, were crucial factors in enabling the diffusion of standardized policy approaches to other territories.

Key actors in migration management, and highly specialized service providers such as the IOM (see Dupeyron, this volume), have developed commercial migration management portfolios based on their organization's understanding of what migration management compromises and represents (see Table 14.1). States and other donors have the possibility to opt-in and buy-in, to become a client of the IOM and benefit from the services and tools they offer, some of which are tailored to assist states' intentions to remotely control and manage migratory processes. While the IOM today usually refers to itself as the "leading inter-governmental organization in the field of migration" (IOM 2015b) and avoiding the term *management*, in the past the organization referred to itself as being a/the leader in migration management, committed to working with states to "manage migration for the benefit of all" (see, e.g., Swing/IOM 2012). Some recent publications and country websites (see e.g. IOM 2014; IOM Georgia 2015) still mention this slogan or use closely-related language. The "benefit of all" notion and "managing-rhetoric" have come under attack in previous years, especially when these discourses were omnipresent; some scholars have started to speak instead of the IOM, for example, as an organization managing migration only for the "benefit of some" (c.f. Georgi 2010).

Though the IOM seems to have abandoned migration management rhetoric, at least to some extent, the "our work" sections of IOM websites, as well as some of its publications, still state that they are active "in the four broad areas of migration management: migration and development, facilitating migration, regulating migration, and addressing forced migration." Furthermore, the IOM mentions cross-cutting activities that are said to include "the promotion of international migration law, policy debate and guidance, protection of migrants' rights, migration health and the gender dimension of migration" (c.f., IOM 2015c). As pointed out by the author in Table 14.1, these areas and activities constitute fields that are also important for other actors and stakeholders that work in similar fields and who have therefore their own vested interests, claims and stakes, including for example: the United Nations Development Programme (UNDP, e.g. migration and development); the ILO (e.g. facilitating and regulating migration, migration law, protection of rights); the UNHCR (e.g. forced migration); the World Health Organization (WHO, migration and health); International Centre for Migration Policy Development (ICMPD, e.g. policy debate, migration regulation); and the Organization for Cooperation and Security in Europe (OSCE, e.g. policy debate, migration regulation). The noted pervasiveness of "migration management" has affected these organizations, even though some of them avoid mentioning the term in their publications or website statements. The ICMPD, for example, states on its website that its goal is to contribute "to good migration governance" (c.f., ICMPD 2015a). Instead of using the term "managing" they use alternative expressions such as "dealing with the current challenges in various fields of migration" (c.f., ICMPD 2015a). Nevertheless, ICMPD still uses the "*m*-word" and conforms to the management framework. For example, it responds by using it to certain calls for funding that make use of the management rhetoric, as do implementation partners of certain ICMPD activities, some of whom in a way 'demand' references to their own management paradigm (e.g., European Commission 2015; Frontex 2008; ICMPD 2015b, 2015c).

Table 14.1 lists a selection of the specific services IOM has tailored to meet the needs of its donors. In the view of the IOM, these tools are best suited to the management – including restrictive management, deterrence and remedies against irregular, unwanted migration – of the different aspects of migration and cross-border mobility. These include assisted voluntary return, lobbied for as a 'better' alternative to forced deportation (Koch 2014), resettlement of refugees, which is also an UNHCR activity, and work on remittances and migration and development. To advertise and 'sell' its solutions (or *tools*) each year the IOM issues its 'Migration Initiatives' (e.g.,

IOM 2015a). This publication outlines the current migration challenges in different world regions and countries, the IOM's tailored strategies, and the IOM's current and intended future responses. The 'Migration Initiatives' outline IOM activities that are already funded (interestingly, failing to mention the funding amounts it has already received and the respective donors) and then adds the additional funding the IOM is still seeking from states, the EU and other entities in order to implement its proposed activities. Donors can find the price tag allotted to each program and then decide to opt for and buy into the specific IOM program activity, or not. Upon receiving funding, the IOM implements its suggested activities in the respective region or country on behalf of its donors. The extent to which the target countries, their governments and (migrant) populations are actually aware of, informed of or consulted in the process of drawing up these strategies, activities and related funding requests is, however, not clarified. Therefore, it remains unclear whether the IOM's "Migration Initiatives" are necessarily perceived by the targeted countries as activities that are necessary and appropriate, as they presumably are by donors.

Arguably, this IOM 'catalogue' is quite useful for all states that seek to outsource and shift migration and border control and management tasks to other countries and regions. Hence IOM migration management delivers, enables and supports the spatial shifting of migration and border policies. IOM migration management, although at first sight seemingly a globally standardized approach, is however in itself quite heterogeneous, and also *spatially* quite heterogeneously applied. The concept and catalogue of IOM 'remote' migration management clearly represent immediate IOM interests, based on the organization's very own perception of necessary activities, geographic targets, specific flows, main mobility routes and particular target – 'risk' or 'beneficiary' – populations. This perception is no doubt shaped by vested interests, including the need to win donor support, expand activities and secure funding and donor interest for future projects, and – given the fact that the IOM is able to find a "market" for its services and 'clients' – matching donor expectations and their own interests in outsourcing and shifting policy tasks and remotely controlling and managing migration with the help of specialized service providers like IOM.

For 2015, the IOM proposed activities for 1.6 billion US$ in total (IOM 2015a: 6–7; here and in the following rounded numbers), with 597 million US$ dedicated to emergency and post-crisis operations; 283 million US$ for assisting migrants, including 'assisted' returns of migrants, asylum-seekers and other persons rejected by states; 256 million US$ for immigration and border management, including the strengthening of security and control measures; and the remainder dedicated to activities in the fields of migration and health (205 million US$), labour migration and development (204 million US$), migration policy development and research (76 million US$), multiregional activities including consultative processes (53 million US$), and migration and environment (8 million US$).

Most of the funding for 2015 was earmarked for IOM activities in West and Central Africa, Asia and the Pacific, the Middle East and North Africa (364/234/204 million US$, IOM 2015a: 6–7). The EU member states, Switzerland and Norway, in contrast, were the 'world region' with the least amount of funding requested and budgeted by IOM for activities within that region: only 89 million US$ out of 1.6 billion US$ funding requests in total, worldwide (IOM 2015a: 6–7). In this context, however, it is important to note that the member states of the EU, taken together, would represent the biggest donor 'state' for the IOM, followed by the US, while the EU as an institution is the biggest multilateral donor to the IOM (European Union Delegation 2014). For Germany, which itself is an important IOM donor *and* client, the IOM asked for contributions totalling 12.95 million US$ (IOM 2015a: 277 and 289). Almost the entirety of this funding (12.2 million US$) was earmarked to support the IOM-assisted voluntary return operations from Germany in order to return rejected asylum-seekers, irregular migrants and other persons back to their home countries. Substantial funding for assisted voluntary returns, as an alternative to state-led deportation, was also foreseen in the cases of the Netherlands (14.9 million US$; total funding requested: 20.4 million US$; IOM 2015a: 277 and 297–298) and Norway (10.7 million US$ of 19.5 million US$; IOM 2015a: 277 and 298–299). In total, almost 52 million US$ of the totally requested amount for this region (89 million US$) was actually earmarked for 'migrant assistance' activities, which are comprised mostly of these AVR programs (IOM 2015a: 277). Hence, the bulk of the funding for these European nations was not actually spent on policy activities actually taking place within these countries, but rather on removing individuals from these European states, reflecting the strong interest of these countries to get support for more restrictive approaches to migration and to outsource parts of their perceived 'migration problems' again *back* to other regions. Unsurprisingly, the IOM requested considerable funding for the neighbouring main transit countries and regions of Europe – North Africa and the Middle East (204 million US$), South-Eastern/Eastern Europe and Central Asia (153 million US$) – especially for so-called "emergency and post-crisis operations" (81 million US$ for North Africa and the Middle East; 27 million US$ for South-Eastern/ Eastern Europe and Central Asia). These crisis operations arguably consist primarily of activities to keep refugee and migrant flows contained within these (transit) countries and at their borders (IOM 2015a: 213 and 233). Other funds were earmarked for "immigration and border management" (31 million US$ for North Africa and the Middle East; 30 million US$ for South-Eastern/ Eastern Europe and Central Asia); strengthening border control measures as a means to prevent flows to other countries and regions, i.e., Europe (IOM 2015a: 213 and 233); and for 'migrant assistance' (30 million US$ for North Africa and the Middle East; 27 million US$ for South-Eastern/Eastern Europe and Central Asia) mainly consisting again in AVR activities, this time focused on the countries receiving returned irregular migrants, asylum-seekers and other persons (IOM 2015a: 213 and 233).

Table 14.1 Migration management according to the International Organization for Migration

Main areas of IOM's migration management	Migration and development	Facilitating migration	Regulating migration	Forced migration
Activities and services in the fields of:	– Migrant diasporas – Remittances – Brain drain and brain gain – Return of qualified nationals	– Family reunification – Labour migration – Recruitment and placement (e.g. of temporary workers) – Pre-departure Information	– Systems for visas, entry & stay – Border management – Technology applications – Assisted voluntary return & reintegration – Counter-trafficking – Counter-smuggling	– Asylum and Refugees – Internally displaced persons – Resettlement – Elections – Post-conflict cituations
(Potential) Overlaps with mandates and activities of e.g.:	– UNDP – World Bank – Financial institutions – Development agencies	– ILO – Private recruitment agencies	– ILO – OSCE – ICMPD – UNODC (UN Office on Drugs and Crime) – EU Frontex – Private Security Corporations	– UNHCR – OECD – Non-governmental relief organizations and charities – Security organizations

Main areas of IOM's migration management	Migration and development	Facilitating migration	Regulating migration	Forced migration

Areas of cross-cutting activities

- Technical Cooperation and Capacity Building
- Data and Research
- Policy Debate and Guidance
- Regional and International Cooperation
- Public Information and Education
- Migration & Health
- Gender Dimension
- Migrant's Rights and International Law
- Integration and Reintegration

(Potential) Overlaps with e.g.:

- UN, ILO, ICMPD, UNHCR, UNDP, UN WOMEN, World Health Organization (WHO), OSCE, EU, EU Frontex, Eurostat
- Non-governmental organizations, think tanks, private corporations, universities, research organizations

Source: author's elaboration of IOM websites (www.iom.int) and publications (IOM 2004: 6)

The second smallest regional funding request for IOM activities projected for 2015 was earmarked for the (very broadly defined) region of 'Central/North America and the Caribbean' (118 million US$; IOM 2015a: 6–7). Activities for the three main immigration/transit countries in this region, the US, Canada and Mexico, comprised only a small part of this funding: Canada: 7/Mexico: 3.7/US: 2.3 million US$ (IOM 2015a: 325), while the bulk of the funding was earmarked for activities in the remaining countries (e.g. Haiti, 48 million US$; Honduras, 7.5 million US$). For the whole region, funding for 'emergency operations' and 'migrant assistance' in the guise of assisted return programs constituted collectively around two thirds of the overall regional funding requested by the IOM. Again, a clear sign of the IOM's intentions to fund and implement primarily activities designed to facilitate and support remote control and management, such as migrant deterrence and return. The 7 million US$ contribution requested for Canada was, as a whole, budgeted to fund return operations and to allow the IOM to provide "voluntary return and reintegration assistance to failed refugee claimants in and around the Greater Toronto Area" (note: only this particular region of Canada; IOM 2015a: 329). This activity was described as follows:

> The Organization informs eligible migrants of their return options based on the programme's criteria. If necessary, migrants are assisted in obtaining travel documents from their respective consulates and embassies. IOM arranges flights and provides transit and/or post-arrival assistance, including medical/non-medical escorts during travel, if necessary. In addition, IOM helps migrants who qualify for in-kind reintegration support with drafting initial reintegration plans. IOM intends to reach out to migrant communities, as well as non-profit and legal organizations that serve them, to raise awareness of the voluntary return process in Canada and in countries of return, and to foster dialogue between migrants and service providers.
>
> (IOM 2015a: 329)

Two million US$ of the total 3.7 million US$ allocated for Mexico (IOM 2015a: 339–342) were also earmarked for return activities, for example to return transit migrants stranded or returned to Mexico from the US and Canada back to their home countries, while, similar to the case of Canada, the entirety of funding requested for the US (2.3 million US$; IOM 2015a: 346) was budgeted only to fund return operations.

For the immediate neighbouring region, South America, the IOM requested for 2015 a total of 177 million US$ in funding (IOM 2015a: 6–7). Nearly the entire funding was requested for only one country, Colombia: 153 million US$, of which 120 million US$ was to benefit IOM activities targeting internally displaced individuals and refugees within Colombia (IOM 2015a: 355). 'Migrant assistance' (that is, return programs) and 'immigration and border management' (10.5/5.3 million US$; IOM 2015a: 355) constituted other priorities projected for this migrant transit and sending region. This could again be interpreted as catering to the interests of the US and Canada in remotely controlling and managing migration in that region with the help of IOM (on this, see the contributions by Dominguez, Gilbert, Koslowski and Labove, this volume).

Status quo and the eventual futures of remote control and management

Since the 1990s, the emergence of new approaches and tools under the umbrella of "migration management," and the increasing availability and influence of specialized providers offering such services, have, taken together, profoundly changed the character of border and migration politics.

In the case of Europe, as a result of these developments and the parallel deepening of European integration and the creation of a common border-less area and joint external border, the majority of EU member states (and the EU as a political entity) have managed to outsource and spatially shift important parts of their migration policy (Zaiotti 2011, 2012; Feldman 2011). Most recent EU discussions, in the context of increasing migrant deaths in the Mediterranean Sea, stress the inevitability of combining expanded border control missions (e.g., by Frontex; on this agency and its activities see den Heijer, this volume) with better capacity-building and management of flows at points and in areas of departure, origin and transit (e.g., European Commission 2015). EU leaders have even started to consider military actions in neighbouring Libya, and a continuation of remote migrant control and management under changed and even military auspices seems therefore very probable (e.g., leaked EU documents distributed by WikiLeaks: Statewatch 2015). The question is whether the EU will opt in the coming months for the strengthening of its own border agency, Frontex; or maybe consider additional new common EU enforcement, control and management agencies; or draw together national (para)military and policy forces; or provide more funding to specialized 'managers' like IOM or the UNHCR; or find the time and political will to explore a strengthening of local leaders and governments in Libya and other countries. Perhaps the EU will opt for a combination of all these measures (making the composition of these measures then crucial), or – worst case – opt instead to wait the current migrant situation out, as was its practice in the past, until media discourses and public attention have calmed down or found new topics.

In comparison, as of yet, the dominant receiving states of migrants in North America, the US and Canada, have demonstrated far less interest than their EU counterparts in the services of the IOM and other international actors, although both countries have cooperated with the IOM and other organizations for many years, and have entrusted them with certain control and management tasks (see Dupeyron, this volume; CBSA 2014; US Department of State 2015) such as the provision of pre-departure information campaigns, the selection of temporary workers, health checks for migrants and their families, migrant transfer and settlement or return, and the resettlement of recognized refugees to Canada and the US, where their care is then taken over by non-governmental organizations, mostly charitable/humanitarian organizations or religious and co-ethnic diaspora associations. The role of international organizations is however still limited and could be further expanded in the future, especially in the light of a continuation of cutbacks of government agencies and government spending on migration policy – such as has happened in Canada (see, e.g. Government of Canada 2012) over the last years – or further reforms of the two countries' migration and border policy, such as the introduction of 'fast track' procedures, which might result in a further privatization and outsourcing of control and management tasks to countries outside the North American "perimeter."

Concerning the remote control and management which is exercised by states in Europe and North America, and is portrayed and discussed in the various contributions to this edited volume, a key question for the future will be to what extent will states further outsource and shift important aspects of their border and migration policy. What are the tasks that states can outsource and shift to other non-state/international or private agencies and other countries and regions? What are the limits, considering the instrumental toolbox of border and migration management, and where should (or will) outsourcing and shifting end? Will, for example, voluntary return programs, like those offered by the IOM, soon completely replace the deportation measures of states and their border agencies? In the case of the EU states, will Frontex perhaps in the coming years take over such activities (as EU assisted voluntary return campaigns), start competing with the IOM and maybe even replace it? And, in turn, will international organizations or private providers 'stop themselves' at one point in this process, that is, are there certain limits – e.g. forced deportation as a last bastion of states' authority – at least for the majority of these service providers? Or will they just continue to carry out anything certain states ask them to do?

These questions, and the contributions to this volume, clearly demonstrate the need to further investigate practices of remote migration control and management. While certain trends have become clear, far too little is known about outsourced and shifted control and management practices: their concrete impacts, their broader implications as well as real-life consequences for individual migrants, and their long-term effects, including the sustainability of these very particular types of foreign intervention in other countries on the basis of outsourcing and long-reaching cooperation with service providers outside/beyond the state. Thinking about the eventual future of remote border control and migration management, and aware of how far already today, geographically speaking, remote control and management reach,[2] another question is: How far can remote control and management be further stretched,[3] and how effective and sustainable can such 'telescoping' of border and migration management in fact be? And, in consequence of this perhaps, if border and migration control/management cannot be extended geographically any further, and proves to be ineffective, does this mean it is finally time to speak about the possibility, and perhaps even the inevitable necessity, of a "migration without borders" scenario (Pécoud and de Guchteneire 2007), globally or at least within geographical areas that are larger than today's geographical constructs of 'EU–Europe' or "US-Canada Perimeter Security"? Thus far, states and economies have been reluctant to agree on and accept greater freedoms of mobility, even as concerns highly skilled migrant workers.[4] As such, it seems unlikely that the near-term future of remote control and management lies in its abolishment and replacement by completely liberalized forms of mobility on the basis of pure market economics. States are, to some extent, currently even more invested than before in keeping control over their borders and migrant flows to their countries. The Schengen Zone, for a long-time considered one of the biggest achievements of the European integration process, is increasingly criticized and its continued existence openly questioned (Euronews 2015; Zeit Online 2015).

In the end, how much more states can outsource, how much farther they can shift policies, and how much longer they can neglect or sit out their 'migration challenges' depends on the public: on how much citizens and migrants know – and *care* – about the nature of both state-driven and privatized border and migration policies. To what extent is the European and North American public really interested in the plight of asylum-seekers and migrants? To what extent do we all care? Are we ashamed of, or protest against the quite obvious neglect of human rights including migrant rights, the daily deaths at sea and along other routes (see den Heijer; Follis; Mountz and Williams, in this volume), the exploitation of other people, increasing global inequality, (civil) wars, climate change and other of the so-called root-causes of refugee and migration flows? This is probably the most challenging question for everyone with a deeper, sustained interest in what states and other actors do in the field of migration. It clearly points to the need of continued and further expanded critical scholarship on remote migration control and management, and a better communication of research results to the public. This book greatly contributes to this scholarship, but can only be the beginning of a continued and intensified scholarly process and active civic engagement.

Notes

1 In its English version: Green Paper on an EU Approach to *Managing Economic Migration*; in the German version: Grünbuch über ein EU-Konzept zur *Verwaltung der Wirtschaftsmigration* (translated: administration of economic migration); in French: Livre vert sur une approche communautaire de la *gestion des migrations économiques* (translated: regulation/governance of economic migration).

2 See for example control and management in sub-Saharan Africa with the help of consultative processes and EU-funded activities carried out by international organizations: e.g., ICMPD 2015b, 2015d.

3 On the geographical "reach" of EU migration and border politics see, for example, the contributions of Dünnwald, Korneev and Leonov, Wolff and Zhyznomirska to this volume.

4 Note for example the more restrictive approaches towards even highly skilled migrants adopted recently by Australia, New Zealand and Canada. Also EU member states, although they have agreed on a joint EU 'blue card,' have in reality not liberalized their migration policies in light of the current Euro and economic crisis, high (youth) unemployment in many EU states and new flows of migrant workers within the Schengen Zone.

Bibliography

Ackleson, J. 2005. Border Security Technologies: Local and Regional Implications. *Review of Policy Research.* Vol. 22, No. 2, pp. 137–155.

Ahmed, B. 2014, *Australia Will Pay Cambodia $35 Million To Take Some Of Its Refugees*, http://thinkprogress.org/world/2014/10/09/3578369/australia-cambodia-refugees/ [24 June 2015].

Amoore, L. 2006. Biometric Borders: Governing Mobilities in the War on Terror. *Political Geography.* Vol. 25, pp. 336–351.

Barron, L. 2014. *Thais Mandate Migrant Worker Health Checks*, www.phnompenhpost.com/national/thais-mandate-migrant-worker-health-checks [24 June 2015].

Bergsten, C. 1994. Managing the World Economy of the Future. In Kenen, P. (ed). *Managing the World Economy.* Washington D.C.: Institute for International Economics, pp. 341–371.

Bhagwati, J. 2003. Borders Beyond Control. *Foreign Affairs.* Vol. 82, No. 1, pp. 98–104.

Castles, S. and Miller, M. 1993. *The Age of Migration. International Population Movements in the Modern World.* London: Palgrave Macmillan.

CBSA (Canada Border Services Agency) 2014. *Evaluation of the Assisted Voluntary Return and Reintegration Pilot Program – Final Report*, www.cbsa-asfc.gc.ca/agency-agence/reports-rapports/ae-ve/2014/avrrpp-pparvr-eng.html [24 June 2015].

Euronews 2015. *Schengen Area-Full Debate,* www.euronews.com/2015/05/27/schengen-area-full-debate/ [24 June 2015].

European Commission 2004. *Green Paper on an EU Approach to Managing Economic Migration. COM (2004) 811 final.* Brussels: Commission of the European Communities.

European Commission 2008. *A Common Immigration Policy for Europe: Principles, Actions and* Tools. *SEC (2008) 2026/SEC (2008) 2027.* Brussels: Commission of the European Communities.

European Commission 2015. *Managing Migration Better in All Aspects: A European Agenda on Migration*, Press Release, http://europa.eu/rapid/press-release_IP-15-4956_en.htm [24 June 2015].

European Union Delegation (Permanent Delegation of the European Union to the UN Office and other International Organizations in Geneva) 2014. *IOM Director General Updates EU Member States on Upcoming High level Dialogue on Migration and Development.* http://eeas.europa.eu/delegations/un_geneva/press_corner/all_news/news/2013/iom_meeting_en.htm [24 June 2015].

Feldman, G. 2011. *The Migration Apparatus. Security, Labor, and Policymaking in the European Union.* Redwood City: Stanford University Press.

Frontex 2008. Frontex-ICMPD-Europol Conference on Mixed Migration Flows. http://frontex.europa.eu/news/frontex-icmpd-europol-conference-on-mixed-migration-flows-MS446D [24 June 2015].

Geiger, M. 2013. The Transformation of Migration Politics: From Migration Control to Disciplining Mobility. In Geiger, M. and Pécoud, A. (eds). *Disciplining the Transnational Mobility of People*. Basingstoke: Palgrave Macmillan, pp. 15–40.

Geiger, M. and Pécoud, A. 2010. The Politics of International Migration Management. In Geiger, M. and Pécoud, A. (eds.) *The Politics of International Migration Management*, Basingstoke: Palgrave Macmillan, pp. 1–20.

Georgi, F. 2010. For the Benefit of Some. The International Organization for Migration and its Global Migration Management. In Geiger, M. and Pécoud, A. (eds.) *The Politics of International Migration Management*, Basingstoke: Palgrave Macmillan, pp. 45–72.

Ghosh, B. 1993. *Movements of People: The Search for a New International Regime. Paper prepared for the Commission on Global Governance*. Geneva: Commission on Global Governance.

Ghosh, B. 2000a. Introduction. In Ghosh, B. (ed.) *Managing Migration. Time for a New International Regime?* Oxford/New York: Oxford University Press, pp. 1–5.

Ghosh, B. 2000b. Towards a New International Regime for Orderly Movements of People. In Ghosh, B. (ed) *Managing Migration. Time for a New International Regime?* Oxford/New York: Oxford University Press, pp. 6–26.

Ghosh, B. 2012. A Snapshot of Reflections on Migration Management. Is Migration Management a Dirty Word? *IMIS-Beiträge*. Vol. 40, pp. 25–31.

Government of Canada, 2012. *Budget 2012. Annex 1: Responsible Spending*, www. budget.gc.ca/2012/plan/anx1-eng.html [24 June 2015].

Grewcock, M. 2013. Australia's Ongoing Border Wars. *Race & Class*. Vol. 54, No. 3, pp. 10–32.

Heller, C. 2014. Perception Management. Deterring Potential Migrants Through Information Campaigns. *Global Media and Communication*. Vol. 10, No. 3, pp. 303–318.

Hernández-León, R. 2013. Conceptualizing the Migration Industry. In Gammeltoft-Hansen, T. and Nyberg-Sørensen, N. (eds.) *The Migration Industry and the Commercialization of International Migration*. London/New York: Routledge, pp. 25–44.

Hess, S. and Kasparek, B. (eds.) 2010. *Grenzregime. Diskurse, Praktiken, Institutionen in Europa*. Berlin: Assoziation A.

Hood, C. 1991. A Public Management for All Seasons. *Public Administration*. Vol. 69, No. 1, pp. 3–19.

House of Commons (Home Affairs Committee) (UK) 2009. *Managing Migration: The Points Based System*, www.publications.parliament.uk/pa/cm200809/cmselect/cmhaff/217/217i.pdf [24 June 2015].

ICMPD 2015a. *Capacity Building*, www.icmpd.org/Capacity-Building.1555.0.html [24 June 2015].

ICMPD 2015b. *Multi-Thematic Programmes 'MIEUX II'*, www.icmpd.org/MIEUX-II.1672.0.html [24 June 2015].

ICMPD 2015c. *Ongoing Projects: Enhancing Georgia's Migration Management (ENIGMMA)*, www.icmpd.org/Ongoing-Projects.1640.0.html [24 June 2015].

ICMPD 2015d. *Participants (Rabat Process)*, www.icmpd.org/Participants.2050.0. html [24 June 2015].

IOM 2004. *Essentials of Migration Management. A Guide for Policy-Makers and Practitioners. Volume I*. Geneva/New York: IOM.

IOM 2014. *Migration Flows in Western Balkan Countries: Transit, Origin and Destination, 2009–2013. IOM Development Fund. Developing Capacities in Migration Management* http://ba.one.un.org/content/dam/unct/bih/news/Migration%20Flows%20in%20Western%20Balkan%20Countries.pdf [24 June 2015].

IOM 2015a. *Migration Initiatives 2015*, http://publications.iom.int/bookstore/index.php?main_page=product_info&cPath=34&products_id=1426 [24 June 2015].

IOM 2015b. *Mission Statement,* www.iom.int/mission [24 June 2015].

IOM 2015c. *Our Work,* www.iom.int/our-work [24 June 2015].

IOM/FOM (Federal Office for Migration, Switzerland) 2005. *International Agenda for Migration Management.* Geneva/Berne: IOM/FOM.

IOM Georgia 2015. *IOM and Georgia – 20 Years Together Managing Migration for the Benefit of All,* http://iom.ge/1/index.php [24 June 2015].

Jones, R. 2012. *Border Walls: Security and the War on Terror in the United States, India, and Israel.* New York: Zed Books.

Kalm, S. 2010. Liberalizing Movements? The Political Rationality of Global Migration Management. In Geiger, M. and Pécoud, A. (eds.). *The Politics of International Migration Management.* Basingstoke: Palgrave Macmillan, pp. 21–44.

Koch, A. 2014. The Politics and Discourse of Migrant Return: The Role of UNHCR and IOM in the Governance of Return. *Journal of Ethnic and Migration Studies.* Vol. 40, No. 6, pp. 905–923.

Loescher, G. and Loescher, A. 1994. *The Global Refugee Crisis.* Santa Barbara: ABC-CLIO.

Loyd, J. and Mountz, A. 2014. Managing Migration, Scaling Sovereignty on Islands. *Island Studies Journal.* Vol. 9, No. 1, pp. 23–42.

Magnet, S. 2011. *When Biometrics Fail: Gender, Race, and the Technology of Identity.* Durham: Duke University Press.

Martin, P., Abella, M. and Kuptsch, C. 2006. *Managing Labor Migration in the Twenty-First Century.* New Haven: Yale University Press.

Mosneaga, A. 2015. Managing International Student Migration: The Practices of Institutional Actors in Denmark. *International Migration.* Vol. 53, No. 1, pp. 14–28.

Muller, B., 2010. Unsafe at Any Speed? Borders, Mobility and Safe Citizenship. *Citizenship Studies.* Vol. 14, No. 1, pp. 75–88.

Overbeek, H. 2002. *Globalisation and Governance: Contradictions of Neo-Liberal Migration Management.* Hamburg: Hamburger Weltwirtschaftsarchiv (HWWA) Discussion Paper 174.

Pécoud, A. 2014. *Depoliticising Migration. Global Governance and International Migration Narratives.* Basingstoke: Palgrave Macmillan.

Pécoud, A. and de Guchteneire, P. (eds.) 2007. *Migration without Borders. Essays on the Free Movement of People.* New York/Oxford: Berghan Books.

Pollit, C. 1990. *Managerialism and the Public Services*: *The Anglo-American Experience.* Cambridge: Basil Blackwell.

Rodriguez, R. 2010. *Migrants for Export: How the Philippine State Brokers Labor to the World.* Minneapolis: University of Minnesota Press.

Rodriguez, R. and Schwenken, H. 2013. Becoming a Migrant at Home. Subjectivation Processes in Migrant-Sending Countries Prior to Departure. *Population, Space and Place.* Vol. 19, No. 4, pp. 375–388.

Salter, M. 2008 (ed.). *Politics at the Airport.* Minneapolis: University of Minnesota Press.

Sassen, S. 1996. *Losing Control? Sovereignty in an Age of Globalisation.* New York: Columbia University Press.

Statewatch 2015. *EU Politico-Military Group advice on the military intervention against 'refugee boats' in Libya and the Southern Central Mediterranean* (WikiLeaks release). www.statewatch.org/news/2015/may/eu-military-refugee-plan-PMG-8824-15.pdf [24 June 2015].

Swing, W. L./IOM 2012. *Managing Migration for the Benefit of All* (Address to the 2012 World Economic Forum Davos by IOM Director General Swing). https://agenda.weforum.org/2012/01/davos-2012-managing-migration-for-the-benefit-of-all/ [24 June 2015].

Taylor, S. 2005a. From Border Control to Migration Management: The Case for a Paradigm Change in the Western Response to Transborder Population Movement. *Social Policy & Administration.* Vol. 39, No. 6, pp. 563–586.

Taylor, S. 2005b. The Pacific Solution or a Pacific Nightmare? The Difference between Burden Shifting and Responsibility Sharing. *Asian-Pacific Law and Policy Journal.* Vol. 6, No. 1, pp. 1–43.

The Commission on Global Governance 1995. *Our Global Neighbourhood.* Oxford: Oxford University Press.

Torpey, J. 2000. *The Invention of the Passport: Surveillance, Citizenship, and the State.* Cambridge: Cambridge University Press.

US Department of State 2015. *International Migration,* www.state.gov/j/prm/migration/ [24 June 2015].

Weiner, M. 1995. *The Global Migration Crisis: Challenges to States and Human Rights.* New York: Longman.

Zaiotti, R. 2011. *Cultures of Border Control: Schengen and the Evolution of European Frontiers.* Chicago: University of Chicago Press.

Zaiotti, R. 2012. Practising Homeland Security across the Atlantic: Practical Learning and Policy Convergence in Europe and North America. *European Security.* Vol. 21, No. 3, pp. 328–346.

Zeit Online (German Newspaper) 2015. *Bayern und Sachsen fordern Einreisekontrollen,* www.zeit.de/politik/deutschland/2015-06/bayern-und-sachsen-fordern-grenzkontrollen [24 June 2015].

Index

9 781138 546493